Post-embryonic Development of the Copepoda

CRUSTACEANA MONOGRAPHS constitutes a series of books on carcinology in its widest sense. Contributions are handled by the Editor-in-Chief and may be submitted through the office of KONINKLIJKE BRILL Academic Publishers N.V., P.O. Box 9000, NL-2300 PA Leiden, The Netherlands.

Published in this series:

CRM 001 - *Stephan G. Bullard*	Larvae of anomuran and brachyuran crabs of North Carolina
CRM 002 - *Spyros Sfenthourakis et al.*	The biology of terrestrial isopods, V
CRM 003 - *Tomislav Karanovic*	Subterranean Copepoda from arid Western Australia
CRM 004 - *Katsushi Sakai*	Callianassoidea of the world (Decapoda, Thalassinidea)
CRM 005 - *Kim Larsen*	Deep-sea Tanaidacea from the Gulf of Mexico
CRM 006 - *Katsushi Sakai*	Upogebiidae of the world (Decapoda, Thalassinidea)
CRM 007 - *Ivana Karanovic*	Candoninae (Ostracoda) from the Pilbara region in Western Australia
CRM 008 - *Frank D. Ferrari & Hans-U. Dahms*	Post-embryonic development of the Copepoda

In preparation (provisional titles):

CRM 00x - *Darren C. Yeo & Peter K.L. Ng*	Southeast Asian freshwater crabs
CRM 00y - *Chang-tai Shih*	Marine Calanoida of the China seas
CRM 00z - *Katsushi Sakai*	Axiidae of the world (Decapoda, Thalassinidea)

Author's addresses:
Dr. F.D. Ferrari, Department of Invertebrate Zoology, MRC-534, National Museum of Natural History, Smithsonian Institution, 4210 Silver Hill Road, Suitland, Maryland 20746, U.S.A.; e-mail: ferrarif@si.edu
Dr. H.-U. Dahms, Institute of Marine Biology, National Taiwan Ocean University, 2 Pei-Ning Road, Keelung, Taiwan 202, Rep. of China; e-mail: hansd@mail.ntou.edu.tw

Manuscript first received 25 September, 2006; final version accepted 18 December, 2006.

Cover: A combination of figs. 30-32, showing adult first swimming legs in three species of calanoid copepods, on which the development of Von Vaupel Klein's Organ, a proposed synapomorphy of the Calanoida, is evident; see pp. 94-96.

Post-embryonic Development of the Copepoda

By

Frank D. Ferrari and Hans-Uwe Dahms

CRUSTACEANA MONOGRAPHS, 8

BRILL

LEIDEN • BOSTON
2007

This book is printed on acid-free paper.

Library of Congress Cataloging-in Publication data

The Library of Congress Cataloging-in-Publication Data is available from the Publisher.

ISSN: 1570-7024
ISBN-13: 978 90 04 15713 2
ISBN-10: 90 04 15713 1

PRINTED IN THE NETHERLANDS

CONTENTS

Authorities with dates of description of all species names mentioned in the main text, captions, and tables, can be found in the Appendix.

PREFACE

Copepod development has been studied for almost 250 years, and published descriptions of the exoskeletal morphology have been the dominant theme for most of that time. Almost half a century transpired between the first description of a post-embryonic developmental stage and a description that included a complete set of all developmental stages of a copepod. Interpretive questions about post-embryonic development based on an incomplete set of stages date from the early nineteenth century, but more recently researchers have discovered that interpretations are more interesting and nuanced if all developmental stages can be incorporated into the analysis.

In this monograph, we focus, whenever possible, on interpretations derived from a complete set of all developmental stages. We diagnose both the nauplius and the copepodid in order to promote such interpretations. We discuss variation in the number of naupliar stages and of copepodid stages among copepods, and also outline variation in the exoskeletal morphology of both the naupliar and copepodid phases of development. Internal anatomy is of interest to us, as are behavior and ecology, although observations from a complete set of developmental stages usually are not available on these topics. We find it interesting that at present stage-specific studies of internal anatomy during the naupliar phase of development are more complete than studies of the copepodid phase, while distributions in space and through time are more apt to be completely known for the copepodid phase of development.

We spend some time discussing variations of the order in which somites are added to the copepod body, and analysing the order segments are added to its limbs during development. One result of particular interest to us is how often the architecture of the body or configuration of the limbs are shaped by failure to form arthrodial membranes, resulting in somite or segment complexes. A model for patterning the body of copepodids has been known for almost a century, and that model is used to help us infer the somite number of the naupliar stages. Observations of limb patterning date from the early twentieth century, but models have been proposed only recently. Because the architecture of the naupliar body and the configuration of its limbs are quite different from the situation for the copepodid body, we analyse in detail the transition between these two phases of development.

Phylogenetic analyses are the coin of the contemporary realm for morphologists, and post-embryonic development has contributed to many phylogenetic hypotheses. We are pleased to discuss them. Finally, we close with suggestions for future studies that seem to us technically possible and intellectually valuable.

We dedicate this work to the memories of Patricia Dudley, late of Columbia University, and Paul Illg, late of the University of Washington, whose thinking about copepod development, and particularly how limbs are patterned, appeared to be so far ahead of their time.

We are pleased to extend our thanks to the following individuals who helped make this work possible. Lana Ong and Molly Kelly Ryan, Smithsonian Institution, prepared the illustrations. David Damkaer of Cocker Creek, Washington, answered questions historical. Mark Grygier, Lake Biwa Museum, collected nauplii of *Leptestheria kawachiensis*; Ted Durbin, University of Rhode Island, cultured developmental stages of *Calanus finmarchicus*; Wim Klein Breteler, Royal Netherlands Institute for Sea Research, cultured stages of *Temora longicornis*; Debbie Steinberg, then at the Bermuda Station for Biological Research, collected copepodids of *Euchirella messinensis*; John Fornshell, Cambridge Scientific Press, collected stages of *Longipedia americana*, *Acrocalanus gibber*, and *Derocheilocaris typica* from waters off Fort Pierce, Florida. Mashiro Dojiri, City of Los Angeles, provided a manuscript copy of his study with Gordon Hendler and Il-hoi Kim on the development of *Caribeopsyllus amphiodiae*; Pedro-Miguel Martínez Arbizu, Deutsches Zentrum für Marine Biodiversitätsforschung, provided a manuscript copy of his work on the phylogenetic position of poecilostome families.

Finally, our special thanks to J. Carel von Vaupel Klein, editor of Crustaceana Monographs, for his extraordinary patience during the preparation of this work.

FRANK D. FERRARI
HANS-UWE DAHMS

AN INTRODUCTION TO COPEPODS AND A BRIEF HISTORY OF STUDIES OF THEIR DEVELOPMENT

Copepods are a speciose group of small crustaceans often placed in the category of class within the Linnaean system. The name Copepoda is derived from the Greek words for 'oar' and 'foot', and refers to the contralateral pairs of thoracic limbs, or swimming legs, that are united by an exoskeletal coupler or interpodal bar. Action of the limb pair, united by the interpodal bar, is linked in a way similar to the sculls of a small boat. An interpodal bar unites the contralateral pairs of some thoracic limbs at least some time during the copepodid phase of development of all copepods. These interpodal bars are a synapomorphy for the Copepoda. The name Copepoda does not refer to the fact that some thoracic limbs are flattened anterioposteriorly and are paddle-like, because this limb configuration is found on many crustaceans.

Currently, more than 12,000 species of copepods are known to science, but that number probably represents less than a quarter of the extant species (Humes, 1994). Copepods vary in size significantly; early naupliar stages of copepods may be less than 0.1 mm in length while the largest adult parasite is over 30 cm. The adult copepod body is made up of a series of somites forming a cephalon, a thorax and an abdomen. Somites of the cephalon do not articulate with one another, and the first thoracic somite always is fused to the cephalon to form a cephalothorax. The remaining six thoracic somites often articulate with one another on the adult body, as do the four abdominal somites. The cephalon bears the limbs of five somites: antenna 1, antenna 2, mandible, maxilla 1 and maxilla 2. Each thoracic somite bears a limb, respectively, the maxilliped, swimming legs 1-4 and limbs 5-6. Three abdominal somites do not bear a limb but the posterior abdominal somite bears an appendage, the caudal ramus.

Copepods live anywhere there is any kind of water. Free-living copepods reside in the open fresh waters of lakes and ponds, as well as in underground aquifers. Water within the sediments of streams and rivers provides a different set of habitats for a different group of free-living copepods. Copepods also have been recovered from the water trapped in parts of terrestrial forest plants and from the surface water of terrestrial leaf-litter. Other free-living copepods successfully inhabit the open water of estuaries, as well as water on and within estuarine sediments. In marine habitats copepods are particularly numerous, and free-swimming copepods are considered the most abundant group of metazoans in open pelagic waters. Marine copepods also are found on and within ocean sediments, and there even are copepods inhabiting the tiny openings within sea-ice. Other kinds of copepods have been successful in adapting to close associations with other aquatic animals and to a lesser extent with aquatic plants. These associations include living immediately around or on a wide variety of stationary or slow-moving invertebrates, as well as living attached to faster moving vertebrates, like fishes. Copepods have been particularly successful in exploiting a wide range of relationships as both external and internal parasites. In fact, parasitism is believed to have evolved from a different free-living ancestor at least twice among copepods.

Successful adaptation to such a wide range of very different habitats has resulted in an unprecedented diversity in the body architecture and limb configuration of copepods. Development of the exoskeletal diversity of copepods is one of the major themes of this monograph, and a brief description of their morphological diversity would challenge any descriptive vocabulary. However, the Linnaean system of classification of copepods, based on differences in body architecture and limb configuration of the exoskeleton, certainly reflects this diversity in a general way. Thus, 1,985 genera and 233 families are a testament to the morphological diversity of the Copepoda.

The post-embryonic development of copepods is divided into two phases, naupliar and copepodid. Each phase, in turn, is divided into a series of stages. A stage represents the period of time during which the exoskeleton of the body does not change architecture and the exoskeleton of the limbs does not change configuration. Instead, the exoskeleton usually changes significantly during the molt between two stages so that a stage represents the period of time between two consecutive molts. Changes in the remaining organs of the body may not be restricted to molts, and some of these organs often appear to develop continuously during one or both phases.

During the naupliar phase of development, there is no direct external expression of somites because the formation of arthrodial membranes between somites is suppressed. However, the naupliar body usually does increase in size from one stage to the next, and a limb bud often may be added during a molt. The addition of a limb bud usually indicates that a somite has been added, although the correlation is not direct. Early nauplii have three well-developed limbs, antenna 1, antenna 2 and the mandible, plus the setose bud of the caudal ramus. Setose buds of some of the limbs between the mandible and caudal ramus are added during the naupliar phase of development.

During the copepodid phase of development, thoracic and abdominal somites, if present, often articulate, and the body usually increases in size and in somite number. There are up to nine well-developed, transformed appendages at the first copepodid stage: antenna 1, antenna 2, mandible, maxilla 1, maxilla 2, maxilliped, swimming legs 1-2 and the caudal ramus. The setose bud of swimming leg 3 also is present. Each remaining thoracic limb is added as a limb bud, one stage later during the copepodid phase than its somite is added to the body. Most limbs also add segment elements during the copepod phase of development. There usually is a significant change in body size and architecture during the molt between the naupliar and copepodid phases.

Published studies of copepod development represent a diverse literature and only a brief chronology of the important descriptive observations and conceptual discoveries is given here. Our understanding of the early history of descriptive works of copepods has benefited significantly from the scholarship of Damkaer (2002). Lange (1756) published the earliest description of the developmental stage of a copepod, a freshwater cyclopid, although Van Leeuwenhoek had provided observations earlier in a letter of 16 October 1699 to Antonio Megliabechi (Fryer, 1998). Lange (1756) illustrated both nauplii and copepodids, and his illustrations are also the earliest for a crustacean nauplius. De Geer (1778) confirmed Van Leeuwenhoek's observations about the size difference between early and later developmental stages of copepods. Ramdohr (1805) described the complete life history of a free-living cyclopid although more species than one were observed in his study. Jurine (1820) documented changes in the population structure of a freshwater cyclopid, "*Monoculus quadricornis*" (probably a species of *Cyclops*). Suriray (1819) illustrated a nauplius that hatched from the embryo of a transformed parasitic copepod. Burmeister (1835) described a chalimus, which Krøyer (1863) later understood to be an immature stage of a parasitic copepod. Wilson (1905) illustrated the complete development of a caligid; nauplius, copepodid, chalimus and adult stages. Dietrich (1915) described and illustrated the sac-like exopod of antenna 2 and the sac-like mandibular palp of the first copepodid of *Cyclops strenuus*; both of these structures are significantly transformed from the segmented configuration on the nauplius.

The history of concepts in copepod development begins a bit later. Von Nordmann (1832) compared the nauplius and the first copepodid of transformed parasites like *Achtheres percarum* and *Tracheliastes polycolpus* to the last naupliar stage and the first copepodid of free-living copepods, which then were known as wingless insects. He concluded that all of these species were the same kind of crustacean. Von Nordmann's (1832) study was the first to apply homologous stages of copepod development to the question of where to place these transformed parasites among a set of systematic categories. Oberg (1906) determined homologies of setae on antenna 1 of *Temora longicornis* between the last nauplius and first copepodid by studying the internal organization of the exoskeleton of the first copepodid as it developed within the last nauplius. Kraefft (1910) interpreted segmentation of the adult limb 5 of *Acartia longiremis* as it appeared from the internal organization of the limb in the fifth copepodid stage, and Lucks (1926) provided a similar interpretation for the swimming leg rami of *Cyclops viridis* [now *Megacyclops viridis*]. Birge & Judey (1908) recognized that the development of *Cyclops bicuspidatus* [now *Acanthocyclops thomasi*] could be interrupted by a long period of quiescence during which the molt of the encased fourth copepodid stage was delayed. Giesbrecht (1913) proposed that during copepodid development, one new somite is added immediately anterior to the anal somite during each molt. Illg (1949) noticed that setae that would be found on the middle segment of the rami of the swimming legs later in development of *Paranthessius columbiae* initially formed on the distal segment complex of the swimming leg rami. The arthrodial membrane that separates these setae on the middle segment from those on the distal complex is formed only later in development. Dudley (1966) figured swimming legs 1-4 of *Notodelphys affinis*, *Pygodelphys aquilonaris*, *Scolecodes huntsmani* and *Doropygus* spp., in which each seta was identified by the copepodid stage in which it first appeared. These important observations were confirmed by Kô (1969d) for the poecilostome, *Ostrincola koe*. Björnberg (1972) used data on naupliar morphology to present the first phylogeny of copepods based on developmental data. Itô (1970) proposed a model for patterning antenna 1 of *Tigriopus japonicus*, and illustrated new and renewed setae on the thoracic limbs of *Harpacticus uniremis* (see Itô, 1971). Uye & Onbé (1975) determined that the duration of the first nauplius of *Pseudodiaptomus marinus* is significantly shorter than the remaining naupliar and copepodid stages so that developmental stages of this species are not of equal duration. Izawa (1987) studied development of several parasitic poecilostomes; the naupliar phase of some species consists of fewer than six stages. Based on the number of limb buds and the number of setae on the bud of the caudal ramus, Izawa (1987) was able to determine the naupliar stages that are progressively suppressed relative to a species with a naupliar phase of six stages. Dahms (1989a) proposed a basic model for patterning antenna 1 during development of representative species from seven families of harpacticoid copepods. Ferrari & Ambler (1992) showed the relationship of the developmental age of setae and the developmental age of arthrodial membranes on swimming leg 3 of *Dioithona oculata*. Ohman et al. (2002) provided stage-specific mortality for all immature stages of *Calanus finmarchicus*.

METHODS AND CONSTRAINTS

This monograph is restricted to the post-embryonic development of the Copepoda, including the terminal adult molt. The morphology of the exoskeleton and the processes that lead to the formation of the structures of the exoskeleton are emphasized, because so much more is known about the variation in the exoskeleton during development than in any other organ of copepods. Development of internal organs, as well as changes in behavior and ecology during development, are discussed but do not receive the same level of attention. The excellent review of diapause by Williams-Howze (1997) obviates its discussion here.

Naupliar stages one through six are termed here NI through NVI; copepodid stages one through six are termed here CI through CVI. The chapters on naupliar and copepodid development are similar in organization. Both begin with a description of the developmental stages of a calanoid copepod, *Phyllodiaptomus annae* or *Ridgewayia klausruetzleri*, respectively, because calanoid limbs usually are made up of more elements, e. g., segments and setae, than comparable limbs of copepods from other orders. However, the two chapters differ in the discussion of developmental variability among the copepods. In the chapter on naupliar development, the variability of both segmentation and setation of limbs is discussed. In the chapter on copepodid development, only variability in limb segmentation is discussed.

We appreciate the power of a phylogenetic hypothesis to structure an argumentation like one about copepod development. However, copepodologists do not have such a system for all copepods or even for a significant number of them. Instead, we believe that there are other contexts in which complex concepts like development can be discussed, and choose here to discuss post-embryonic development in straightforward, comparative terms so that our arguments will proceed from description through comparison to analysis. We usually compare differences among species, and avoid comparisons of differences among genera, families or orders whenever possible. We often use familial and ordinal names of copepods to provide a framework for our comparisons, but without anticipating hypotheses about descendent relationships among species in these categories. We avoid comparing differences among higher taxonomic categories because we have serious doubts about the diagnoses of many orders, families, and genera of copepods. These doubts center on the issue of character analysis; there are only a few published studies that include careful analyses of the character states accompanying the diagnoses of orders, families, and genera (positive examples of the careful analysis of character states can be found in Park, 1995, 2000; Willen, 2000; Seifried, 2003). As a result of these doubts, our choices of ordinal taxon names may differ from contemporary lists. We accept Calanoida, Cyclopoida, Harpacticoida, Misophrioida and Siphonostomatoida without discussion. Very little is known of monstrilloid development and nothing of the development of mormonilloids and gelyelloids, so adequacy of the diagnoses of Monstrilloida, Mormonilloida and Gelyelloida does not concern us here. We believe that the polyarthrans should be removed from the Harpacticoida (see Dahms, 2004b) because neither a larval synapomorphy (Dahms, 2004b) nor an adult synapomorphy (Tiemann, 1984) has been identified for polyarthran plus oligoarthran harpacticoids. Although no formal taxon has been proposed for polyarthrans, we refer to them here by that name. We recognize the Thaumatopsylloida of Ho et al. (2003) because the bud of the fourth swimming leg and apparently the sixth thoracic somite are present at the first copepodid stage (M. Dojiri, pers. comm., e-mail 26 October 2005). This is a significant difference from the known body architecture of the first copepodid of all other copepods on which the bud of the fourth swimming leg and the sixth thoracic somite do not appear until the second copepodid stage. Because the architecture of the first copepodid varies so little among the remaining orders of copepods, and has been described as phylotypic (Ferrari, 2003), the thaumatopsylloids may have branched off early in the lineage of copepods. We do not find interesting the published opinion (Boxshall & Halsey, 2004) removing the families of Poecilostomatoida to Cyclopoida but refer to them here as poecilostomes out of deference to the unpublished analysis of P.-M. Martínez-Arbizu (in litt.). We remain uncomfortable with the Platycopioida, because no analysis exists of the character states used to diagnose this order. Synapomorphies for the Platycopiidae have been proposed, e. g., a second dorsal seta on the proximal exopodal segment of swimming legs 2-4, but whether these synapomorphies are equivalent in number or degree of transformation to those of other copepod orders remains to be determined. Here we refer to these copepods by their family name, Platycopiidae.

These are interesting times for biological nomenclature. A proposed phylocode (De Queiroz & Contino, 2001) is objective, but its unrestrained structure of dichotomies lacks organizational simplicity, and it is disruptive of historical precedent, perhaps as it must be. The Linnaean system has the inertia of history, and a limited number of categories results in an organizational simplicity. However, the Linnaean system does not require an understanding of relationships of taxa within ranks, and it lacks a methodology for ensuring that different taxa belonging to the same category have been derived in ways that are evolutionarily similar, i. e., that features which define different taxa belonging to the same category are derived by equivalent processes. We continue to rely on the Linnaean system here, despite its limitations, because an alternative system applicable to copepods has not been proposed.

A glossary is provided for some of the terms used here. Literature citations in the text are meant to exemplify, rather than exhaust, the set of species that express the structures or processes under discussion. The bibliography is divided into three sections: literature on development that is cited in the text; all remaining literature we could find in which one aspect or another of the post-embryonic development of copepods has been considered; and citations of publications that do not consider copepod development but which help to clarify certain points in the text. Among the last section are citations of publications in which limb patterning of other crustaceans has been studied; we included these because so little is known about this important process among any group of crustaceans. In the bibliography and citation sections, the titles of cited literature are complete although transliterated words and the capitalization of adjectives may vary.

We include original observations of the antenna 2 of the copepods, *Calanus finmarchicus* and *Longipedia americana*, the mystacocaridan, *Derocheilocaris typica*, and the spinicaudatan branchiopod, *Leptestheria kawachiensis*, because these help advance a diagnosis of the nauplius and constrain the naupliar phase of development of crustaceans. Naupliar development of antenna 2 of the two copepods also provides observations about how its exopod is patterned. We include original observations of swimming leg 1 of the calanoids, *Temora longicornis*, *Acrocalanus gibber*, and *Euchirella messinensis* to show how an hypothesis of limb patterning can be used to determine homologies of segments. In all of these cases, the limbs were cleared and dissected in lactic acid, stained by adding a solution of chlorazol black E dissolved in 70% ethanol / 30% fresh water to the limbs in lactic acid. Stained limbs were examined in glycerol with bright-field or differential interference optics. Drawings were made with a camera lucida and then digitized; resulting electronic files were edited and illustrations were produced from edited electronic files.

Finally, a word about the terms "appendage", "limb" and "swimming leg" as used here. As indicated in the glossary, we consider an appendage to be a paired, proximodistal extension of a somite, which pair is symmetrical about the dorsoventral axis of the somite. A limb is a paired appendage with a propodal/ramal configuration, which is found on the five cephalic somites or the seven thoracic somites. The caudal ramus is not considered a limb because it appears to lack a propodal/ramal configuration. Paired limbs on thoracic somites 2-5 are here called swimming legs 1-4 because on many copepods they function to propel the copepod through the water, usually in characteristic jumping/swimming movement. These limbs may function in ways other than swimming in some copepods, but will be termed swimming legs in all species discussed here. The paired limbs on thoracic somites 6-7 are here called limbs 5-6. An interpodal bar often is not observed between the contralateral pair of limb 5 of many copepods, and an interpodal bar has only been suggested for limb 6 of a few copepods. In many publications, limb 5 of gymnoplean copepods is called a swimming leg because its configuration on a few gymnopleans resembles swimming legs 1-4; in all species here it will be called limb 5.

THE NAUPLIUS AND NAUPLIAR DEVELOPMENT

Naupliar stages of copepods can be diagnosed by the presence of a thin, attenuate arthrite originating on the coxa of the protopod of antenna 2 (fig. 1B-F for *Calanus finmarchicus* or fig. 2A-F for *Longipedia americana*). The naupliar arthrite is an articulating segmental structure which is moved by a pair of muscles (figs. 1B, F, 2A, F). The muscle pair originates on the dorsal wall of the protopod, and appears to attach anteriorly or posteriorly to the base of the arthrite. Fahrenbach (1962) correctly understood that this structure on antenna 2 of *Diarthrodes cystoecus* was moved by muscles, but he identified it as a gnathobase. The term 'gnathobase' traditionally has been used to describe a non-articulating, ventral extension of the coxa, including the coxal endite, usually of the mandible. The gnathobase of the copepod mandible is present only in post-naupliar developmental stages of copepods, with the exception of NIV-NVI of most species of calanoids in which it is present, along with the naupliar arthrite of antenna 2. The coxal endite on the maxilla 1 of copepodids of *Euryte longicauda* also has been described as a gnathobase (Ferrari & Ivanenko, 2005).

The mandibular gnathobase of copepodids is a well-studied structure that moves food through the mouth opening. The naupliar arthrite on antenna 2 of copepods appears to have a similar function, and has been referred to as a masticatory process in some descriptive publications (e. g., Dudley, 1966; Björnberg, 1972; Dahms & Bresciani, 1993), perhaps reflecting this presumed function. The arthrite also has been mistakenly identified as a seta in many descriptive publications of copepods, although in harpacticoids its morphology is quite complex (see examples in Dahms, 1990c). However, the naupliar arthrite is not a seta, because it is moved by muscles. Nor is it a segment of the limb, because it does not appear to be located along a proximodistal axis, and because the nuclei of epidermal cells cannot be observed within the arthrite (unpubl. obs. of *Artemia salina*) as they can be in limb segments (unpubl. obs. of *Dioithona oculata*). The structure of the naupliar arthrite of copepods can be better understood from the corresponding structure on nauplii of the mystacocaridan, *Derocheilocaris typica* (fig. 3A) or the spinicaudatan branchiopod, *Leptestheria kawachiensis* (fig. 3B). The naupliar arthrite of these crustaceans is a bifurcate structure originating on a

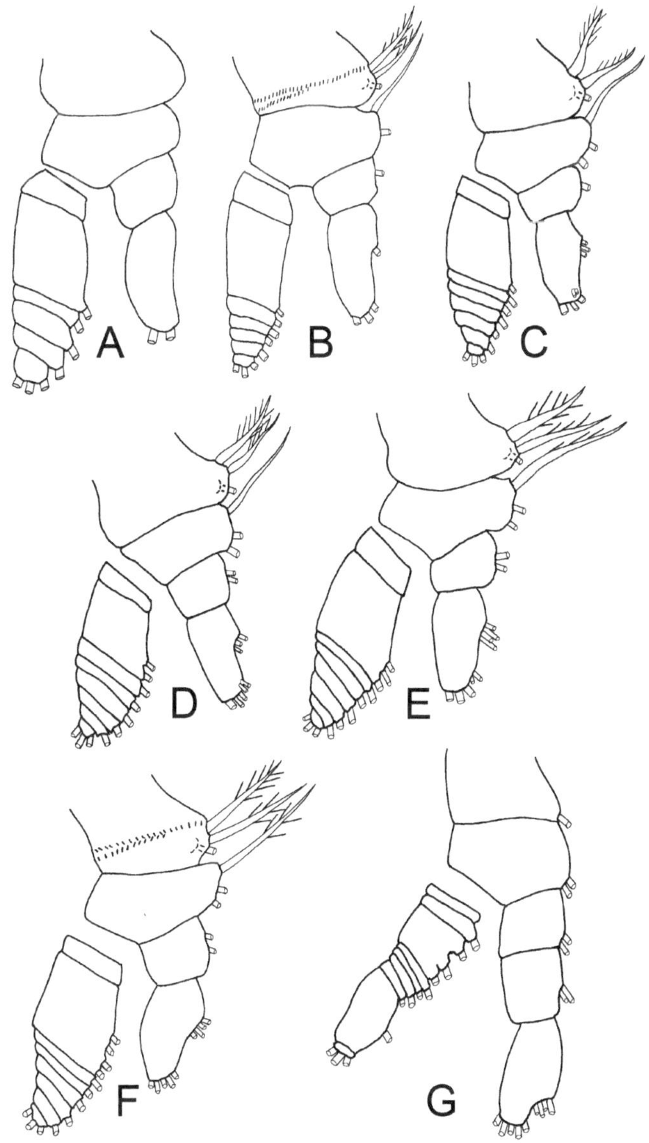

Fig. 1. Antenna 2 of *Calanus finmarchicus*. A, NI; B, NII; C, NIII; D, NIV; E, NV; F, NVI; G, CI. Exopod detached; images not to scale; broken lines within coxa of B and F indicate location of muscles terminating on naupliar arthrite.

ventral, quadrate extension of the coxa; attachment of the muscles is more easily observed in this configuration.

A naupliar arthrite may be absent from antenna 2 of species of parasitic copepods with a naupliar phase consisting of free-swimming stages without a

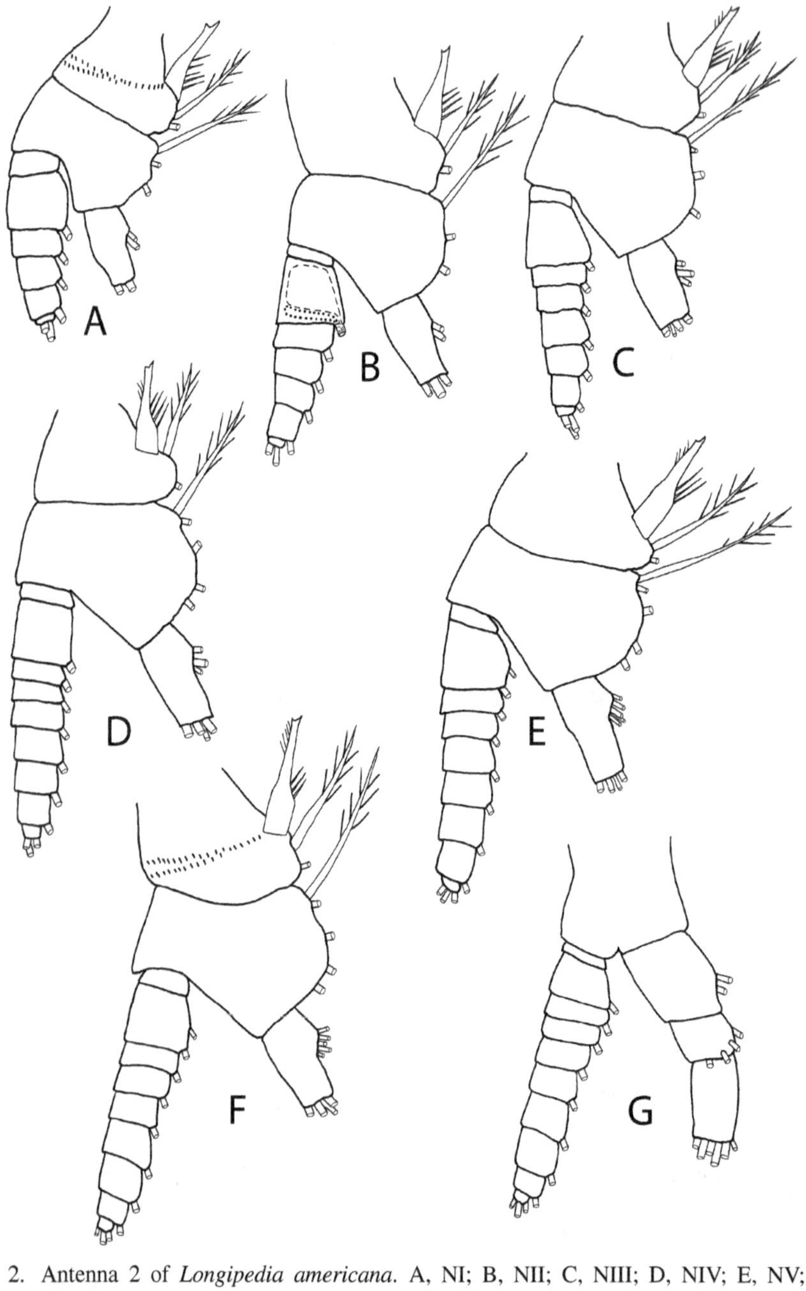

Fig. 2. Antenna 2 of *Longipedia americana*. A, NI; B, NII; C, NIII; D, NIV; E, NV; F, NVI; G, CI. Images not to scale; broken lines within coxa of A and F indicate location of muscles terminating on naupliar arthrite; dotted lines within elongate 2nd exopodal segment of B are transposed from another specimen and show the configuration of the exoskeleton of the following stage as an elongate (proximal) and a short (distal) segment, each segment with a ventral seta.

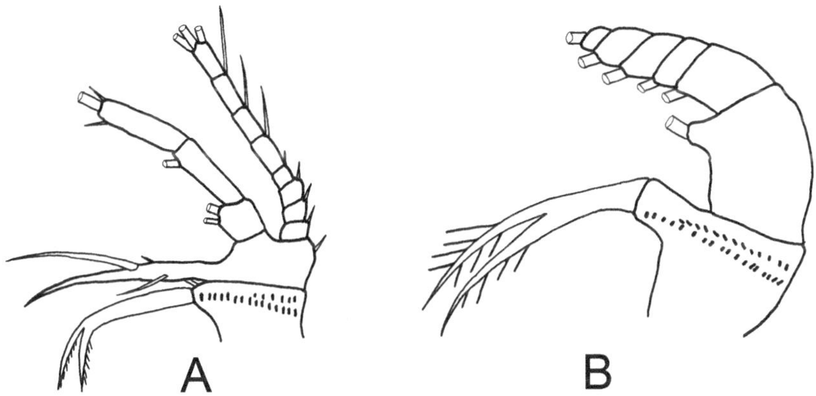

Fig. 3. A, *Derocheilocaris typica*, antenna 2 of first metanauplius; B, *Leptestheria kawachiensis*, antenna 2 of third metanauplius. Images not to scale; broken lines within coxa indicate location of muscles terminating on naupliar arthrite.

mouth opening, e. g., *Caligus elongatus* (see Piasecki, 1996) or *Scottomyzon gibberum* (see Ivanenko et al., 2001). However, a naupliar arthrite is present on the only free-living nauplius of *Monstrilla hamatapex*, which has no mouth, gut or anus, and so presumably does not feed (Grygier & Ohtsuka, 1995); the arthrite of this species may serve to attach the nauplius to its host. A naupliar arthrite also may be absent from antenna 2 of copepods whose early naupliar stages lack a mouth, e. g., NI of *Calanus finmarchicus* (fig. 1A) or NI of *Pseudodiaptomus marinus* (see Uye et al., 1983), although the arthrite is present in the naupliar stages in which a mouth is present, e. g., NII-NVI of *Calanus finmarchicus* (fig. 1B-F). Loss of the naupliar arthrite, after an initial appearance on antenna 2 of early developmental stages, usually marks the end of the naupliar phase of development. The well-developed naupliar arthrite on NI-NV of the tachidiid, *Tachidius discipes* and the harpacticid, *Harpacticus uniremis*, described as a masticatory process in these harpacticoids, is reduced at NVI (Dahms, 1990c).

For copepods, the diagnosis of a nauplius can be further refined by the following two attributes: somites of the body are not separated by arthrodial membranes during thc naupliar phase; post-mandibular appendages maxilla 1, maxilla 2, the maxilliped, swimming legs 1-2, and the caudal ramus, if present, are expressed as unarticulated limb buds throughout the naupliar phase. Thoracic somites 2-5 of the first copepodid of most copepods usually are separated by arthrodial membranes. The presence of at least some of the following transformed limbs: maxilla 1, maxilla 2, the maxilliped, swimming legs 1-2 or caudal ramus, also characterizes copepodid stages. For copepods, then, the presence of the above reconfigured limbs, the presence of thoracic somites separated by an arthrodial membrane, and the presence of an interpodal bar uniting the contralateral pair of swimming legs 1-2 correlates well with the loss of the naupliar arthrite and marks the termination of the naupliar phase of development.

An example of naupliar development of a copepod with six naupliar stages is the calanoid, *Phyllodiaptomus annae*; the following description of its development is derived from Dahms & Fernando (1993b). A calanoid copepod was chosen because the buds of maxilla 1 through swimming leg 2 are added during the naupliar phase. The nauplius of *Phyllodiaptomus annae* is broadly oval and flattened dorsoventrally; its length is about twice its width in the early stages. A red naupliar eye is located between the bases of the paired antenna 1 although the color and shape of the eye are lost soon after clearing; the eye is not figured here. Changes in the form of the body and appendages are shown in figs. 4-8.

Nauplius I (fig. 4A): body oval and elongate. Bud of caudal ramus with a spinulose seta posterioventrally. Labrum and ventral body wall well-developed, but unornamented. Antenna 1 3-segmented (fig. 5A); proximal segment with 1 spinulose seta at the distoventral edge; middle segment with 2 ventral setae of similar size, at midlength and distally; distal segment with 3 terminal spinulose setae; ventral seta largest, dorsal margin with denticles; terminal aesthetasc absent. Coxa of antenna 2 with arthrite (fig. 6A); basis with 3 ventral setae, proximal longest. Exopod 6-segmented; 1st segment unarmed, 4 following segments each with 1 distoventral seta; terminal segment with 2 setae. Endopod 1-segmented with 1 ventral seta at mid-length and 2 terminal setae. Coxa of mandible with 1 ventral seta at mid-length; basis with 2 ventral setae at mid-length (fig. 7A). Mandibular exopod 4-segmented; distal segment with 2 terminal setae and remaining segments each with 1 distoventral seta. Endopod 1-segmented with 2 terminal setae plus 2 ventral groups of setae, a proximal group of 2 setae and a group of 2 setae at mid-length.

Nauplius II (fig. 4B): differing from NI as follows: labrum with tiny spinules over proximal third; ventral body wall with 3 rows of spinules. Bud of caudal ramus asymmetrical, left slightly larger, but with a row of denticles only on right ramus; seta on right caudal ramus in dorsal position, seta on left in posterior position. Antenna 1 with 1 small aesthetasc on knob between 2 terminal setae (fig. 5B). Ventral terminal seta smaller than that of NI. Coxa of antenna 2 with a spinulose seta anterioventrally (fig. 6B); arthrite of coxa spinulose along inner edge, with long spinules at base. Distal segment of exopod with 3 terminal setae. Endopod with 2 ventral setae and a 3rd terminal

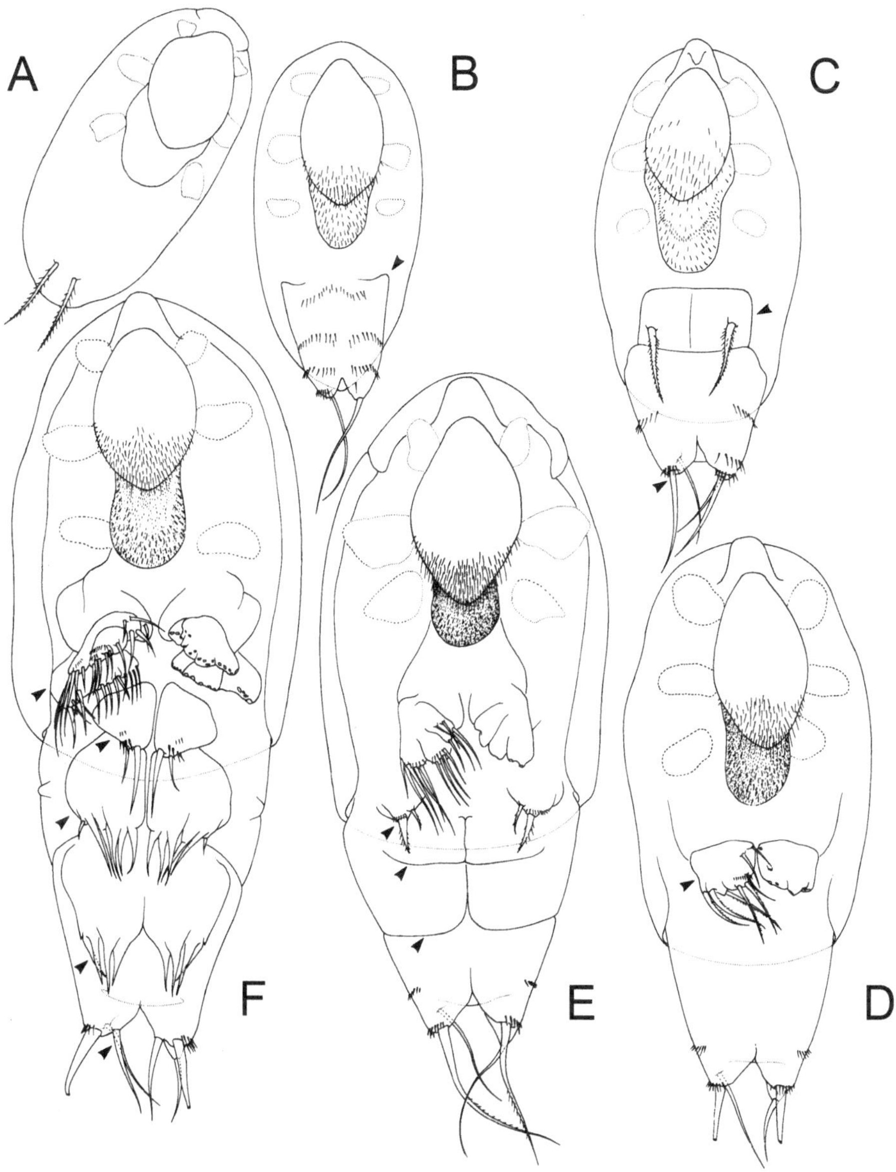

Fig. 4. *Phyllodiaptomus annae*, habitus, ventral (modified from Dahms & Fernando, 1993b). A, NI; B, NII; C, NIII; D, NIV; E, NV; F, NVI. Arrowheads to new structures.

seta. Mandible with posteriodistal seta on the basis (fig. 7B); 1-segmented endopod with proximal group of 3 setae and with 3 setae terminally. Exopod with 2 setae on proximal segment.

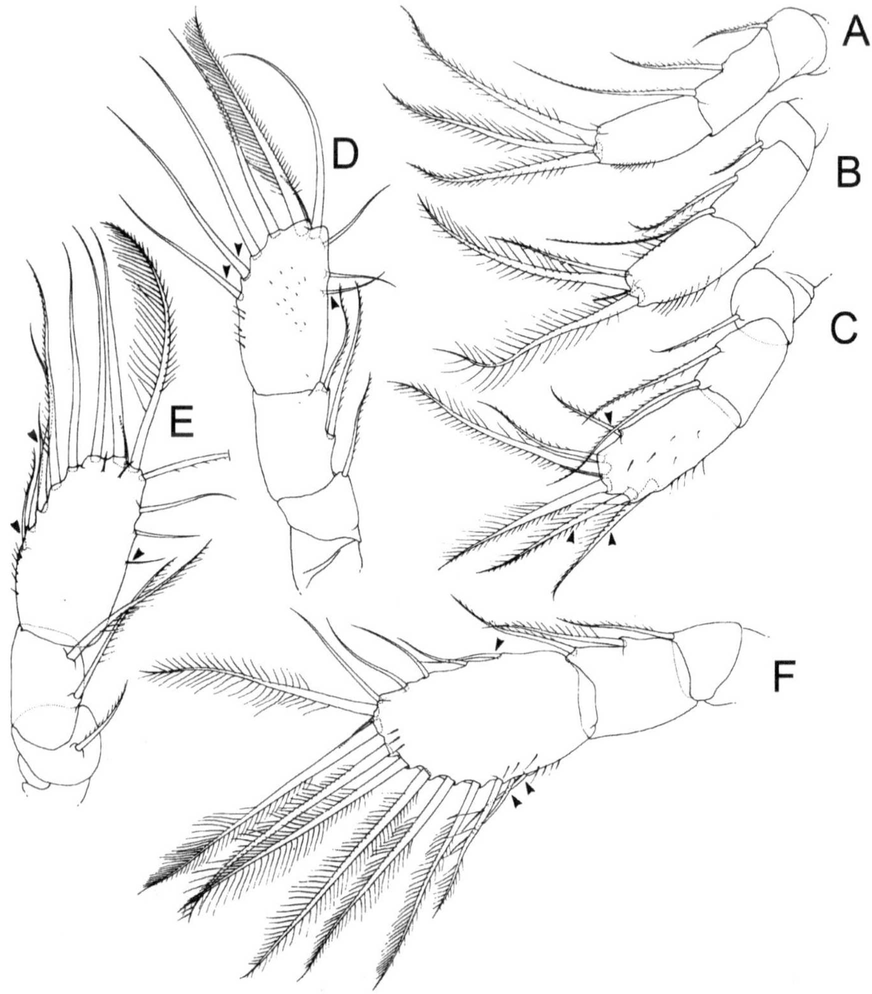

Fig. 5. *Phyllodiaptomus annae*, antenna 1 (modified from Dahms & Fernando, 1993b). A, NI; B, NII; C, NIII; D, NIV; E, NV; F, NVI. Arrowheads to new structures.

Nauplius III (fig. 4C): differing from NII as follows: anterior row of spinules on ventral body wall absent; 2nd row absent on right side; 3rd row present on left side. Caudal ramus with terminal seta added on right side and on left side; the former terminal seta on left now on dorsal lobe. Distal segment of antenna 1 with 1 short seta on distoventral margin and 2 well-sclerotized setae dorsally, as well as a few denticles (fig. 5C). Antenna 2 with 2 coxal setae distoventrally; 4 ventral setae on basis (fig. 6C). Endopod with 4 terminal setae and 3 setae in mid-ventral group; incomplete arthrodial membrane distal to mid-ventral group of 3 setae. Exopod with 3 setae on elongate (2nd

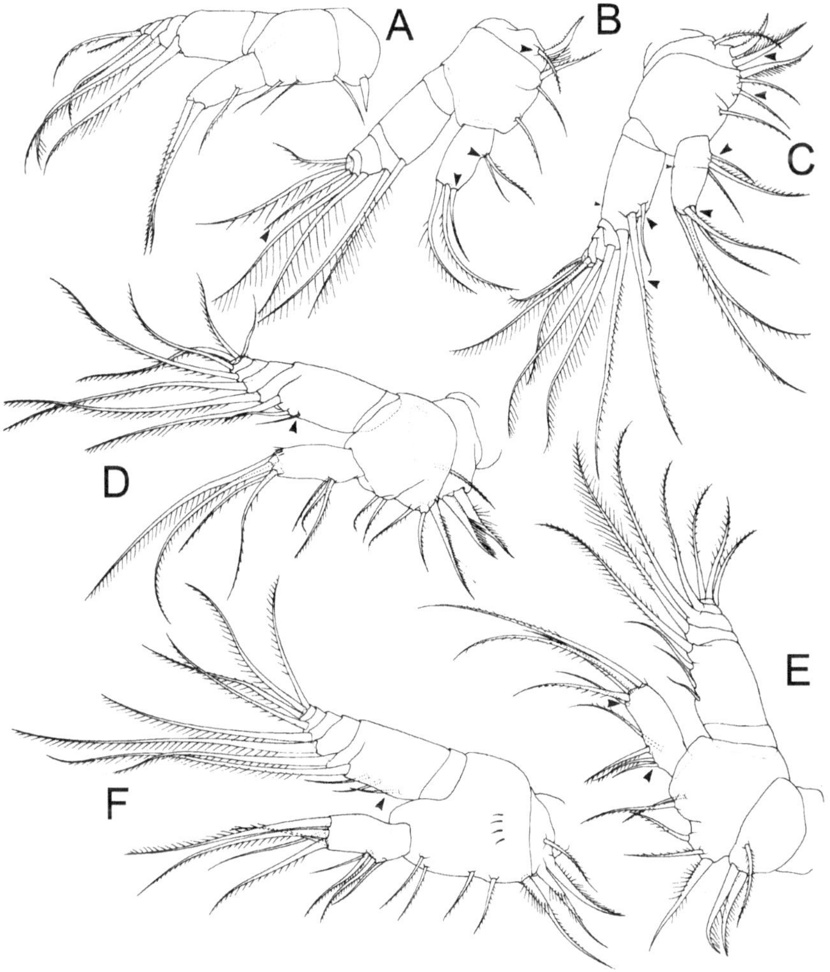

Fig. 6. *Phyllodiaptomus annae*, antenna 2 (modified from Dahms & Fernando, 1993b). A, NI; B, NII; C, NIII; D, NIV; E, NV; F, NVI. Arrowheads to new structures.

to proximal) segment; ventrally an incomplete arthrodial membrane distal to middle seta of that segment. Basis of mandible with 2 setae on posteriodistal group (fig. 7C). Unilobe bud of maxilla 1 with 1 spinulose seta (fig. 4C).

Nauplius IV (fig. 4D): differing from NIII as follows: antenna 1 with 3 new setae on distal segment, 1 on proximoventral face and 2 on proximodorsal face (fig. 5D). Exopod of antenna 2 with 1 seta added proximally to the elongate (2nd to proximal) segment (fig. 6D). Coxa of mandible bearing a large ventral gnathobase with a proximal seta, and 2 thin and 1 thick ventral attenuations; 1 distoventral seta on coxa (fig. 7D). Basis with 6 ventral setae.

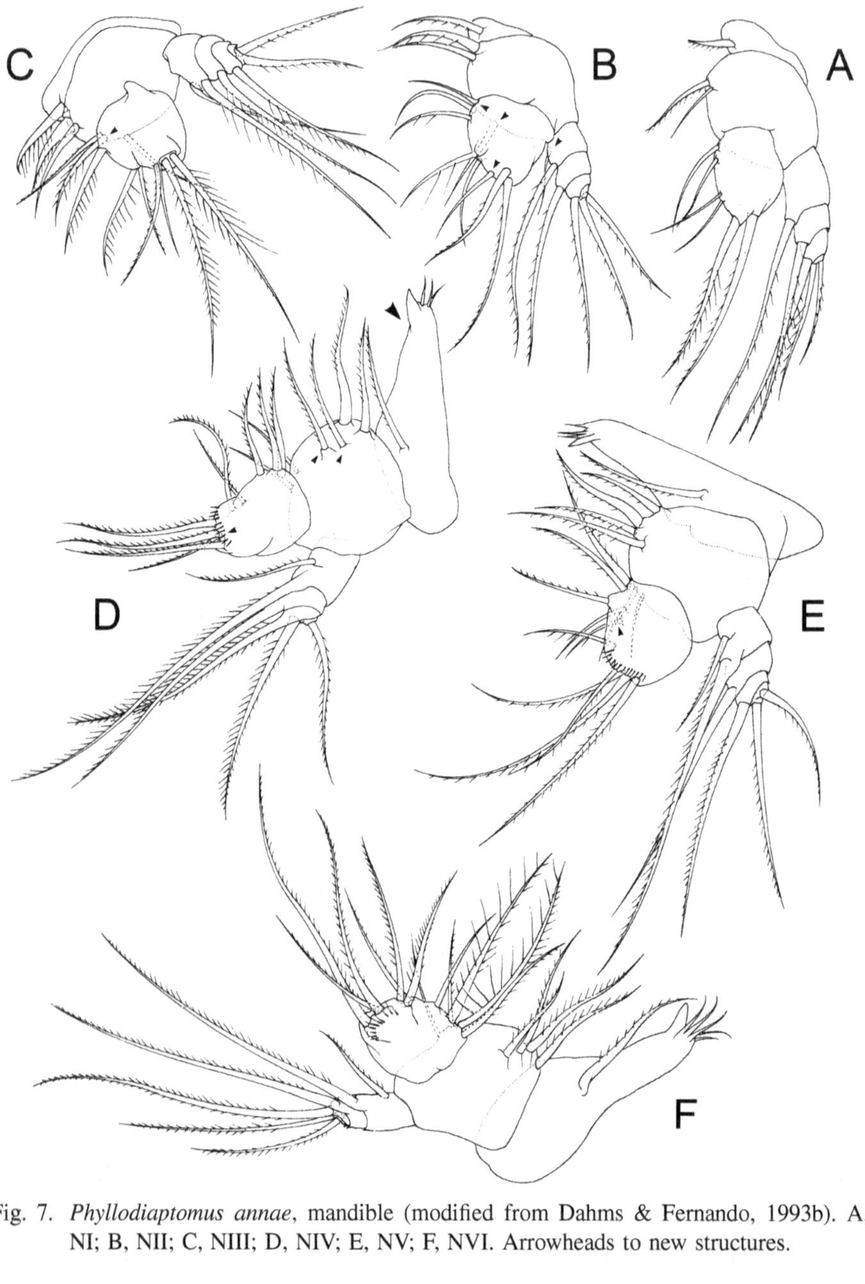

Fig. 7. *Phyllodiaptomus annae*, mandible (modified from Dahms & Fernando, 1993b). A, NI; B, NII; C, NIII; D, NIV; E, NV; F, NVI. Arrowheads to new structures.

Endopod with 4 terminal setae. Maxilla 1 multi-lobe, with 1 seta each on 3 distodorsal lobes, 3 setae and a row of denticles on the distoventral lobe, and 1 posterior and 1 ventral setae (figs. 4D, 8A).

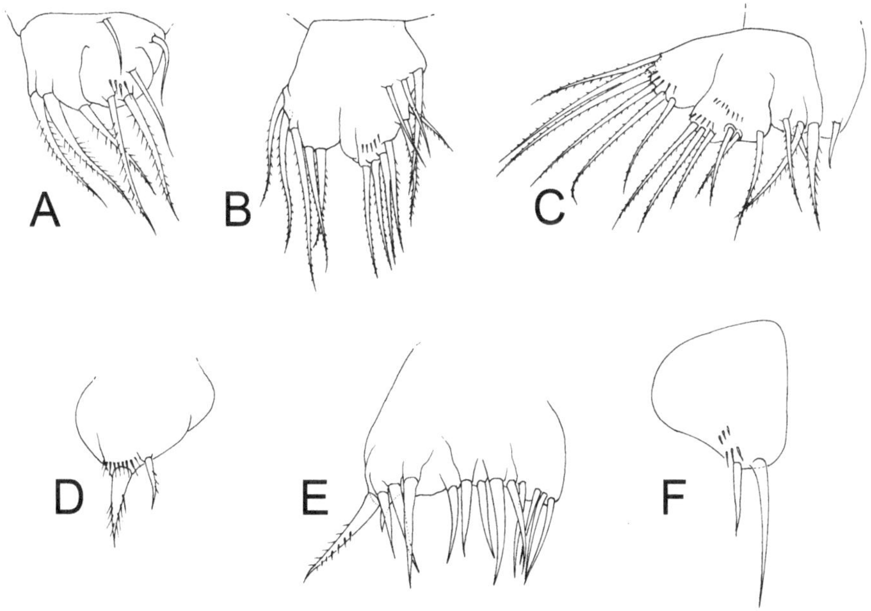

Fig. 8. *Phyllodiaptomus annae* (modified from Dahms & Fernando, 1993b) maxilla 1, A, NIV; B, NV; C, NVI; maxilla 2, D, NV; E, NVI; maxilliped, F, NVI.

Nauplius V (fig. 4E): differing from NIV as follows: antenna 1 with 3 new setae on distal segment, 1 on proximoventral face and 2 on proximodorsal face (fig. 5E). Endopod of antenna 2 with 4 setae in proximal group of setae at midlength and 1 inner seta added to terminal group (fig. 6E). Mandibular endopod with a terminal group of 5 setae (fig. 7E). Maxilla 1 with 5 setae on dorsal lobe, 4 setae on distal lobe and 5 setae ventrally (figs. 4E, 8B). Maxilla 2 a unilobe bud with 2 setae, dorsal larger and with a row of denticles at its base (figs. 4D, 8D).

Nauplius VI (fig. 4F): differing from NV as follows. Bud of caudal ramus with 3 setae (fig. 4F). Antenna 1 with 3 new setae on distal segment, 1 on proximoventral face and 2 on proximodorsal face (15 setae in total plus 1 aesthetasc) (fig. 5F). Exopod of antenna 2 with 1 new seta proximally on elongate segment, 5 setae in total (fig. 6F). Mandible with one more attenuation on the ventral face of gnathobase (fig. 7F). Maxilla 1 with 6 setae on the dorsal lobe, 6 setae on a terminal lobe, 3 setae on ventral lobe, and 1 ventral seta (fig. 8C). Maxilla 2 with 5 setae on terminal lobe and 9 setae on an indistinct series of ventral lobes with an attenuate process near the base (fig. 8E). Maxilliped a unilobe bud with 2 setae and 2 groups of 3 denticles each (fig. 8F). Swimming leg 1 a bilobe bud with 3 setae on the dorsal lobe and 3 setae on the ventral lobe (fig. 4F). Swimming leg 2 a bilobe bud with 3 setae on the dorsal lobe and 2 setae on the ventral lobe (fig. 4F).

Variation in the number of naupliar stages

Copepods develop through at most six naupliar stages; no study of copepod development has reported more than six naupliar stages. As a general rule, six naupliar stages have been reported for most free-living copepods. The number of naupliar stages in the Harpacticoida, a large order of mostly free-living species, is always six (Dahms, 1990c). Gurney (1932) reported six stages without exception for the Calanoida, which are exclusively free-living, and Elgmork & Langeland (1970) assumed six naupliar stages for all Cyclopoida. Izawa (1987) suggested at most six stages for poecilostomes, which are often associated with or parasitic on other invertebrates, and Itoh & Nishida (1997) reported six naupliar stages for the primitive poecilostome, *Hemicyclops japonicus*.

Among calanoids, reports of fewer than six naupliar stages are rare and may result from errors in sampling field-collected populations. For example, only 5 naupliar stages were reported for *Pseudodiaptomus euryhalinus* (see Johnson, 1948), *P. coronatus* (see Jacobs, 1961; Grice, 1969), *P. aurivilli* (see Ummerkutty, 1964), *P. ardjuna* (see Alvarez & Kewalramani, 1970), *P. acutus* (see Björnberg, 1972; Fanta, 1982), *P. richardi* (see Cicchino, 1975), and *P. binghami* (see Goswami, 1978a). In general, the NI of calanoids has a caudal ramus of symmetrical buds, each of which bears only 1 seta. For those species of *Pseudodiaptomus* that were reported with five naupliar stages, NI with the above configuration of the caudal ramus was the stage consistently absent and assumed not to be expressed among these species of *Pseudodiaptomus*. However, careful culturing of *Pseudodiaptomus marinus* (see Uye & Onbé, 1975) revealed a duration time of only a few minutes for its first naupliar stage, which had a caudal ramus of symmetrical buds with only 1 seta. Duration times for the remaining naupliar stages, NII-NVI were 1-3 days. It seems likely, then, that all of the above species of *Pseudodiaptomus* have six naupliar stages, but because the duration of NI is significantly less than that of the remaining stages, NI was not observed. If there are calanoids with fewer than six naupliar stages, they most likely may be found among those species living closely associated with the benthos. For example, Matthews (1964) cultured from eggs only four naupliar stages of the benthopelagic calanoid, *Chiridius armatus*.

Five naupliar stages have been reported for some free-living cyclopoids of the family Cyclopidae, e. g., *Cyclops strenuus* by Dietrich (1915), *Megacyclops viridis* [as *Cyclops viridis*], by Lucks (1926), *Speocyclops racovitzai* by Lescher-Moutoué (1966), *Eucyclops serrulatus* by Auvray & Dussart (1966) and *Ectocyclops rubescens* by Carvalho (1971). However, six nauplii are known for other free-living Cyclopidae, e. g., *Halicyclops neglectus* by Candeias (1966), *Apocyclops royi* by Alvarez-Valderhaug & Kewalramani (1979) or Chang & Lei (1993), *Bryocyclops caroli* by Bjornberg (1984), *Macrocyclops albidus* by Defaye (1984), *Mesocyclops aequatorialis* by Dahms & Fernando (1992), *Thermocyclops consimilis* by Dahms & Fernando (1992), *Paracyclops fimbriatus* by Karaytug & Boxshall (1996), as well as for other free-living cyclopoids like *Cyclopina longifera* [probably *Cyclopina longifurca* Sewell, 1924, now *Paracyclopina longifurca* see Smirnov (1935)] by Goswami (1977a), *Cyclopina schneideri* by Grainger & Mohammed (1991), *Dioithona oculata* by Ferrari & Ambler (1992). The issue of the number of naupliar stages for Cyclopidae has been discussed by Elgmork & Langeland (1970), who summarized the literature and identified the primary difficulty as distinguishing NIV and NV. Elgmork & Langeland (1970) used the pattern and shape of setules on setae of the caudal ramus to separate NIV and NV of *Cyclops scutifer*. However, these two stages also differ in the number of setae on the bud of maxilla 1, which increases during the molt from NIV to NV of many free-living cyclopoids, and this attribute also can be used to separate these two naupliar stages.

Among commensal and parasitic poecilostomes, six nauplii are known for species of Lichomolgidae, e. g., *Lichomolgus canui* by Costanzo (1969) or *Zygomolgus poucheti* by Calafiore & Costanzo (1992), and species of Sabelliphilidae, e. g., *Modiolicola insignis* by Costanzo (1984) or *Herrmannella rostrata* by Costanzo & Calafiore (1985). However, fewer than six naupliar stages have been reported by Izawa (1973, 1975a, 1986b) for cultured poecilostomes like *Sarcotaces pacificus*, *Colobomatus pupa*, *Anchistrotos pleuronichthydis*, *Neanthessius renicolis* and *Pseudacanthocanthopsis apogonis*. An important correlation between egg size and the number of naupliar stages was noted (Izawa, 1987) for these poecilostomes. Species whose eggs are usually less than 120 μm, and have little stored lipid material, pass through six naupliar stages. Larger-sized eggs have greater amounts of stored lipids, and the species with larger-sized eggs have a lower number of naupliar stages. More importantly for these poecilostomes, the particular naupliar stages that were not expressed could be formalized (Izawa, 1987) in the following way, relative to development of a poecilostome with a six-stage naupliar phase (table I): if five stages are expressed, then NII of a poecilostome with a six-stage phase is suppressed; if four stages are expressed, NII and NIII are suppressed; if three stages are expressed, NII-NIV are not present; if only two stages were present, then NII-NV are suppressed. This last category successfully predicts the situation of another poecilostome with two nauplii, the gastrodelphyid, *Sabellacheres illgi* (cf. Dudley, 1964), so that this developmental formula (Izawa, 1987) has predictive power beyond the species studied. As a general analytical procedure, if fewer than six nauplii are present, identification of the missing naupliar stage or stages usually depends on finding differences in the exoskeleton corresponding to two non-consecutive stages.

Not all parasitic copepods with fewer than six naupliar stages follow the above formula for missing naupliar stages. Cyclopoids and siphonostomatoids provide several exceptions. As noted previously, consecutive naupliar stages usually differ in the morphology of the exoskeleton. However, differences in the exoskeleton may not always be expressed between consecutive naupliar stages of cyclopoids and siphonostomatoids. For example, Dudley (1966) cultured the notodelphyid cyclopoids, *Pygodelphys aquilonaris*, *Notodelphys affinis* and *Doropygus seclusus* through five naupliar stages. Two of these stages correspond to particular stages of a six-stage phase (for examples of a six-stage phase, see Dahms & Fernando, 1992; Ferrari & Ambler, 1992). Dudley's (1966) first nauplius, without a unisetose bud of maxilla 1 and with one seta on the bud of the caudal ramus, appears to correspond to NI, and her fifth nauplius, with setose bilobe buds of swimming legs 1-2, appears to correspond to NVI. The second, third and fourth naupliar stages described by Dudley all appear identical, with a unisetose bud of maxilla 1 and six setae on the bud of the caudal ramus. These nauplii do not appear to correspond to any stage of a cyclopoid with a six-stage naupliar phase; among those cyclopoids, a bud of maxilla 1 with one seta appears only in

TABLE I

Stages missing from a naupliar phase of six stages when only 5, 4, 3, or 2 naupliar stages are present

NI	NII	NIII	NIV	NV	NVI
	5				
	4	4			
	3	3	3		
	2	2	2	2	

NII or NIII, but the caudal ramus of these stages bears one seta at NII or at most three setae at NIII.

There are no siphonostomatoid copepods from field capture or from cultures for which six naupliar stages have been reported. The largest number, four naupliar stages, was observed for *Scottomyzon gibberum* cultured outside of its starfish host (Ivanenko et al., 2001). *Scottomyzon gibberum* is a less derived species having branched off early during the evolution of siphonostomatoids (V. N. Ivanenko, pers. comm.). The first and second nauplii of *S. gibberum* do not differ in segmentation or setation of their limbs (Ivanenko et al., 2001). However, the second nauplius does have tubercles on antenna 2 and the mandible; these tubercles are absent from the first nauplius.

Among caligid fish parasites, the two naupliar stages of *Caligus spinosus* were separated by setation of antenna 1 (Izawa, 1969). The second of two naupliar stages of *C. elongatus* was separated by the presence of a ventral process that was considered homologous to the basis of the maxilliped of CI (Piaseki, 1996). Both nauplii of *C. spinosus* and *C. elongatus* were similar to NI of copepods with six naupliar stages, i. e., without a unisetose bud of maxilla 1 and with one seta on the bud of the caudal ramus.

Some species of caligids have two naupliar stages that are identical in the morphology of their exoskeleton. Molting between these two, identical naupliar stages was observed for *Caligus centrodonti* by Gurney (1934a) and for *C. clemensi* by Kabata (1972). Similar observations of two identical nauplii are reported for the related *Lepeophtheirus salmonis* by Johnson & Albright (1991) and for *L. dissimulatus* by Lewis (1963). The attribute of two identical naupliar stages for caligids with only two naupliar stages is also shared with the thaumatopsylloid, *Caribeopsyllus amphiodiae* (cf. M. Dojiri, pers. comm., e-mail 26 October 2005). Nauplii of *C. amphiodiae* were cultured outside of their brittle star host; the 2nd and 3rd of three nauplii differed in size but not in appendage morphology.

It should be noted, however, that the culturing protocol for caligid-like parasitic copepods often includes detaching from a female her embryo sac with its embryos, and maintaining the embryo sac separately until nauplii have hatched. Chemical communication from the female to the embryos in the sac has not been investigated directly, or even indirectly by observing embryo release from sacs detached from females for differing periods of time. An interruption of maternal communication may affect the number of expressed naupliar stages.

No free-living naupliar stage has been observed for a number of siphonostomatoids belonging to the families Pennellidae (see Bennet, 1962; Perkins,

1983; Izawa, 1997) and Nicothoidae (see Gurney, 1930a; Heron & Damkaer, 1986), and CI is assumed to be the stage released from the embryo sac of these species. Perkins (1983) inferred that the pennellid, *Cardiodectes medusaeus* had no naupliar stage of development because only copepodids were observed in the embryo sac. Izawa (1997) was unable to hatch a nauplius from cultured eggs of *Peniculisa shiinoi*. Heron & Damkaer (1986) concluded that the nicothoid, *Hansenulus trebax* hatches as a copepodid rather than as a nauplius, because embryos, nauplii and copepodids were observed in the embryo sac, but copepodids were the only free-living stage observed on the host. Swimming legs 1-2 of these copepodids have 1-segmented rami, and this configuration for these limbs is indicative of the first copepodid stage of copepods.

The only investigation of a presumably free-living species of Misophrioida is that by Gurney (1933a) of *Misophria pallida*. He reported a naupliar stage that hatched from an embryo in an embryo sac carried by a female. The nauplius, presumably NI, molted directly to the first copepodid.

Variation in the order of appearance of limb buds

Maxilla 1 is present at NII as a unilobe bud with one seta in all polyarthrans and in some harpacticoids, cyclopoids and poecilostomes (e. g., *Harpacticus uniremis* (see Itô, 1971), *Paraleptastacus brevicaudatus* (see Dahms, 1990a), *Parastenhelia megarostrum* (see Dahms & Hicks, 1996), *Dioithona oculata* (see Ferrari & Ambler, 1992), *Hemicyclops japonicus* (see Itoh & Nishida, 1997), *Taeniacanthus lagocephali* (see Izawa, 1986a), *Tegobomolochus nasicola* (see Izawa, 1986b) and *Panaietis yamagutii* (see Izawa, 1986b)). The bud of maxilla 1 of these copepods remains unchanged with one seta at NIII; it is multi-lobe and bears more than one seta from NIV to NVI. However, in other harpacticoids, e. g., *Paramphiascella fulvofasciata* (see Rosenfield & Coull, 1974), *Amphiascus undosus* (see McMillan, 1991) or *Stenhelia palustris* (see Dahms & Bresciani, 1993), the unilobe bud of maxilla 1 with one seta does not appear until NIV; this is also the case for the poecilostomes, *Oncaea media* (see Malt, 1982), *Neoergasilus japonicus* (see Urawa et al., 1980a) and *Pseudomyicola ostreae* (see Nakamura et al., 1979). Do et al. (1984) report a unilobe bud of maxilla 1 with one seta on NV of *Pseudomyicola spinosus*. For the cyclopoids, *Apocyclops dengizicus* (see Alvarez-Valderhaug & Kewalramani, 1979) and *A. royi* (see Chang & Lei, 1993), the seta on the bud of maxilla 1 initially is present at NIII. Maxilla 1 does not appear to be present as a simple, setose bud during the early naupliar phase of siphonostomatoid development (Ivanenko et al., 2001). Instead, this limb first appears as a complex lobe along with swimming legs 1-2 during the last naupliar stage, and this last nauplius appears to correspond to NVI of a copepod with a six-stage naupliar phase. For all known calanoids, maxilla 1, as a unilobe bud with one seta, appears at NIII. At NIV, maxilla 1 is multi-lobe and bears more than one seta; a conformation similar to the same stage of cyclopoids and harpacticoids.

Maxilla 2 does not appear as a setose bud during the naupliar phase of species of polyarthrans, harpacticoids, cyclopoids, poecilostomes, or siphonostomatoids. Its appearance at NV is unique to calanoids. Nor does a unilobe setose bud of the maxilliped appear during the naupliar phase of species of polyarthrans, harpacticoids, cyclopoids, poecilostomes, or siphonostomatoids. Its appearance with two terminal setae at NVI also is unique to calanoids.

The bilobe setose buds of swimming legs 1-2 first appear at NVI of copepods with six naupliar stages. The appearance of these limb buds at this stage is considered to be conserved among copepods (Izawa, 1987), and the presence of a stage with these limb buds was used by that author to identify the last naupliar stage of any poecilostome, and thus to align the naupliar stages of poecilostome copepods with five or fewer naupliar stages (Izawa, 1987), as explained above. The thaumatopsylloid, *Caribeopsyllus amphiodiae* is one of only a few copepods that does not express the buds of swimming legs 1-2 during the naupliar phase of development (M. Dojiri, pers. comm., e-mail 26 October 2005).

The data of first appearances, as summarized in table II, can be applied with some predictive power to nauplii of copepods, particularly calanoids, for which fewer than six stages were observed. For example, among the four naupliar stages of *Chiridius armatus* reported by Matthews (1964), there is a stage with buds of swimming legs 1 and 2; this stage probably corresponds to NVI. Two earlier stages lack the setose bud of maxilla 1 and most likely correspond to NI and NII. Maxilla 1 of the remaining stage is a multilobe

TABLE II

First appearance of setose buds of copepod limbs during the naupliar phase of development

Maxilla 1:	NII [harpacticoids and cyclopoids]; NIII [calanoids]; NVI [siphonostomatoids]
Maxilla 2:	NV [calanoids]
Maxilliped:	NVI [calanoids]
Swimming leg 1:	NVI [calanoids, harpacticoids, cyclopoids, poecilostomes, siphonostomatoids]
Swimming leg 2:	NVI [calanoids, harpacticoids, cyclopoids, poecilostomes, siphonostomatoids]

bud with 2 setae. This kind of limb bud for maxilla 1 usually is found in NIV and NV of calanoids. However, the bud of maxilla 2, usually present on NV and NVI, is not present on the stage of *Chiridius armatus* in question, so this stage most likely corresponds to NIV.

Differences among the six naupliar stages of copepods include changes in the number of limb buds, in the number of segments of limb rami, or in the setation of limb segments including the caudal ramus. A key cannot be written to identify unequivocally the naupliar stages of all copepods, due to the degree of this variation and to the absence of one or more stages during the naupliar phase of development. However, the key below is useful for copepods in which all six naupliar stages are present, and particularly for free-living animals:

Key to Copepod Nauplii I-VI

Three transformed limbs, bud of caudal ramus with 1 pair of setae .... NI
Bud of maxilla 1 a simple lobe with 1 seta or posterior part of body distinctly narrower than anterior part .... NII
Bud of caudal ramus with more than 1 pair of setae .... NIII
Mandibular gnathobase present and/or bud of maxilla 1 multi-lobe with no more than 6 setae .... NIV
Bud of maxilla 2 present or bud of maxilla 1 multi-lobe with at least 7 setae .... NV
Bud of swimming legs 1-2 present .... NVI

Variation in transformed appendages

During the naupliar phase of development, antenna 1 of copepods usually is 3-segmented at NI with the proximal segment unarmed. However, among the polyarthrans, antenna 1 of NI may be 6-segmented in *Longipedia minor*, or 5-segmented on NII-NVI of *L. minor* and 5-segmented on NI-NVI of *Canuella perplexa* (see Dahms, 1990c). The unarmed, proximal segment may be missing in harpacticoids, e. g., *Parategastes sphaericus*, *Pseudotachidius* sp. (see Dahms, 1990c) or the arthrodial membrane between the proximal and middle segment may be poorly formed in some poecilostomes, e. g., *Pseudacanthocanthopsis apogonis* (see Izawa, 1986b). The arthrodial membrane between the middle and distal segments of some harpacticoids may also be poorly formed, e. g., *Parathalestris harpactoides* (see Dahms, 1990c). Setae may be added to the distal segment of antenna 1 of many copepods from NIII to NVI, although the addition of setae in sets of one proximoventral and two proximodorsal setae is limited to calanoids and to polyarthrans like *Longipedia americana* (see Onbé, 1984). Examples of the addition of different sets of ventral setae to the distal segment of antenna 1 have been observed for harpacticoids like *Stenhelia palustris* (see Dahms & Bresciani, 1993) and *Drescheriella glacialis* (see Dahms, 1987a), or for cyclopoids like *Mesocyclops edax* (see Dahms & Fernando, 1995) and *Dioithona oculata* (see Ferrari & Ambler, 1992). Setal additions to the distal segment of antenna 1 of poecilostomes have been generalized by Izawa (1987, figs. 4, 5). The notodelphyid cyclopoids lose the proximoventral seta and the mid-ventral seta of the middle segment at NIII, e. g., *Scolecodes huntsmani*, at NIV, e. g., *Doropygopsis longicauda*, and at NV, e. g., *Pygodelphys aquilonaris* (see Dudley, 1966).

There are many examples of the addition of setae to the protopod of antenna 2. Of particular interest is the addition of the seta distal and adjacent to the naupliar arthrite. This seta has been reported at NI of *Paraleptastacus brevicaudatus* by Dahms (1990a) and *Dioithona oculata* by Ferrari & Ambler (1992), or first observed at NII for *Macrocyclops fuscus* by Dahms & Fernando (1994) and *Mesocyclops edax* by Dahms & Fernando (1995), or presented at NIII of *Longipedia americana* by Onbé (1984), *Mesocyclops* cf. *thermocyclopoides* by Dahms & Fernando (1993a), *M. aequatorialis similis* by Dahms & Fernando (1992), *Thermocyclops consimilis* by Dahms & Fernando (1992), and *T. decipiens* by Dahms & Fernando (1993a). The addition of ventral setae to the middle and distal setal groups of the endopod of antenna 2 has been reported for species of calanoids, harpacticoids, cyclopoids and poecilostomes, and this addition of setae to the endopod may alternate between the middle and distal setal groups from one stage to the next. Siphonostomatoids do not add setae to the endopod of antenna 2.

The addition of complete arthrodial membranes to the exopod of antenna 2 occurs only in calanoids, e. g., *Calanus finmarchicus* (fig. 1), polyarthrans, e. g., *Longipedia americana* (fig. 2), and some cyclopoids, e. g., *Mesocyclops edax* (see Dahms & Fernando, 1995) or *M. leuckarti* (see Dahms & Fernando, 1993c). The addition of arthrodial membranes to this ramus is discussed in detail in the chapter "Patterning the appendages of copepods" (see below). For all other copepods, there is no change in the number of arthrodial membranes in the exopod of antenna 2 throughout the naupliar phase of development.

Setae may be added to the exopod of antenna 2 during the naupliar phase; for details see p. 74 ff. One example is the addition of a third seta to the crown group of setae on the distal segment. The addition of this seta occurs at different stages: NII of *Tigriopus japonicus* (see Itô, 1970); NIII of

Scutellidium hippolytes (see Dahms, 1993b), *Stenhelia palustris* (see Dahms & Bresciani, 1993), *Longipedia americana* (see fig. 2), *Dioithona oculata* (see Ferrari & Ambler, 1992), or *Neanthessius renicolis* (see Izawa, 1986b, in the 2nd of 5 nauplii = NIII); NIV of *Paraleptastacus brevicaudatus* (see Dahms, 1990c) or *Tegobomolochus nasicola* (see Izawa, 1986b, in the 3rd of 5 nauplii = NIV); NV of *Tisbe gracilis* (see Dahms & Bergmans, 1988); NVI of *Parastenhelia megarostrum* (see Dahms & Hicks, 1996). These observations suggest that if three crown setae are present on the terminal segment of the exopod of a copepod with fewer than six naupliar stages, like *Notodelphys affinis* or *Pygodelphys aquilonaris* (see Dudley, 1966), a missing stage may be NI, because that stage is expected to bear only two crown setae.

One seta or more may also be added proximally and ventrally to the proximal segment of the exopod of antenna 2 of cyclopoids, e. g., *Mesocyclops edax* (see Dahms & Fernando, 1995), *M. aequatorialis similis* (see Dahms & Fernando, 1992), *Thermocyclops consimilis* (see Dahms & Fernando, 1992), and *Dioithona oculata* (see Ferrari & Ambler, 1992), or poecilostomes like *Taeniacanthus lagocephali* (see Izawa, 1986a), *Philoblenna arabica* (see Izawa, 1986b), or *Doridicola sepiae* (see Izawa, 1986b). These setal additions appear to follow the pattern of setal additions expressed by calanoids (see further below).

Species of calanoids are the only copepods for which a gnathobase is present on the mandibular coxa during the naupliar phase of development. The mandibular gnathobase initially is presented at NIV. In all other copepods, there is no coxal gnathobase on the mandible of any naupliar stage. The addition of setae to the mandible may include a second, proximoventral seta to the proximal exopodal segment, a situation that appears to be similar to the addition of a seta to the proximal exopodal segment of antenna 2. Stage-specific variation of the addition of this seta includes NII of *Longipedia americana* (see fig. 2), *Scutellidium hippolytes* (see Dahms, 1993a) and *Hemicyclops japonicus* (see Itoh & Nishida, 1997); on the 2nd of 5 nauplii = NIII of *Panaietis yamagutii* (see Izawa, 1986b); on the 2nd of 3 nauplii = NIV of *Philoblenna arabica* (see Izawa, 1986b); and on the 2nd of 2 nauplii = NVI of *Doridicola sepiae* (see Izawa, 1986b).

Variation in setation of limb buds

The bud of maxilla 1 is unilobe, and bears a single seta at NII and NIII in many species of harpacticoids, cyclopoids, and poecilostomes, and at NIII in species of calanoids. Setae are added to the bud of maxilla 1 of most of these copepods beginning with the molt to NIV when the bud of maxilla 1 is multi-lobe. The bud of swimming leg 1, presented at NVI, may bear up to four setae on its presumptive exopod and up to three setae on its presumptive endopod (Ferrari, 2000). In contrast, the bud of swimming leg 2, also presented at NVI, may only bear up to three setae on its presumptive exopod, and only up to two setae on its presumptive endopod. Furthermore, significant variation in the numbers of setae on these limb buds has been documented (Ferrari, 2000). Setae are added to the bud of the caudal ramus of most species of copepods beginning at NII; however, some species of siphonostomatoids do not add setae to the bud of this appendage (Ivanenko et al., 2001).

Internal anatomy

Relative to the naupliar exoskeleton, much less is known about the internal anatomy of the nauplius, because there are fewer observations of internal development and because many of the reports do not include observations of all naupliar stages (e. g., Claus, 1858b, 1863; Grobben, 1881). Exceptions are copepods for which only one naupliar stage is known (Nordmann, 1832, 1864; Claus, 1858a, 1861), and the works of Fanta (1973, 1976, 1982), who described aspects of the internal anatomy for all stages of three copepods with a naupliar phase of six stages, *Pseudodiaptomus acutus*, *Euterpina acutifrons* and *Oithona ovalis*, representing calanoids, harpacticoids, and cyclopoids, respectively.

In general, a cuticle-lined esophagus runs anteriorly and dorsally from the mouth before turning posteriorly to end in a cone-like protrusion into the midgut. The midgut, with glandular cells, is covered by smooth, longitudinal muscles and is divided by a valve into a spherical anterior part and a cylindrical posterior part. Cellular architecture changes between the anterior and posterior parts of the midgut. The cuticle-lined hindgut is well-muscled and ends at the posterior anus. The mouth is open at NI of *Diarthrodes cystoecus* (see Fahrenbach, 1962), *Euterpina acutifrons* (see Fanta, 1972), *Oithona ovalis* (see Fanta, 1976) and *O. davisae* (see Uchima & Hirano, 1986), NII of *Pseudodiaptomus acutus* (see Fanta, 1982) and NIII of *Calanus finmarchicus* (see Lowe, 1935). Although the mouth of *Doropygus seclusus* is open at NI, the stomodaeum, or esophagus, does not open into the midgut until NIII (Dudley, 1966); the posterior end of the proctodaeum, or hindgut, is not open to the anus until NII (Dudley, 1966). A dorsal diverticulum of the midgut forms at NIII of *Euterpina acutifrons* (see Fanta, 1972). The tubular foregut of *Lernaea cyprinacea* develops at NII but does not become funnel-shaped until NIII, during which stage the midgut begins to form (Benedetti et al., 1992). NIII is the last naupliar stage reported for *L. cyprinacea* after which a complete gut is present at the first copepodid. Wax esters are the main lipid component of all nauplii of *Euchaeta japonica* (see Lee et al., 1974) but were not detected in nauplii of *Calanus helgolandicus* (see Lee et al., 1972).

The antennary gland, an excretory gland of the nauplius that is functionally comparable to the maxillary gland of copepodids, has been described by Fahrenbach (1962: 349) as "dorsal lateral to the basis" with the excretory pore opening "on the posterior side of the antenna at the level of the exopod" for *Diarthrodes cystoecus*. Excretions are stored as "urinary concretions" within the body of *Oithona ovalis* (see Fanta, 1976) and *Pseudodiaptomus acutus* (see Fanta, 1982). No labrum or labral glands form in notodelphyids during the naupliar phase of development (Dudley, 1966).

The naupliar nervous system includes a large dorsal "brain" and a pair of thick circumesophageal connectives that unite ventrally as a subesophageal ganglion. From this ganglion, paired ventral nerves emerge, continuing to the posterior end of the body. Anteriorly, neither protocerebrum, deutero-cerebrum or tritocerebrum can be distinguished in *Diarthrodes cystoecus* (see Fahrenbach, 1962) but a protocerebrum that is divided into two lobes has been observed in *Euterpina acutifrons* (see Fanta, 1972) and *Oithona ovalis* (see Fanta, 1976). Other changes in the nervous system during naupliar development include: appearance of the ganglion of antenna 1 and antenna 2; regression of the glomeruli of antenna 2 (these glomeruli are not reported in the first copepodid); thickening anteriorly of paired ventral nerves of *Diarthrodes cystoecus* (see Fahrenbach, 1962), which progressively unite posteriorly. A study of the innervation of setae of the appendages at NV of *Eucalanus pileatus* suggests that some setae are mechanoreceptors, while others may play a role in both mechanoreception and chemoreception (Bundy & Paffenhöfer, 1997).

Nauplii do not have longitudinal, striated muscles, apparently because the body is not composed of movable parts. There are oblique, striated, extrinsic muscles originating on the dorsal body wall and inserting in the protopod of the limbs, and striated, intrinsic muscles within the limbs of *Euterpina acutifrons* (see Fanta, 1972) and *Oithona ovalis* (see Fanta, 1976). The number of oblique, extrinsic muscles to the appendages increases, with variation, as naupliar development progresses. The number for *Euterpina acutifrons* increases from two to antenna 1, two to antenna 2, and two to the mandible at NI-NIII, to two muscles to antenna 1, three to antenna 2, and three to the mandible at NIV-NV, and finally to four muscles to antenna 1, seven to antenna 2, and four to the mandible at NVI. The changes in extrinsic muscles of *Oithona ovalis* differ: two to antenna 1, two to antenna 2, and two to the mandible at NI-NII; two to antenna 1, three to antenna 2, and three to the mandible at NIII; three to antenna 1, three to antenna 2, three to the mandible and two to maxilla 1 at NIV; and finally three to antenna 1, four to antenna 2, three to the mandible and two to maxilla 1 at NV-NVI. For the calanoid, *Pseudodiaptomus acutus* (see Fanta, 1982) NII has two extrinsic muscles to antenna 1, three to antenna 2, and two to the mandible; NIII-NVI three to antenna 1, four to antenna 2, and two to the mandible.

Ecology

The occurrence throughout the world's oceans of nauplii that were not differentiated to species or stage has been discussed (Sazhina, 1985). Studies of the distributional ecology of nauplii usually are concerned with the vertical structure of pelagic species; such studies often are not stage-specific or species-specific. Nauplii that were not differentiated to species or stage are more likely to be found within the thermocline of stratified waters (Incze et al., 1996). Diel changes were detected in vertical structure of nauplii identified only to the order in Calanoida, Cyclopoida and Harpacticoida (cf. Ferrari et al., 2003).

Some studies of nauplii identified to species, but not necessarily to stage, are available. Nauplii of the limnetic *Leptodiaptomus novamexicanus* [reported as *Diaptomus novamexicanus*] occasionally appear to undergo reverse vertical migration (Redfield & Goldman, 1980); however, migration parameters in general did not change with the changes in abundance of individuals in a cubic meter of water. Nauplii of *Acartia clausi*, categorized as NI-NIV or as NV-NVI, were found throughout a 4-m water column with the NV-NVI group more likely to be collected close to the substrate during the day (Landry, 1978a). Feeding nauplii of *Calanus finmarchicus* (stages NIII-NVI) were found closer to the surface than the non-feeding stages NI-NII (Durbin et al., 2000a).

Swimming has been studied for nauplii of only a few species, primarily free-living planktonic calanoids and cyclopoids. There have been no stage-specific comparisons of swimming. Species of *Eucalanus* use antenna 2 and the mandible to swim (Paffenhöfer & Lewis, 1989) while other species of calanoids, like *Centropages typicus* and *Calanus finmarchicus*, use all three naupliar appendages (Björnberg, 1986a; Titelman & Kiørboe, 2003a). The amount of time spent moving these appendages may vary among species of the same genus, e. g., *Eucalanus pileatus* or *E. crassus* and *E. hyalinus* (see Paffenhöfer & Lewis, 1989). Swimming behavior appears to vary significantly among different naupliar stages of the same species although two basic categories of swimming have been generalized. Long periods of sinking punctuated by a brief series of fast jumps have been contrasted with slower, almost continuous swimming (Gauld, 1959; Gerritsen, 1978; Buskey, 1994; Paffenhöfer et al., 1996; Titelman & Kiørboe, 2003a, b). These two categories are not restricted to species or to a naupliar stage, but may be dependent on the motility of prey. Particular swimming behaviors may predispose nauplii to attacks by predators (Buskey et al., 1993), and so swimming behaviors may represent an adaptive balance between feeding efficiency and predator avoidance (Titelman & Kiørboe, 2003b). There is evidence that larger, i. e., older, naupliar stages respond to smaller deformations of their adjacent fluid field than do smaller, i. e., younger, stages (Green et al., 2003). Thus, older naupliar stages should be able to detect disturbances by predators more easily than younger nauplii.

Antenna 2 and the mandible produce a feeding current in calanoid nauplii (Storch, 1928) or more specifically the asymmetrical motion of the rami of antenna 2 does (Paffenhöfer & Lewis, 1989). The tips of setae of some calanoid mouthparts may be chemoreceptors (Friedman & Strickler, 1975). Feeding may result from continuous swimming movements or may take place only during jump-skip movements. A quite different naupliar feeding mechanism has been described for harpacticoid copepods living in the fronds of large marine algae (Harding, 1954; Green, 1958; Fahrenbach, 1962). Movement of the mandibles brings the oral area of the nauplius against the wall of a plant cell. Adduction of the "gnathobase" (the naupliar arthrite) of antenna 2 at about 4 strokes per second rasps at the wall of the cell. When the cell wall is broken, the "gnathobase" is used to push the cell contents into the esophagus.

Naupliar bioenergetics, including measurements of growth rate as determined by uptake of carbon and nitrogen or of respiration, have been studied only for nauplii of a few free-living copepods, e. g., *Calanus pacificus* (see Fernández, 1979) and *Eudiaptomus graciloides* (see Hamburger & Boetius, 1987). Naupliar growth rate was affected positively by increasing temperature and negatively by decreasing carbon content of their food, e. g., *Calanus helgolandicus* and *Pseudocalanus elongatus* (see Green et al., 1991), and

Calanus chilensis (see Torres & Escribano, 2003), as well as the stage at which feeding began. For example, the pattern for species of *Calanus* (e. g., Hygum et al., 2000) includes a negative growth rate for the non-feeding NI-NII, followed by a long duration for NIII, which is the first feeding stage, and short but identical stage duration for NIV-NVI. A similar pattern has been reported for the unrelated calanoid, *Metridia pacifica* (see Pinchuk & Paul, 1998).

Patterns of higher instantaneous survival for NI-NII, relative to later naupliar stages, have been reported (Eiane & Ohman, 2004) for the calanoids, *Calanus finmarchicus*, *Pseudocalanus elongatus* and the cyclopoid, *Oithona similis*. The absolute value of the rates of survival differed among the three species, perhaps reflecting differences in swimming behaviors. Instantaneous survival rates also differed within the same stage of the same species from different localities (Ohman et al., 2004), as well as among species of the same genus (Ohman & Wood, 1996; Eiane et al., 2002; Ohman et al., 2002). Naupliar survival of older stages of *Calanus pacificus* appeared to be less variable than that of younger naupliar stages (Mullin, 1995). Nauplii of the metridinids, *Pleuromamma gracilis* and *Metridia longa* have been observed to bioluminesce (Evstigneev, 1982, 1984; Lapota & Losee, 1984; Lapota et al., 1988).

Development times and suggested interpretations have been surveyed for free-living, planktonic copepods (Hart, 1990; Sabatini & Kiørboe, 1994; Kumar & Ramamohama Rao, 1998; Peterson, 2001; Hirst et al., 2003). As might be expected, the rate of copepod post-embryonic development often is dependent on temperature and food, although *Cyclops vicinus* was able to compensate its developmental time through acclimation to temperature (Munro, 1974). With unlimited food and at a constant temperature, developmental time from NI to NVI of many copepods is independent of body size. Developmental times for species of marine calanoid nauplii are shorter than developmental times for species of freshwater calanoids (Peterson, 2001). Development of *Acartia tonsa* is inhibited by exposure to a series of compounds some of which may antagonize ecdysone (Anderson et al., 2001).

Isochronal development hypothesizes that the absolute time spent at each stage is identical for a species developing at the same temperature. While attractive for its predictive power, the rule may not hold generally even for free-living, pelagic copepods. Data from Uye & Onbé (1975) on the short duration of NI of *Pseudodiaptomus marinus* (noted above) compromise the rule, and there also is evidence for the short duration of NI for other calanoids, a prolonged duration of NII, and a short duration of NVI (Peterson, 2001). Some indications of the lack of predictability for the isochronal rule can be found in studies of different species of *Calanus*. Development of *Calanus pacificus* was described as isochronal (Fernández, 1979) but *Calanus australis* was isochronal only for NIV-CIII (Peterson & Painting, 1990). Based on these studies, the rule of isochronal development can be expected to have poor predictive power for many host-associated copepods, especially those with fewer than six naupliar stages. An intergeneric equiproportional rule (Hart, 1990) predicts that the duration of a life history stage takes up a constant proportion of postembryonic development in all species within the same genus, but clear support for this rule has not been established.

THE COPEPODID PHASE OF DEVELOPMENT

The copepodid stages of copepods usually have thoracic somites separated by an arthrodial membrane, the post-mandibular appendages usually are transformed appendages, an interpodal bar unites the contralateral pairs of swimming legs, but a naupliar arthrite is absent from the coxa of antenna 2. The following description of body and appendages (figs. 9-20) for the six copepodid stages of the gymnoplean, *Ridgewayia klausruetzleri* is modified from Ferrari (1995) and serves as an introduction to the copepodid phase of development. A description of only the body of the six copepodid stages of the podoplean, *Dioithona oculata* and the thaumatopsylloid, *Caribeopsyllus amphiodiae* is presented in order to compare the development of the three basic adult architectures of copepods. For purposes of interpretation, thoracic somites usually bear a limb or a limb bud. The posterior abdominal somite bears an appendage, the caudal ramus, but the other abdominal somites do not bear a limb bud or a limb.

Please note that the abdominal somites are numbered herein according to their order of presentation in the developmental sequence: **not** according to their linear position along the anterioposterior axis of the body.

Copepodids of *Ridgewayia klausruetzleri*

CI: Body divided into broad anterior section and narrow posterior section (fig. 9A, I). Anterior section a long anterior part and 3 smaller articulating parts; posterior section of 2 articulating parts. Antenna 1 of 10 articulating segments (fig. 10A). Antenna 2 with coxa, basis, 2-segmented endopod and 9-segmented exopod (as for CVI female in fig. 11A, except setation of endopod fig. 11F). Mandible with coxa including ventral gnathobase, basis, 2-segmented endopod and 5-segmented exopod (as for CVI female in fig. 12A, except setation of endopod fig. 12D). Maxilla 1 with praecoxal endite, coxa with exite and endite, basis with exite and 2 endites, 1-segmented exopod, and 2-segmented endopod (as for CVI female in fig. 13A, except setation in fig. 13F). Maxilla 2 with syncoxa including 1 praecoxal endite and 1 coxal endite, basis with 2 endites, ramus with 2 lobes and distal section poorly differentiated (as for CVI female in fig. 14A, except setation of endopod in fig. 14B for CII). Maxilliped with syncoxa of 3 praecoxal endites and 1 coxal endite, basis with 2 endites poorly differentiated proximally, and 2-segmented endopod (fig. 15A). Swimming leg 1 with coxa,

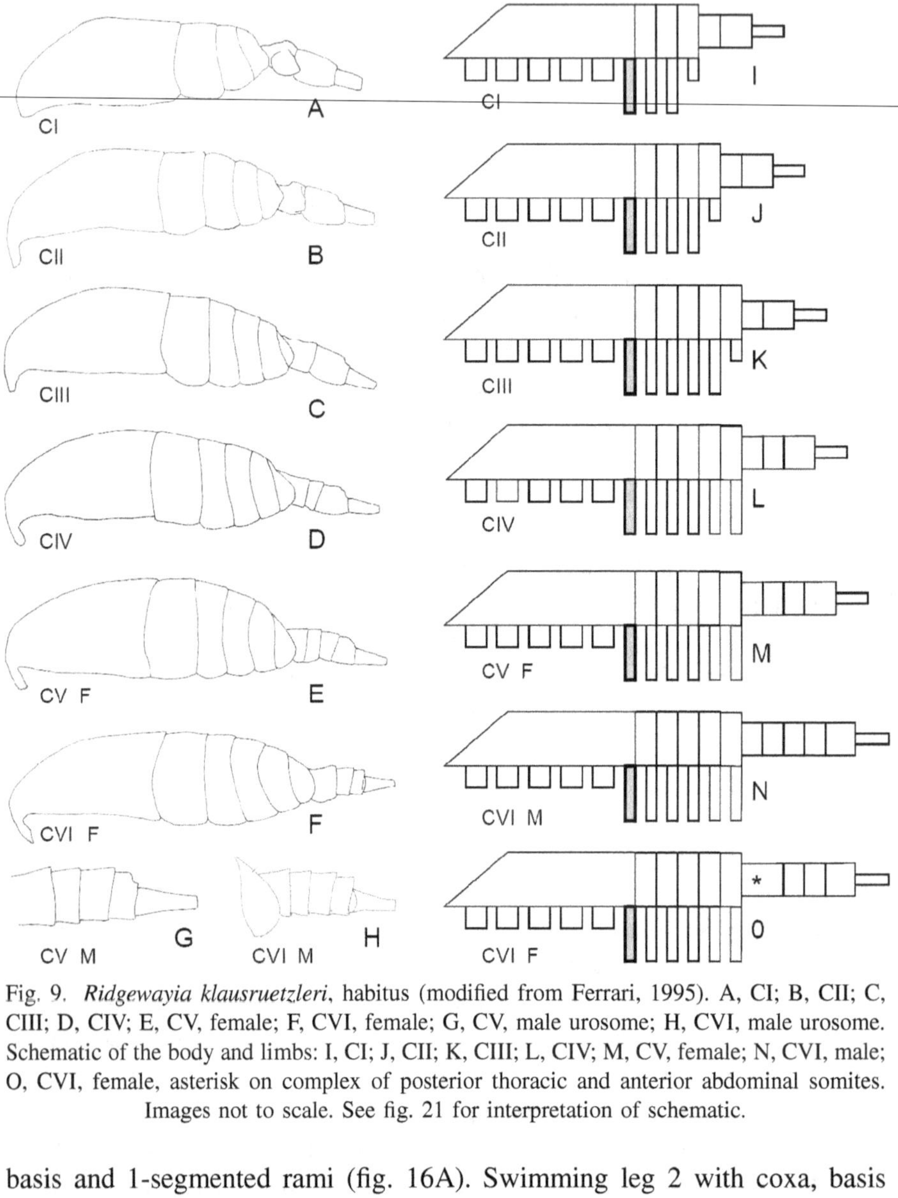

Fig. 9. *Ridgewayia klausruetzleri*, habitus (modified from Ferrari, 1995). A, CI; B, CII; C, CIII; D, CIV; E, CV, female; F, CVI, female; G, CV, male urosome; H, CVI, male urosome. Schematic of the body and limbs: I, CI; J, CII; K, CIII; L, CIV; M, CV, female; N, CVI, male; O, CVI, female, asterisk on complex of posterior thoracic and anterior abdominal somites. Images not to scale. See fig. 21 for interpretation of schematic.

basis and 1-segmented rami (fig. 16A). Swimming leg 2 with coxa, basis and 1-segmented rami (fig. 17A). Swimming leg 3 as bilobe bud (fig. 18A).

CII: Differs from CI as follows. Anterior section with 4 smaller articulating parts (fig. 9B, J). Antenna 1 of 17 articulating segments (fig. 10B). Antenna 2, as for CVI female except setation of endopod (fig. 11E). Maxilla 1 as for CVI female except setation of endopod (fig. 13E). Maxilla 2 as for CVI female (fig. 14A). Maxilliped with 3-segmented endopod (fig. 15B). Swimming leg

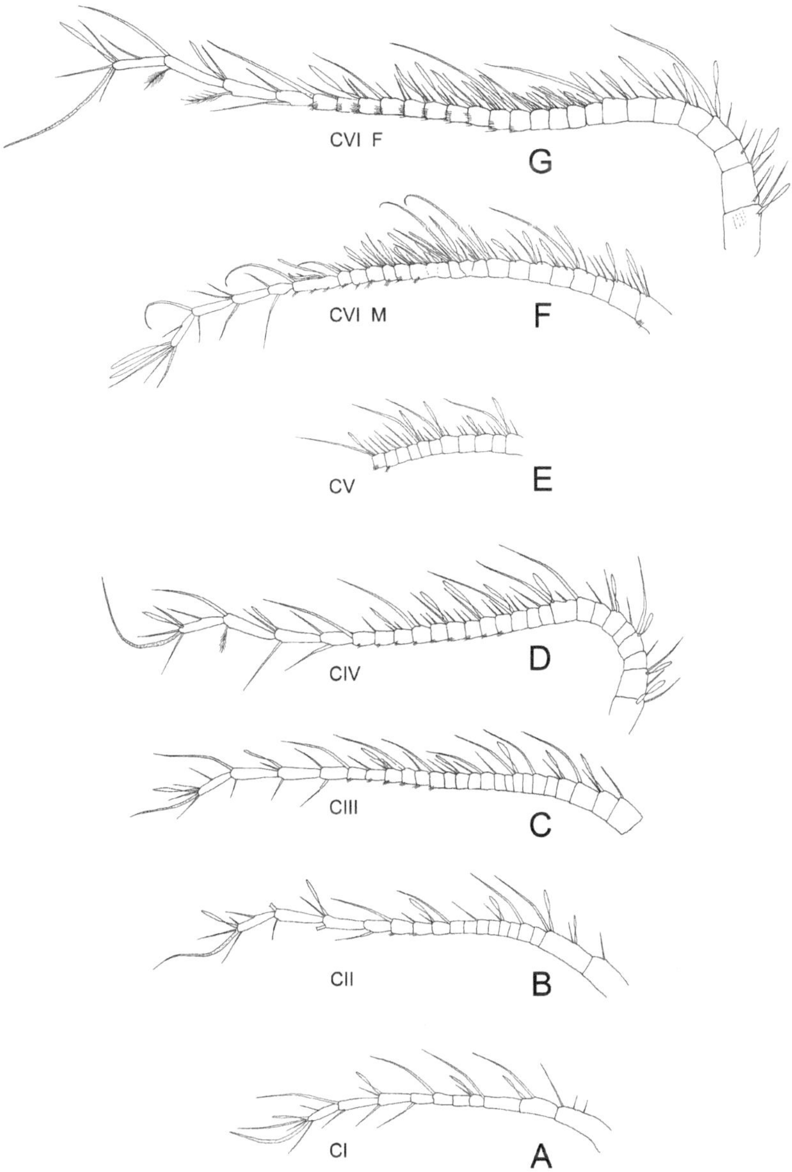

Fig. 10. *Ridgewayia klausruetzleri*, antenna 1 (modified from Ferrari, 1995). A, CI; B, CII; C, CIII; D, CIV; E, CV, segments 4-14; F, CVI, male; G, CVI, female. Images not to scale.

1 with 2-segmented rami (fig. 16B). Swimming leg 2 with 2-segmented rami (fig. 17B). Swimming leg 3 with coxa, basis and 1-segmented rami (fig. 18B). Swimming leg 4 as bilobe bud (fig. 19A).

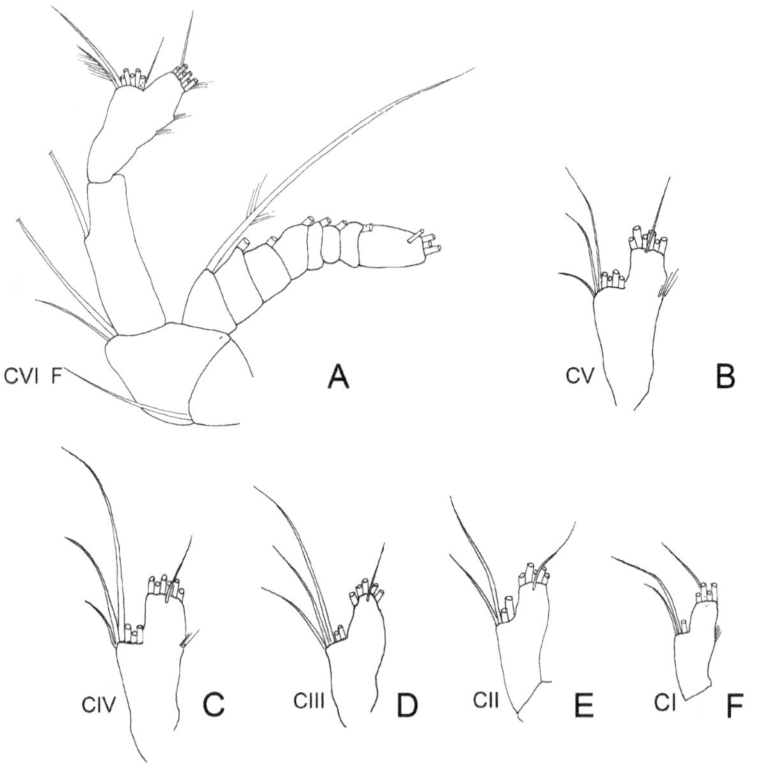

Fig. 11. *Ridgewayia klausruetzleri*, antenna 2 (modified from Ferrari, 1995). A, CVI female; B, CV, setation of terminal endopodal segment; C, CIV, setation of terminal endopodal segment; D, CIII, setation of terminal endopodal segment; E, CII, setation of terminal endopodal segment; F, CI, setation of terminal endopodal segment. Images not to scale.

CIII: Differs from CII as follows. Anterior section with 5 smaller articulating parts (fig. 9C, K). Antenna 1 of 24 articulating segments (fig. 10C). Antenna 2 as for CVI female except setation of endopod (fig. 11D). Mandible as for CVI female except setation of endopod (fig. 12C). Maxilla 1 as for CVI female except setation (fig. 13D). Maxilliped with 4-segmented endopod (fig. 15C). Swimming leg 3 with 2-segmented rami (fig. 18C). Swimming leg 4 with coxa, basis and 1-segmented rami (fig. 19B). Limb 5 as bilobe bud (fig. 20A).

CIV female: Differs from CIII as follows. Posterior section of 3 articulating parts (fig. 9D, L). Antenna 1 of 25 articulating segments (fig. 10D). Antenna 2 as for CVI female except setation of endopod (fig. 11C). Mandible as for CVI female except setation of endopod (fig. 12B). Maxilla 1 as for CVI female except setation (fig. 13C). Maxilliped with 5-segmented endopod (fig. 15D).

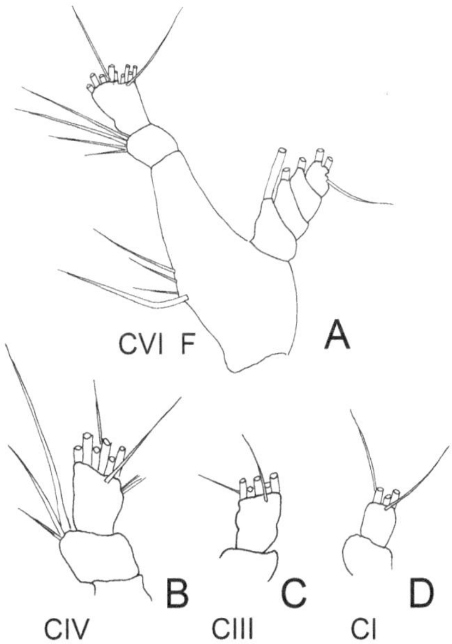

Fig. 12. *Ridgewayia klausruetzleri*, mandibular palp (modified from Ferrari, 1995). A, CVI female; B, CIV, setation of endopod; C, CIII, setation of terminal endopodal segment; D, CI, setation of terminal endopodal segment. Images not to scale.

Swimming legs 1-3 setation as shown (figs. 16C, 17C, 18D). Swimming leg 4 with 2-segmented rami (fig. 19C). Limb 5 with coxa, basis and 1-segmented rami (fig. 20B).

CV female: Differs from CIV female as follows. Posterior section of 4 articulating parts (fig. 9E, M). Antenna 1 of 26 articulating segments (fig. 10E). Antenna 2 as for CVI female except setation of endopod (fig. 11B). Maxilla 1 as for CVI female except setation (fig. 13B). Maxilliped different setation (fig. 15E). Swimming legs 1-4 with 3-segmented rami (as for CVI female figs. 17E, 18E, 19C, 20D). Limb 5 both exopods 2-segmented, left endopod 2-segmented (fig. 20C), right endopod 1-segmented.

CVI female: Posterior section of 4 articulating parts (fig. 9F, O). Antenna 1 of 26 articulating segments (fig. 10F). Antenna 2 (fig. 11A). Mandible (fig. 12C). Maxilla 1 (fig. 13A). Maxilla 2 (fig. 14A). Maxilliped (fig. 15F). Swimming leg 1 (fig. 16D). Swimming leg 2 (fig. 17D). Swimming leg 3 (fig. 18D). Swimming leg 4 (fig. 19D). Limb 5 with 3-segmented exopod and 2-segmented endopod fig. 20D).

CIV male: body and appendage segments do not differ from CIV female.

CV male: Differs from CIV male as follows: limb 5, right endopod 1-segmented.

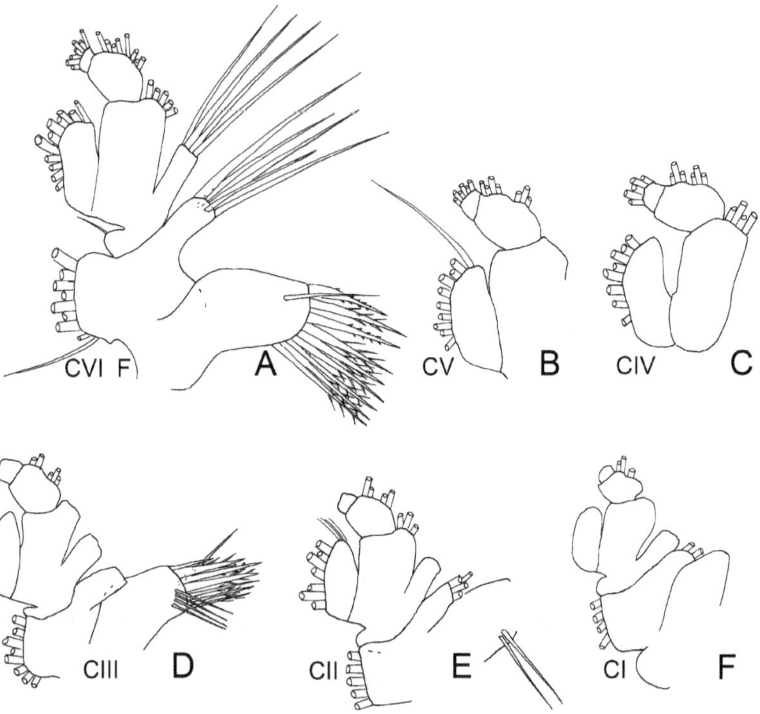

Fig. 13. *Ridgewayia klausruetzleri*, maxilla 1 (modified from Ferrari, 1995). A, CVI female; B, CV, setation of exopod and endopod; C, CIV, setation of distal endite of basis, exopod and endopod; D, CIII, setation of praecoxal endite and exite, and proximal and middle sets of the endopod; E, CII, setation of posterior face of praecoxal endite and of exite, coxal endite, distal endite of basis, proximal and middle sets of the endopod and of exopod; F, CI, setation of praecoxal exite, coxal endite, proximal and middle sets of the endopod. Images not to scale.

CVI male: Differs from CV male as follows: posterior section of 5 articulating parts (fig. 9H, O). Limb 5 right exopod 2-segmented, right endopod 1-segmented (fig. 20 E), left exopod 3-segmented, left endopod 2-segmented (fig. 20F).

Copepodids of *Dioithona oculata*

The body of a podoplean copepod like *Dioithona oculata* differs from the body of a gymnoplean in the following ways:

CI: Body divided into broad anterior section and narrow posterior section (fig. 21A). Anterior section of long anterior part and 3 smaller articulating parts; posterior section of 2 articulating parts.

CII: Anterior section of 4 smaller articulating parts (fig. 21B).

CIII: Posterior section of 3 articulating parts (fig. 21C).

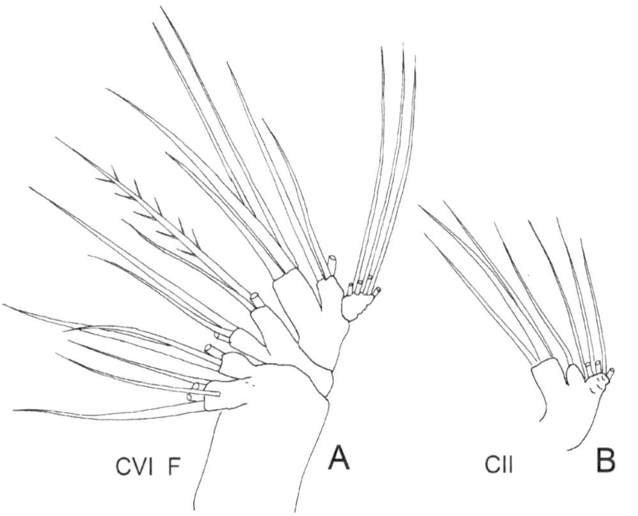

Fig. 14. *Ridgewayia klausruetzleri*, maxilla 2 (modified from Ferrari, 1995). A, CVI female; B, CII, setation of ramus. Images not to scale.

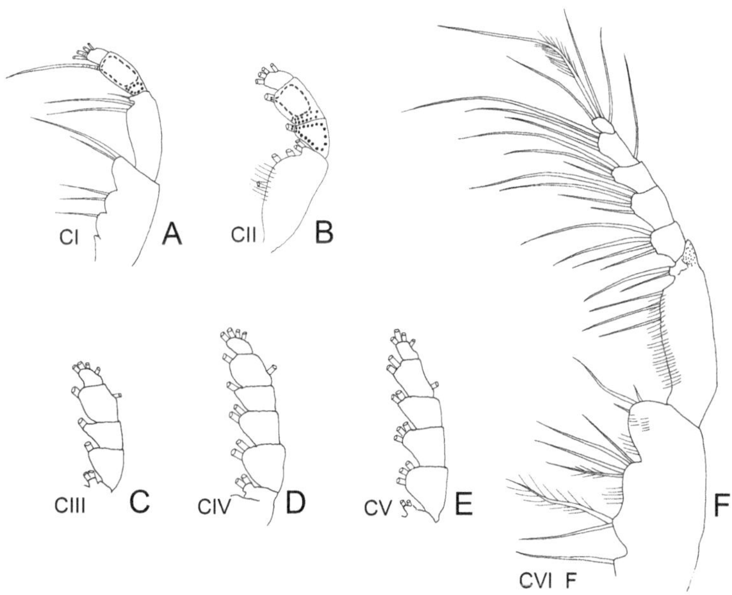

Fig. 15. *Ridgewayia klausruetzleri*, maxilliped (modified from Ferrari, 1995). A, CI; B, CII, basis + ramus; C, CIII, distal endite of basis + ramus; D, CIV, distal endite of basis + ramus; E, CV, distal endite of basis + ramus; F, CVI female. Dotted lines within penultimate endopodal segment of A, and antepenultimate and penultimate endopodal segments of B, indicate configuration of the exoskeleton of the following stage; a short, proximal, new segment is formed within the penultimate segment. Images not to scale.

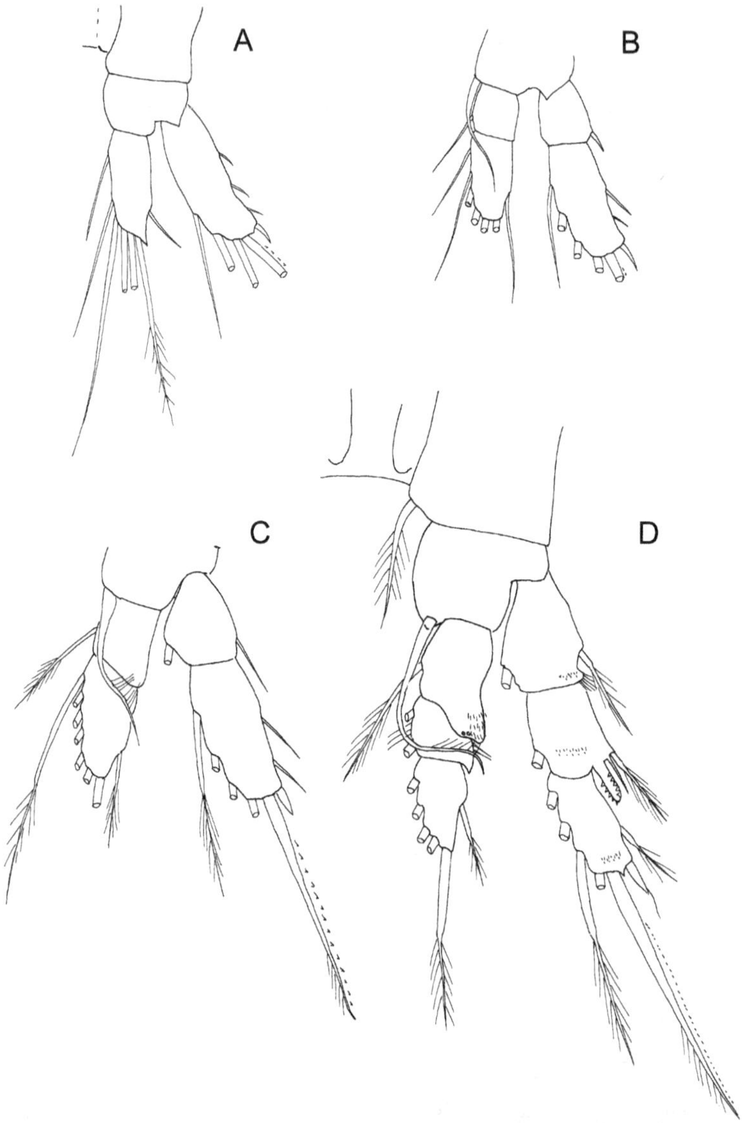

Fig. 16. *Ridgewayia klausruetzleri*, swimming leg 1 (modified from Ferrari, 1995). A, CI; B, CII, exopod and endopod; C, CIV female, exopod and endopod; D, CVI female. Images not to scale.

CIV: Posterior section of 4 articulating parts (fig. 21D).

CV: Posterior section of 5 articulating parts (fig. 21E).

CVI male: Posterior section of 6 articulating parts (fig. 21F).

CVI female: Anterior section similar to CVI male; posterior section of 5 articulating parts (fig. 21G).

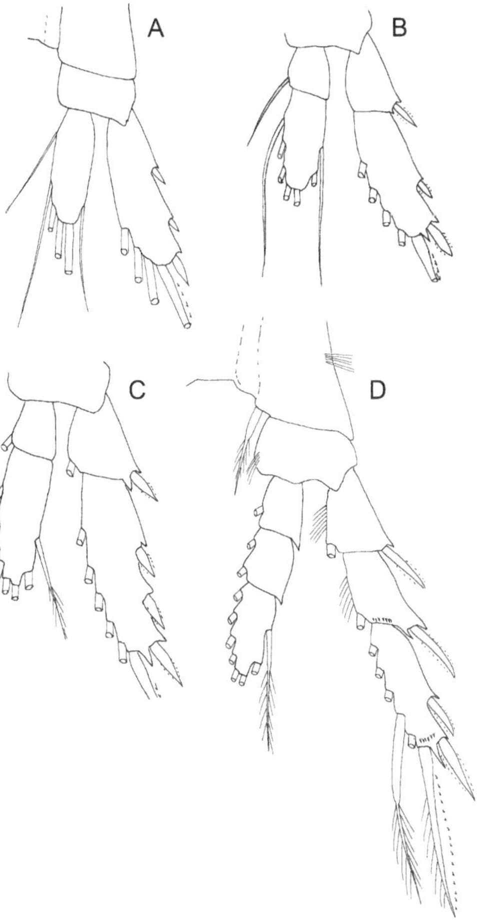

Fig. 17. *Ridgewayia klausruetzleri*, swimming leg 2 (modified from Ferrari, 1995). A, CI; B, CII, exopod and endopod; C, CIV female, exopod and endopod; D, CVI female. Images not to scale.

Copepodids of *Caribeopsyllus amphiodiae*

Development of a thaumatopsylloid is exemplified by *Caribeopsyllus amphiodiae* (M. Dojiri, from pers. comm., e-mail 26 October 2005) and differs from both the gynmoplcan and podoplean as follows:

CI: Body divided into broad anterior section and narrow posterior section (fig. 22A). Anterior section of long anterior part and 2 smaller articulating parts; posterior section of 2 articulating parts.

CII: Posterior section of 3 articulating parts (fig. 22B).

CIII: Does not differ from CII (fig. 22C).

CIV: Posterior section of 4 articulating parts (fig. 22D).

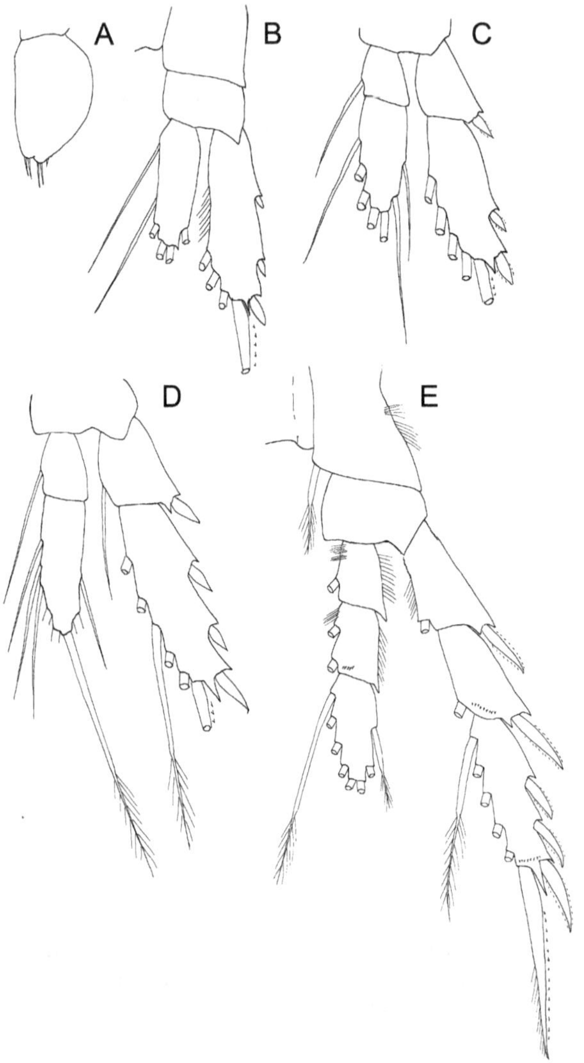

Fig. 18. *Ridgewayia klausruetzleri*, swimming leg 3 (modified from Ferrari, 1995). A, CI; B, CII; C, CIII, exopod and endopod; D, CIV female, exopod and endopod; E, CVI female. Images not to scale.

CV: Does not differ from CIV (fig. 22E).

CVI male: Posterior section of 5 articulating parts (fig. 22F).

CVI female: Anterior section similar to CVI male; posterior section of 4 articulating parts (fig. 22G).

The anterior section of the body of the gymnoplean, *Ridgewayia klausruetzleri* at CI is a long cephalothorax of 5 cephalic somites plus the first thoracic

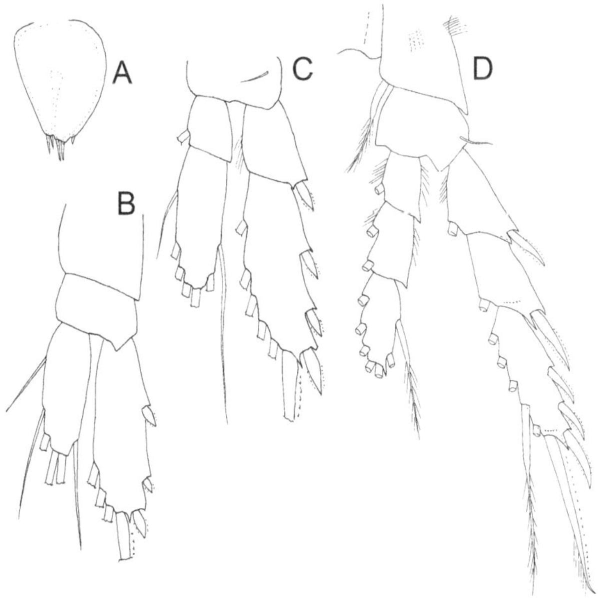

Fig. 19. *Ridgewayia klausruetzleri*, swimming leg 4 (modified from Ferrari, 1995). A, CII; B, CIII; C, CIV female, exopod and endopod; D, CVI female. Dotted lines within A show the configuration of the exoskeleton of the following stage. Images not to scale.

somite, which does not articulate anteriorly; the following three somites, articulating anteriorly and posteriorly, are the second, third and fourth thoracic somites (fig. 9A, I). The posterior section consists of the fifth thoracic somite and the posterior abdominal somite. At CII, the anterior section consists of the articulating second, third, fourth and fifth thoracic somites; the posterior section consists of the sixth thoracic somite and the posterior abdominal somite (fig. 9B, J). At CIII, the anterior section includes the articulating second, third, fourth, fifth and sixth thoracic somites; the posterior section consists of the seventh thoracic somite and the posterior abdominal somite (fig. 9C, K). At CIV, the anterior section is unchanged from CIII, and the posterior section consists of the seventh thoracic somite, the anterior [or second, i. e., in development] abdominal somite, and the posterior [or first, i. e., in development] abdominal somite (fig. 9D, L). At CV, the anterior section remains unchanged, and the posterior section consists of the seventh thoracic somite, the second abdominal somite, a middle [or third] abdominal somite, and the posterior abdominal somite (fig. 9E, M). At CVI, the anterior section of the male remains unchanged; the posterior section consists of the seventh thoracic somite, the second abdominal somite, two middle [the third and fourth] abdominal somites, and the posterior abdominal somite (fig. 9H, N).

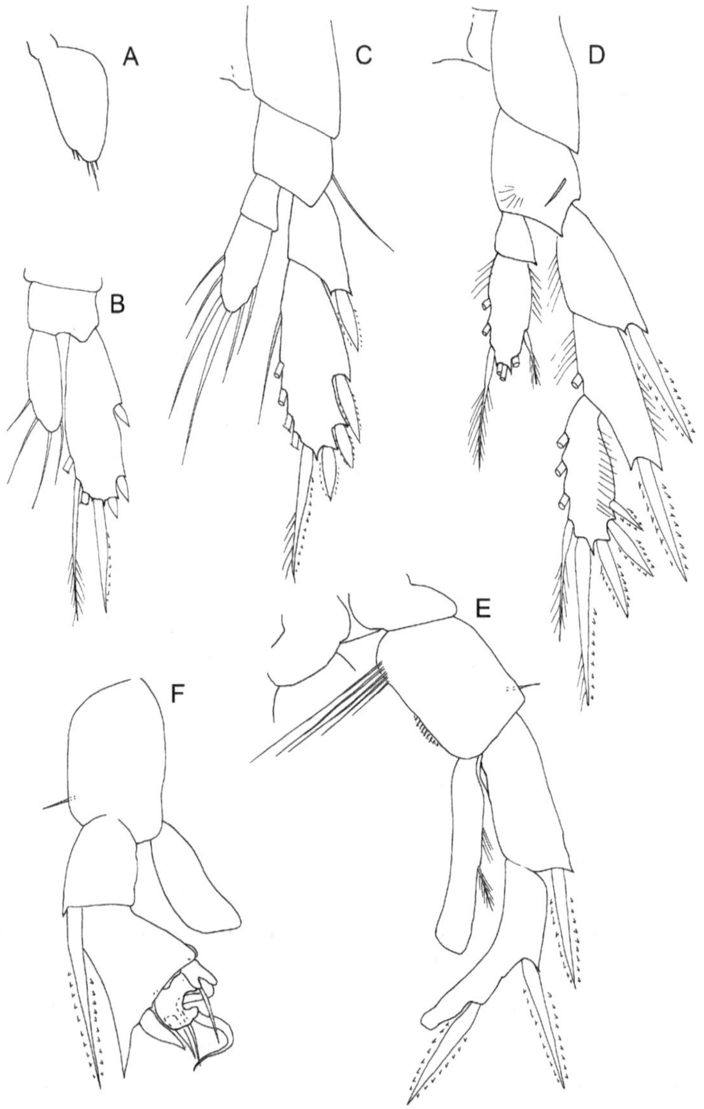

Fig. 20. *Ridgewayia klausruetzleri*, limb 5 (modified from Ferrari, 1995). A, CIII; B, CIV female, basis, exopod and endopod; C, CV, female; D, CVI female; E, CVI male, right limb; F, CVI male, left limb. Images not to scale.

The posterior section of the CVI female has a somite complex consisting of the seventh thoracic somite unarticulated with the second (anterior) abdominal somite, plus the third and the fourth abdominal somites, and the posterior abdominal somite (fig. 9F, O). At CIII and later stages, the anterior section of the body of the gymnoplean corresponds to the adult prosome; the

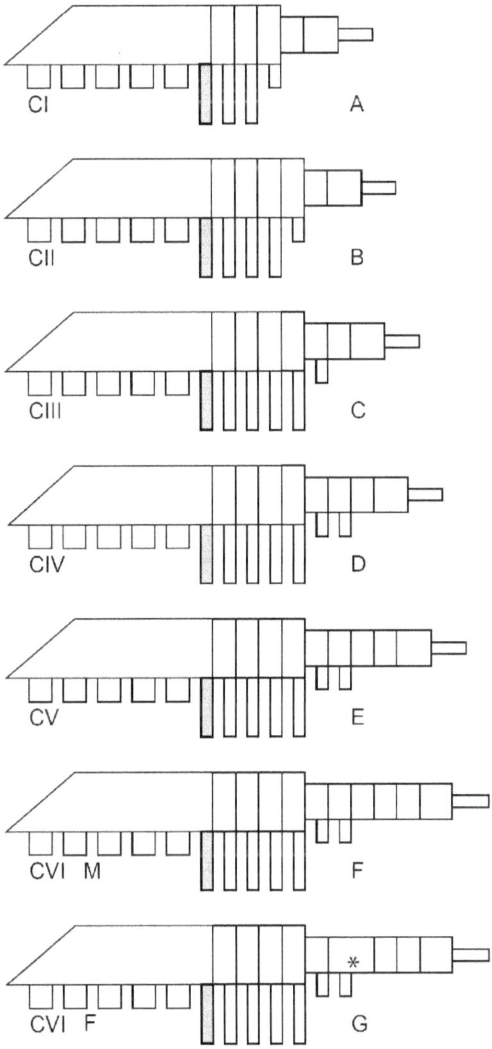

Fig. 21. *Dioithona oculata*, schematic of the body and limbs of copepodids (modified from Ferrari & Ambler, 1992). A, CI; B, CII; C, CIII; D, CIV; E, CV; F, CVI male; G, CVI female, asterisk on complex of posterior thoracic and anterior abdominal somites. Square is cephalic limb; long, vertical rectangle is transformed thoracic limb; dark, first thoracic limb is maxilliped; short, vertical rectangle is thoracic limb bud; horizontal rectangle is caudal ramus.

remaining somites of the posterior part of the body include some or all of those comprising the adult urosome.

The body of *Dioithona oculata* differs from that of *Ridgewayia klausruetzleri* at CIII-CVI (fig. 9C-H, K-O vs fig. 21C-G). At CIII, the anterior section is unchanged from CI-CII and consists of a long cephalothorax of

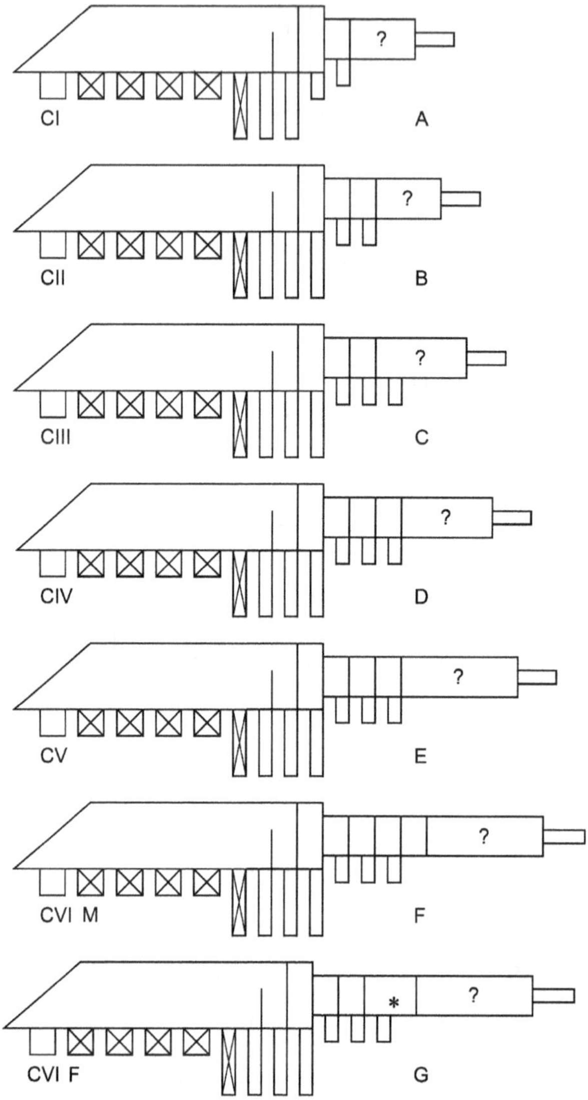

Fig. 22. Schematic of the body and limbs of *Caribeopsyllus amphiodiae* copepodids (modified from M. Dojiri, pers. comm., e-mail 26 October 2005). A, CI; B, CII; C, CIII; D, CIV; E, CV; F, CVI, male; G, CVI, female, asterisk on complex of posterior thoracic and anterior abdominal somites. X denotes limbs that are never present; see fig. 21 for interpretation of morphology; question mark indicates somite composition is unclear.

five cephalic somites plus the first thoracic somite, which is unarticulated anteriorly; the following four articulating somites are the second, third, fourth and fifth thoracic somites (fig. 21C). The posterior section consists of the sixth thoracic somite, the seventh thoracic somite, and the posterior abdominal somite. At CIV, the anterior section remains unchanged, and the posterior section consists of the sixth and the seventh thoracic somites, the anterior (the second) abdominal somite, and the posterior abdominal somite (fig. 21D). At CV, the anterior section remains unchanged, and the posterior section consists of the sixth and seventh thoracic somites, the second abdominal somite, a middle (or third) abdominal somite, and the posterior abdominal somite (fig. 21E). At CVI, the anterior section of the male remains unchanged, and the posterior section consists of the sixth and seventh thoracic somites, the second abdominal somite, two middle (the third and fourth) abdominal somites, and the posterior abdominal somite (fig. 21F); there are no complexes on the posterior section of the male. The posterior section of the CVI female is composed of the sixth thoracic somite followed by a somite complex of the seventh thoracic somite unarticulated with the second abdominal somite, the third and the fourth abdominal somites, and the posterior abdominal somite (fig. 21G). At CII and later stages, the anterior section of the podoplean body corresponds to the adult prosome; the remaining somites on the posterior part of the body include some or all of those comprising the adult urosome.

The body of *Caribeopsyllus amphiodiae* (fig. 22A-G) differs from that of *Ridgewayia klausruetzleri* (fig. 9A-O) and *Dioithona oculata* (fig. 21A-G) throughout copepodid development. The anterior section of the body of *Caribeopsyllus amphiodiae* at CI is a long cephalothorax of five cephalic somites without limbs, except for A1, plus a limbless first thoracic somite and the second thoracic somite, which is unarticulated dorsally with the third thoracic somite. The following free somite, articulating anteriorly and posteriorly, is the fourth thoracic somite (fig. 22A). The posterior section of the body consists of the fifth thoracic somite and the posterior abdominal somite; it also may include the sixth thoracic somite (whose limb bud is presented at CII), which does not articulate posteriorly with the posterior abdominal somite. At CII, the anterior section remains unchanged; the posterior section consists of the fifth and sixth thoracic somites plus the seventh thoracic somite (whose limb bud is presented at CIII), which does not articulate posteriorly with the posterior abdominal somite (fig. 22B). The anterior section of CIII remains unchanged; the posterior section consists of the fifth and sixth thoracic somites plus the seventh thoracic somite, which does not articulate posteriorly with the posterior abdominal somite; it also may include the second abdominal somite unarticulated between the seventh thoracic somite and the posterior abdominal somite (fig. 22C). At CIV, the anterior section remains unchanged, and the posterior section consists of the fifth, sixth and seventh thoracic somites, and the posterior abdominal somite; the posterior section also may include the second and third abdominal somites, which are unarticulated between the seventh thoracic somite and the posterior abdominal somite (fig. 22D). The anterior section of CV remains unchanged, and the posterior section consists of the fifth, sixth and seventh thoracic somites, and the posterior abdominal somite; it may also include the second, third and fourth abdominal somites unarticulated between the seventh thoracic somite and the posterior abdominal somite (fig. 22E). At CVI, the anterior section of the body of the male remains unchanged; the posterior section consists of the fifth, sixth and seventh thoracic somites, the anterior (or second) abdominal somite, and the posterior abdominal somite; the posterior section also may include the third, fourth and fifth abdominal somites unarticulated between the second abdominal somite and the posterior abdominal somite (fig. 22G). The posterior section of CVI female consists of the fifth and sixth thoracic somites, the seventh thoracic somite, which does not articulate with the anterior (or second) abdominal somite, and the posterior abdominal somite; the posterior section also may include the third, fourth and fifth abdominal somites, unarticulated between the second abdominal somite and the posterior abdominal somite (fig. 22F).

Caribeopsyllus amphiodiae differs from gymnopleans and podopleans in two ways. The sixth thoracic somite is present at CI, while the sixth thoracic somite of gymnopleans and podopleans is presented at CII. As a result, *C. amphiodiae*, throughout its copepodid phase, bears one more somite than the comparable copepodid stages of gymnopleans and podopleans. Second, the anterior section of the body of *C. amphiodiae* corresponds to the adult prosome at CI; the remaining somites include some or all of those comprising the adult urosome. The anterior section of the body of gymnopleans and podopleans does not correspond to the adult prosome at CI.

In summary, the anterior section of the body corresponds to the adult prosome of *Ridgewayia klausruetzleri* at CIII, to the adult prosome of *Dioithona oculata* at CII, and to the adult prosome of *Caribeopsyllus amphiodiae* at CI. These differences result in a different number of thoracic somites incorporated into the urosome of the adult body as follows: one thoracic somite to the urosome of the adult body of *Ridgewayia klausruetzleri*; two thoracic somites to the urosome of *Dioithona oculata*; three thoracic somites to the urosome of *Caribeopsyllus amphiodiae*. In addition, *C. amphiodiae* bears one somite more than *Ridgewayia klausruetzleri* or *Dioithona oculata* throughout the copepodid phase of its development.

Variation in the number of copepodid stages

Most copepods, including all free-living species from a variety of different habitats, as well as many different kinds of parasitic copepods, pass through five immature copepodid stages before a terminal molt to the adult CVI takes place. Examples with six copepodid stages include free-living marine calanoids like *Ridgewayia klausruetzleri* (see Ferrari, 1995) and *Platycopia orientalis* (see Ohtsuka & Boxshall, 1994), free-living freshwater calanoids like *Megadiaptomus hebes* (see Ranga Reddy & Rama Devi, 1985), free-living marine harpacticoids like *Thalestris longimana* (see Dahms, 1990b), free-living freshwater harpacticoids like *Canthocamptus mirabilis* (see Itô & Takashiro, 1981), free-living marine cyclopoids like *Dioithona oculata* (see Ferrari & Ambler, 1992), free-living freshwater cyclopoids like *Macrocyclops albidus* (see Defaye, 1984), marine cyclopoids like *Euryte longicauda*, which is associated with an invertebrate (see Ferrari & Ivanenko, 2005), free-living marine poecilostomes like *Hemicyclops japonicus* (see Itoh & Nishida, 1995), marine poecilostomes like *Doridicola longicauda* (see Costanzo et al., 1994) and *Midicola spinosus* (see Do et al., 1984, as *Pseudomyicola spinosus*) living in invertebrates, marine poecilostomes like *Taeniacanthus lagocephali* (see Izawa, 1986a) parasitic on fish, marine siphonostomatoids like *Asterocheres lilljeborgi* (see Ivanenko & Ferrari, 2003) associated with an invertebrate, marine siphonostomatoids like *Scottomyzon gibberum* (see Ivanenko et al., 2001) symbiotic on an invertebrate, and marine siphonostomatoids like *Caligus elongatus* (see Piasecki, 1996) parasitic on fish.

Fewer than six copepodid stages have been reported for a number of parasitic copopods, but often it is difficult to determine whether these fewer stages are a natural phenomenon or whether the missing stages may simply remain uncollected from the field or suppressed during laboratory culture. If there are copepods that pass through fewer than six copepodid stages, a likely group would include those copepods with a naupliar-like adult body, and with fewer than seven thoracic appendages, e. g., nicothoids like *Hansenulus trebax* (see Heron & Damkaer, 1986), the ventriculinid, *Heliogabalus phascolia* (see Lützen, 1968), or parasitic copepods of uncertain family or ordinal placement like *Allantogynus delamarei* (see Changeux, 1961) or *Selioides bocqueti* (see Carton, 1964). Copepods for which more than six copepodid stages have been reported belong to the Caligidae and are discussed in the following section.

An exhaustive key to separate different copepodid stages of all copepods cannot be constructed due to the presence of somite complexes and the degree of variation in appendage segmentation. However, the persistence of a 3-segmented exopod on swimming leg 4 of most free-living copepods, as well as many copepods associated with invertebrate hosts, permits the following diagnoses of six copepodid stages:

KEY TO COPEPOD COPEPODIDS I-VI

Two transformed swimming legs, rami of both 1-segmented CI
Three transformed swimming legs, rami of third 1-segmented CII
Four transformed swimming legs, rami of fourth 1-segmented CIII
Four transformed swimming legs, exopod of fourth 2-segmented CIV
Four transformed swimming legs, exopod of fourth 3-segmented CV
Copulatory and oviducal or genital openings present CVI

The above key breaks down for free-living copepods when the addition of the arthrodial membrane to swimming leg 4 is delayed until the molt to CVI as it is, for example, among the monophyletic lineage of cyclopids that includes *Thermocyclops decipiens*, *Mesocyclops edax* and *Diacyclops thomasi* (see Ferrari, 1998).

Stage correspondence of copepods with chalimus stages

Caligid-like copepods (e. g., Caligidae, Cecropidae, Euryphoridae, Lernaeopodidae, Lernaeoceridae and Pennellidae) exhibit a unique variation in the copepodid phase of development. Developmental stages comparable to copepodid stages II-V are often attached to a fish host directly by antenna 2 or by a frontal filament that is held with the maxilla 2 (Benz, 1989). These stages may express a modified morphology, and they are called chalimus stages. Swimming legs of these chalimus stages often do not add arthrodial membranes or setae during molts between two successive stages. In some species, the swimming legs may be reduced in size and morphology to the extent that these limbs appear bud-like on the chalimus. These secondary buds of swimming legs 1-2 have been reconfigured from transformed limbs of the first copepodid, while the transformation of swimming legs 3-4 to a secondary limb bud is often less dramatic. All four swimming legs are reconfigured to transformed limbs during the terminal adult molt to CVI.

Caligid-like copepods molt to a first copepodid from a last nauplius that may be the only nauplius, or the second of two nauplii. This last nauplius does not appear to correspond to NVI of copepods with six naupliar stages;

rather it appears to be an NI. Like other parasitic copepods and free-living copepods, the first copepodid of caligids is a free-swimming stage, and like many other parasites it is also dispersive and infective. Body architecture of the first copepodid of caligid-like copepods is very similar to that of all other copepods, with the exception of the thaumatopsylloids. The body includes a cephalon with five appendages, five thoracic somites and a posterior abdominal somite. Swimming legs 1-2 are transformed limbs with unarticulated rami, and swimming leg 3 is a bud; the posterior abdominal somite bears a caudal ramus. This is a remarkably conserved morphology among copepods, and caligid-like copepods at CI reinforce the concept of the first copepodid as the phylotypic stage of copepods (Ferrari, 2003). The first copepodid of caligid-like copepods usually molts to chalimus 1, which usually attaches to a fish host. A second, unattached copepodid has been reported for the lernaeocerids, *Lernaeocera branchialis* by Sproston (1942) and *Lernaeenicus sprattae* by Schram (1979). This second copepodid may attach to the host using A2 or Mx2, and is transformed into a chalimus 1 that is attached by a frontal filament to the host fish. However, molting has not been observed between the second copepodid stage and the first chalimus stage of these two species (Schram, 1979) so that development of these lernaeocerids may be interpreted as including only one copepodid stage and a polymorphic first chalimus.

Chalimus 1 of caligid-like copepods resembles CII of other copepods in the number and kind of somites: a cephalon with five limbs, six thoracic somites and a posterior abdominal somite. Swimming legs 1-3 often may appear similar to transformed limbs and swimming leg 4 is a bud. Three more molts result in chalimus stages 2-4, respectively. Chalimus 4 of caligids like *Caligus elongatus* has been reported to molt directly to an adult (see Piasecki, 1996); this also is the situation for lernaeopodids like *Salmincola californiensis* (see Kabata & Cousens, 1973) and lernaeocerids like *Lernaeenicus sprattae* (see Schram, 1979). However, a pre-adult stage has been reported for lernaeopodids like *Neobrachiella robusta* (see Kabata, 1986) and caligids like *Lepeophtheirus salmonis* (see Johnson & Albright, 1991). The pre-adult stage usually is followed by the adult, although a second pre-adult has been reported to occur just before the adult stage of caligids like *Caligus clemensi* and *Lepeophtheirus pectoralis* (see Kabata, 1972; Boxshall, 1974a). However, there are no direct observations of molting between the two pre-adults, or between a pre-adult and an adult.

A copepodid phase with more than four chalimus stages has been reported for caligids like *Caligus epidemicus* (see Lin et al., 1996) and *C.*

multispinosus (see Lin et al., 1997). However, chalimus 5-6 differ from each other and from chalimus 4 only in the shape of the prosome, and there are no direct observations of molting between chalimus 4-5 or between chalimus 5-6. Furthermore, among caligid-like copepods, copulation has been observed only between the free-swimming female and the free-swimming male of the lernaeocerid, *Lernaeenicus sprattae*. If copulation between a free-swimming male and a free-swimming female is the usual situation for caligid-like copepods, then sperm in spermatophores attached to a free-swimming pre-adult female or sperm in the cuticular seminal receptacle of the free-swimming pre-adult female would be lost during any subsequent molt and so would be unavailable to fertilize eggs of the adult female. This loss would require remating by the molted female and wasted reproductive effort of the initial copulating male. One or more pre-adults reported for caligid-like copepods may simply represent different morphs of the adult stage, and chalimus 5-6 may simply be different morphs of chalimus 4. These morphs may result from the continued expansion of an exoskeleton that is initially soft and poorly-sclerotized after molting. A continued expansion of parts of the exoskeleton has been reported for the caligid-like *Salmincola californiensis* (see Kabata & Cousens, 1973) and *Neobrachiella robusta* (see Kabata, 1986), as well as for other siphonostomatoids like *Scottomyzon gibberum* (see Röttger, 1969; Ivanenko et al., 2001). Because there have been no observations of molting between a first and second copepodid, between a first and second pre-adult, or between a pre-adult and an adult, a copepodid phase for caligid-like copepods of one copepodid stage, chalimus stages 1-4, and one adult stage aligns well with CI-VI of other copepods. This seems to be the most likely situation for these parasitic copepods.

Addition of appendages

At CI, the transformed limbs of many copepods like *Ridgewayia klausruetzleri* are antenna 1, antenna 2, mandible, maxilla 1, maxilla 2, the maxilliped, all of which originate on the cephalothorax, and swimming leg 1, swimming leg 2, plus the bud of swimming leg 3 each originating from consecutive articulating thoracic somites. The bud of swimming leg 4, originating on the articulating fifth thoracic somite, is added at CII, and the bud of limb 5, originating on the articulating sixth thoracic somite is added at CIII. No limb buds are added during the molts to CIV-CVI, and there are no differences between females and males of *R. klausruetzleri* in the addition of appendages.

During CI-III, there is no difference between *Dioithona oculata* and *R. klausruetzleri* in limb number including limb buds. However, the bud of limb 6, originating from the articulating seventh thoracic somite, is added in *D. oculata* during the molt to CIV. No limbs are added during molts to CV-CVI, and there are no differences between males and females in the addition of appendages of *D. oculata*. *Caribeopsyllus amphiodiae* differs most notably from *R. klausruetzleri* and *D. oculata* in the absence of antenna 2, mandible, maxilla 1, maxilla 2, and the maxilliped. More subtle differences include a setose bud of leg 4 present on CI and the initial appearance of the buds of limbs 5-6 at CII-CIII, respectively. In addition, limbs 5-6 are not initially presented on an articulating sixth or seventh thoracic somite, respectively. Rather, limbs 5-6 first appear on the posterior somite complex that includes the sixth thoracic somite or the seventh thoracic somite unarticulated with the posterior abdominal somite.

Literature reports of the development of copepod swimming legs 1-4 and limbs 5-6 have been surveyed and analysed extensively (Ferrari, 1988). During the copepodid phase of development, there is little variation in the stage at which a limb bud initially is presented, or in the order of appearance of limb buds, which is exclusively anterior to posterior (Ferrari, 1988). The bud of limb 5 of *Lamproglena chinensis* has been reported to be presented initially at CII and the bud of limb 6 initially at CIII (see Kuang, 1962), rather than at CIII and CIV as is the situation for other copepods. However, these unusual observations may be misinterpretations of limb bud morphology, because a similar configuration was not reported for related species (Grabda, 1963; Kuang, 1980).

The presentation of the bud of limb 5 may be delayed until CIV, e. g., *Balaenophilus unisetus* (see Aurivillius, 1879), until CV, e. g., *Mytilicola intestinalis* (see Costanzo, 1959), or CVI, e. g., *Monstrilla helgolandica* (see Pelseneer, 1914). In a similar fashion, the initial presence of the bud of limb 6 may be delayed until CV, e. g., *Oncaea media* (see Malt, 1982), or until CVI, e. g., *Zaus robustus* (see Itô, 1976). Species of *Acartia* and *Candacia* fail to express the bud of limb 5 at CIII, although the transformed limb is present at CIV (Ferrari & Ueda, 2005), and limb 5 has been reported as suppressed in *Sabellacheres illgi* (see Dudley, 1964) and *Porcellidium fimbriatum* (see Bocquet, 1948). Ferrari (1988) proposed that the development of the genital plate of female gymnoplean copepods was not part of the bud of limb 6 of calanoids, but recent scanning electron micrographs of the genital somite complex of several different species of *Pseudodiaptomus* by Walter et al.

(2002, figs. 9A-D, 10A-B) suggest that the genital flap plus operculum may indeed be homologous to the bud of limb 6.

Variation in transformed appendages

The morphology of antenna 1 during the copepodid phase of development has been reported extensively. Comparative development of representative species from six copepod orders has been described (Boxshall & Huys, 1998). Development of antenna 1 also has been compared among six families within the Harpacticoida (cf. Dahms, 1989a) and within 29 genera of the cyclopoid family Cyclopidae (cf. Schutze et al., 2000), as well as within three genera of the cyclopoid family Notodelphyidae (cf. Dudley, 1966). The number of segments of antenna 1 increases with increasing copepodid development (Dudley, 1966; Dahms, 1989a; Boxshall & Huys, 1998; Schutze et al., 2000). Female segmentation usually is complete by CV but males often undergo important changes during the molt to CVI (Dahms, 1989a). A stable terminal section of eight segments on antenna 1 is established early in development (Dahms, 1989a; Boxshall & Huys, 1998; Schutze et al., 2000), or if segmentation is not stable then setation of this section of the limb is stable (Dudley, 1966). Variation in segment number of antenna 1 usually results from a failure to express an arthrodial membrane separating two segments later in development (Dahms, 1989a; Boxshall & Huys, 1998; Schutze et al., 2000). Less often, variation results from the secondary loss of an arthrodial membrane that was expressed earlier in development between two segments (Dudley, 1966). This latter case usually occurs late in the development of male copepodids (Dudley, 1966; Dahms, 1989a).

The number of ramal segments of antenna 2 does not change throughout the copepodid phase of development of calanoids and polyarthrans (Dahms, 1993b; Ferrari, 1995), and ramal segmentation does not change during copepodid development of species of harpacticoids and siphonostomatoids, although the segment number of their exopod is reduced from that of the naupliar phase (Dahms, 1993b). At CI of cyclopoids and poecilostomes, the exopod of antenna 2 is a small, poorly-sclerotized, wrinkled structure with 1-3 setae (Ferrari & Ambler, 1992; Huys & Böttger-Schnack, 1994; Ferrari & Ivanenko, 2005). The wrinkled structure is lost at CII, although one or two setae, presumably terminal ramal setae, may remain throughout the copepodid phase of cyclopoids like *Dioithona oculata* and *Euryte longicauda* (see Ferrari & Ambler, 1992; Ferrari & Ivanenko, 2005). However, harpacticoids like *Macrosetella gracilis* (see Huys & Böttger-Schnack, 1994) and poecilostomes like *Anchistrotos pleuronichthydis* or *Critomolgus anthopleurus* (see Izawa, 1986; Kim, 2003) do not retain these ramal setae after CI.

The segmental configuration of the mandible does not change throughout the copepodid phase of development of most copepods, although the exopod of cyclopoids like *Dioithona oculata* (see Ferrari & Ambler, 1992) or harpacticoids like *Drescheriella glacialis* (see Dahms, 1987a) may have fewer segments than that of calanoids like *Ridgewayia klausruetzleri* (see Ferrari, 1995). At CI, the basis plus rami (or palp) of the mandible of cyclopid copepods is a poorly-sclerotized, often bifurcate, wrinkled structure with terminal setae, e. g., *Euryte longicauda* (see Ferrari & Ivanenko, 2005). This wrinkled structure is lost at CII while the terminal, ramal setae are retained throughout the copepodid phase. The mandible of the poecilostomes, *Hemicyclops ctenidis* (see Kim & Ho, 1992) or *Ergasilus hypomesi* (see Kim, 2004) and the siphonostomatoids, *Dermatomyzon nigripes* or *Asterocheres lilljeborgi* (see Ivanenko & Ferrari, 2003) does not change during the copepodid phase. However, the corresponding parts of the mandible of poecilostomes and siphonostomatoids are not well-understood.

The segmental configuration of maxilla 1 does not change through the copepodid phase of development, although there may be fewer segments and protopodal endites in cyclopoids like *Dioithona oculata* (see Ferrari & Ambler, 1992) or harpacticoids like *Drescheriella glacialis* (see Dahms, 1987a) relative to segments or endite numbers for calanoids like *Ridgewayia klausruetzleri* (see Ferrari, 1995). The corresponding parts of poecilostomes and siphonostomatoids (see Kim & Ho, 1992; Ivanenko & Ferrari, 2003; Kim, 2004) remain to be analysed. A poorly-sclerotized, terminal structure without setae has been reported on maxilla 1 of CI of some poecilostomes like *Midicola spinosus* [as *Pseudomyicola spinosus*] (see Do et al., 1984) or *Critomolgus anthopleurus* (see Kim, 2003). This wrinkled structure is lost at CII, and can be interpreted as a serial homolog of the mandibular palp of cyclopid copepods at CI of *Euryte longicauda* (see Ferrari & Ivanenko, 2005).

The segmental configuration of maxilla 2 does not change throughout the copepodid phase of development of copepods, although there may be fewer segments and protopodal endites in cyclopoids like *Dioithona oculata* (see Ferrari & Ambler, 1992) or harpacticoids like *Tisbe gracilis* (see Dahms & Bergmans, 1988) relative to segments or endite numbers for calanoids like *Ridgewayia klausruetzleri* (see Ferrari, 1995). The corresponding parts of poecilostomes like *Hemicyclops ctenidis* (see Kim & Ho, 1992) or *Ergasilus hypomesi* (see Kim, 2004) or siphonostomatoids like *Dermatomyzon nigripes* or *Asterocheres lilljeborgi* (see Ivanenko & Ferrari, 2003) are not well-understood.

Changes in the configuration of the maxilliped during the copepodid phase of development are confined to the incremental but significant addition of segments and/or setae to the endopod of most calanoids like *Ridgewayia klausruetzleri* (see Ferrari, 1995), polyarthrans like *Longipedia americana* or *Coullana canadensis* (see Ferrari & Dahms, 1998), some cyclopoids like *Oithona similis* (see Ferrari & Ivanenko, 2001) and siphonostomatoids like *Dermatomyzon nigripes* or *Asterocheres lilljeborgi* (see Ivanenko & Ferrari, 2003). In several studies, segmental homologies have been proposed (Ferrari & Dahms, 1998; Ferrari & Ivanenko, 2001). For other cyclopoids like *Pygodelphys aquilonaris*, there is no change in segmentation (Dudley, 1966). For some siphonostomatoids, like *Caligus epidemicus*, segmentation does not change although the shape of particular segments may change significantly (Lin et al., 1996). Among poecilostomes, changes in the maxilliped include its suppression at CI for *Ergasilus hypomesi* (see Kim, 2004) or a reconfiguration of the limb during the molt to CII for *Conchyliurus quintus* and *Critomolgus anthopleurus* (see Kim, 1994, 2003). A significant reconfiguration of the male maxilliped takes place during the molt to CVI of *Midicola spinosus* (see Do et al., 1984), *Hemicyclops ctenidis* (see Kim & Ho, 1992) and *Ergasilus hypomesi* (see Kim, 2004).

The swimming legs of copepods almost always undergo important changes during the copepodid phase of development, and there is a significant variation in these changes among species (Ferrari, 1988). To summarize, most changes involve truncation of development of a coordinated pattern of swimming leg development called the general pattern (Ferrari, 1988). The general pattern has been hypothesized to be ancestral for copepods because it is represented among so many copepods, including those species assumed to be basal in many of the copepod orders. Truncation may occur at several different steps during limb development resulting in significant developmental variability among species. In a few species, a change in swimming leg configuration may result from a delay in the formation of a particular limb element, e. g., the arthrodial membrane that separates two segments (Ferrari, 1998). An unusual variability is expressed in caligid-like siphonostomatoids. Swimming legs 1-2 of these copepods at CI are similar to those of most copepods at CI. However, during the four chalimus stages, corresponding to CII-CV, swimming legs often do not add arthrodial membranes or setae, and in some species the limb is transformed in such a way that it appears as a secondary limb bud (Ferrari, 1988). The molt to CVI, identified as either a pre-adult or an adult, results in adult swimming legs 1-4 whose morphology is remarkably similar to the swimming legs of copepods that do not develop through a set of chalimus stages.

The segmental configuration of limbs 5-6 of many copepods does not change from the limb bud step during the copepodid phase of development (Ferrari, 1988). The major exception is limb 5 of calanoids for which significant variability may be expressed in the segmentation of both exopod and endopod of both females and males. In males, this variability is thought to be an adaptation to some aspect of its performance during copulation (e. g., Blades & Youngbluth, 1980), and this variability may be of particular importance for spermatophore transfer to conspecific females.

Limb 5 of the female and male of calanids like *Neocalanus tonsus* develops much like swimming legs 1-4 (see Campbell, 1934, as *Calanus tonsus*), but the male limb 5 differs slightly from that of the female by adding a group of denticles or sensilla to the middle and distal exopodal segments of the limb on the side of the genital opening, during the terminal molt to a CVI adult. These denticles or sensilla presumably aid in manipulating the spermatophore during mating. Similar denticles or sensilla on the exopod of the fifth limb on the side of the genital opening can be found in most calanoids, and appear to be a synapomorphy for the order Calanoida. The male limb 5 of many calanoids expresses more than the simple morphological variability exhibited by *Neocalanus tonsus*. However, some of this variability appears to result from a reduction in the number of some exopodal segments and some or all endopodal segments by truncation during development (Ferrari & Ueda, 2005). Limb 5 of female calanoids also exhibits significant variability; in most cases this variability appears to result from the suppression of development of the endopodal segments, and the diminution of exopodal size and truncation of exopodal segmentation. The resulting configuration often includes a terminal seta or pointed attenuation of the distal segment (Heron & Bowman, 1971; Ferrari & Ueda, 2005). The extent to which this structure may be used to remove selected spermatophores from the female's genital complex or aid in removal of unwanted sperm from the seminal receptacles of the female has not been investigated.

Internal anatomy

Less is known about the internal anatomy of immature copepodid stages than about the naupliar stages of many copepods. In the notodelphyids,

Notodelphys affinis, *Pygodelphys aquilonaris* and *Doropygus longicauda*, a labrum is present although the midgut of CI is blocked by large yolk globules in its anterior part, and the foregut and hindgut are not open. At CII, yolk has dissipated from the midgut, which now has a lumen; the foregut and the hindgut are open (Dudley, 1966). A functional midgut valve is found at the level of the posterior section of the body, as well as a functional anal valve at CII. No differences were found in the cells of the foregut, midgut or hindgut of CI-CVI of *Lernaea cyprinacea*, although there are differences between the free adult female and the embedded adult female in length and cellular zonation of the midgut (Sabatini et al., 1987). At CI, extrinsic muscles of the functioning appendages of notodelphyids are striated but those of the bud of swimming leg 3 and of the longitudinal muscles of the posterior section of the body are not striated (Dudley, 1966). A ventral nerve from a neuropile at the level of swimming leg 3 at CI continues to the end of the body. Setae of the maxilliped of copepodids of *Temora stylifera* possess dendrites that suggest both mechanosensory and chemosensory functions (Paffenhöfer & Loyd, 1999).

Total lipid content of *Calanus finmarchicus* along with dry weight increase exponentially from CI to CV, but then only slightly during the molt to CVI (Kattner & Krause, 1987). Wax esters make up a greater proportion of total lipids as the copepodid phase of development proceeds, although there is a proportional decrease in CVI females as wax esters are allocated to egg production. Copepodids of the sympatric Southern Ocean calanoids, *Calanus propinquus*, *Calanoides acutus* and *Rhincalanus gigas* show an increase in lipid storage with development, and *Calanus propinquus* and *Calanoides acutus* also increase the carbon-chain lengths of stored lipids during development (Kattner et al., 1994). Copepodids of both *Calanoides acutus* and *Rhincalanus gigas* accumulate wax esters but of different carbon-chain lengths; copepodids of *Calanus propinquus* accumulate triacyloglycerols (Kattner et al., 1994). Wax esters are also the main lipid component of CI-CVI of *Euchaeta japonica* (see Lee et al., 1974) and CIII-CVI of *Calanus helgolandicus* (see Lee et al., 1972). The oil sac of CV *Calanus finmarchicus* takes up a proportionately larger space in specimens with larger prosome length, perhaps because the other internal organs do not vary with prosome length while the oil sac may vary (Miller et al., 2000). Gonadal tissue of *Calanus finmarchicus* initially has been detected at CV (Crain & Miller, 2000). The size and shape of the gonad of CV females goes through significant changes prior to and during the molt to CVI, and cell size of the presumptive testis and ovary is a better predictor of sex than is gonad size (Crain & Miller, 2000).

Functional morphology, swimming, and feeding behavior

Copepodid stages of calanoid copepods like *Temora stylifera* and *Centropages velificatus* move continuously through the water while they generate a feeding current. The first copepodid of *Eucalanus pileatus* is active only about half of the time (Paffenhöfer et al., 1996) and copepodids of *Acartia tonsa* have been characterized as intermittent swimmers (Buskey, 1994). Swimming speeds of the cyclopoid *Cyclops scutifer* increase as copepodid development proceeds (Gerritsen, 1978) but speeds of the calanoid *Acartia tonsa* do not appear to increase among early copepodids (Buskey, 1994). Copepodid stages of *Eurytemora affinis* appear to limit particle intake to a subset of the food particle sizes presented to them (Allan et al., 1977).

Seasonal cycles, vertical distribution, vertical migration

Copepodids of many species of planktonic calanoids from both marine and freshwater habitats assort vertically in the water column during the day. In general, this assortment has a well-characterized pattern. Adults are found deepest with progressively younger copepodid stages progressively closer to the water surface for the lagoonal *Acartia clausi* (see Landry, 1978a), for the freshwater *Leptodiaptomus novamexicanus* [reported as *Diaptomus novamexicanus*] (see Redfield & Goldman, 1980), for the oceanic *Pleuromamma xiphias* (see Ferrari, 1985) or for the oceanic *Calanus pacificus* and *Metridia lucens* (see Osgood & Frost, 1994b). For *Calanus finmarchicus*, this assortment also has been expressed as the proportion of specimens of a specific stage collected from a particular depth stratum (Dale & Kaartvedt, 2000). During the day, the degree to which vertically assorting copepodid stages of *C. finmarchicus* are separated may exhibit differences among spatially separated groups. Copepodids in the Atlantic Water and copepodids in the transition water between Polar Water and Atlantic Water express such differences (Dale & Kaartvedt, 2000).

Some variation in this general pattern of vertical assortment has been reported among pelagic calanoids (e.g., Yamaguchi et al., 2004). CI of *Paraeuchaeta norvegica* may be found slightly deeper than CII although the remaining copepodid stages assort deeper with increasing age (Fleddum et al., 2001). Both *Calanus helgolandicus* and *C. glacialis* may express an inverse vertical assortment, with younger stages deeper than older ones (Williams & Conway, 1980; Unstad & Tande, 1991), a pattern also expressed by *Gaidius variabilis* (see Yamaguchi et al., 2004). Planktonic cyclopoid species like *Dioithona oculata*, which form swarms during the day, seem to assort horizontally rather than vertically. CII-CVI of *D. oculata* make up the swarm, and CI, as well as the nauplii, are found outside the swarms (Ambler et al., 1991); all copepodids disperse horizontally at night (Ferrari et al., 2003). Daytime assortment of *D. oculata* appears to be similar to assortment of some parasitic cyclopoids, e. g., *Pachypygus gibber*, where CI is the free-swimming dispersive and infective stage of this notodelphyid parasite, while CII-CVI are found in the host (Hipeau-Jacquotte, 1978).

Copepodids of many species of copepods have been reported to undertake diel migrations. The most common migration studied is a diel vertical migration in which copepodid stages found deeper in the water column during the day migrate toward the surface at night (e. g., Landry, 1978; Redfield & Goldman, 1980; Ferrari, 1985). However, copepodids of *Metridia lucens* may occasionally perform reverse vertical migrations (Osgood & Frost, 1994a). The horizontal migrations of the swarming copepodids of *Dioithona oculata* are not as distinctive and have been described simply as dispersive, although a migration signal can be characterized from strata-specific sampling (Ambler et al., 1991; Ferrari et al., 2003).

For immature copepodids, migratory parameters such as percent migratory participation and migratory amplitude, often are not as pronounced as those for adult males; adult females often express intermediate parametric values, e. g., copepodids of the limnetic *Leptodiaptomus novamexicanus* (see Redfield & Goldman, 1980) or copepodids of the pelagic, marine *Calanus finmarchicus* (see Dale & Kaartvedt, 2000). Migratory participation and amplitude do not change with an increase in density of copepodids of *L. novamexicanus* in the late summer. However, migration parameters for copepodids of *Paraeuchaeta norvegica* may be controlled by food availability (Fleddum et al., 2001).

Most seasonal studies of copepodids have been carried out on marine planktonic copepods, usually calanoids like *Acartia clausi* or *Metridia pacifica* (see Landry, 1978; Batchelder, 1985) and less often on freshwater planktonic cyclopoids such as *Cyclops strenuus strenuus* or *Mesocyclops leuckarti* (see Elgmork, 1959; Alekseev, 1982) or harpacticoids like *Microsetella norvegica*, which is associated with unicellular, marine phytoplankton (Uye et al., 2002). Some of the most detailed studies are of diapausing marine calanoids like *Calanus finmarchicus*, *Calanus agulhensis*, *Calanus chilensis* or *Eucalanus inermis* (cf. e. g., Gislason & Astthorsson, 1996; Escribano et al., 1998; Gaard, 2000; Huggett & Richardson, 2000; Tande et al., 2000; Hidalgo et al., 2005). Generally, these calanoids become quiescent in waters well below 100 m as CIV or CV and then return to surface waters as CVI, or as eggs, with the onset of seasonal primary production, which often is triggered by upwelling. One or more generations follow before the late season descent to the depth of quiescence by CIV or CV; the descent usually is preceded by the diminution of seasonal primary production. Variations on this simple model include a distribution of developmental stages of *Calanus finmarchicus* after ascent, which correlates with the direction of water movement (Gislason et al., 2000), or differences in vertical assortment and diel vertical migration of *Calanus agulhensis* in food-rich areas as opposed to food-poor areas (Huggett & Richardson, 2000). Interannual differences in the abundance of copepodids of *Calanus finmarchicus* do not show a clear relationship with water temperature, because the annual effects of advection of cold water containing *C. finmarchicus* cannot be separated from effects of interannual changes in water temperature (Tande et al., 2000).

Development times, mortality, etc.

Isochronal development has been proposed for the copepodids of many free-living calanoid copepods, but it does not appear to be an effective predictor as a general rule for development for all copepods (Hart, 1990; Peterson, 2001). For example, CV has been reported as the stage of longest duration for three species of *Cyclops* (cf. Zánkai, 1987). Copepodids of *Cyclops vicinus* appear to be able to acclimate to temperature changes in a way that calanoids like *Eudiaptomus gracilis* cannot (Munro, 1974), and their stage durations are affected. CI is regarded as the dispersive stage for many parasites, and is of longer duration than those stages immediately following CI. The poecilostome, *Hemicyclops gomsoensis* is a good example of a copepod with a long duration of CI (Itoh, 2003). In general, the rate of increase in body size diminishes with older copepodid stages.

General statements about copepodid mortality from predation are available but usually these are restricted to planktonic, marine species. For example, copepodids of *Acartia clausi* appeared to be less affected by predation than the nauplii (Landry, 1978). Studies of stage-specific mortality are much more restricted. Stage-specific mortality declines to a constant level after CI for *Calanus finmarchicus*, before increasing at CV-CVI (Ohman et al., 2002). In contrast, stage-specific mortality of co-occurring species of *Pseudocalanus* is more uniform during copepodid development. The negligible stage-specific mortality of *Oithona similis* was assumed to reflect a relatively immotile feeding strategy (Eiane & Ohman, 2004). Data from contrasting marine habitats over a broad geographical range indicate that stage-specific mortality for *Calanus finmarchicus* varies appreciably (Ohman et al., 2004).

PATTERNING THE COPEPOD BODY

The variability in body architecture described in the previous chapter for the gymnoplean, *Ridgewayia klausruetzleri*, the podoplean, *Dioithona oculata* and the thaumatopsylloid, *Caribeopsyllus amphiodiae* represents fundamental differences in the general body architecture of copepods. Most of the remaining variation in the association of somites along the anterioposterior axis of the body results from one of two processes: the formation of somite complexes that result from the failure of an arthrodial membrane to form between two somites; or the suspension of the addition of somites to the body. These two processes are discussed below. The transformation of the shape of individual somites during development is a process that will not be discussed here.

A basic understanding of how the body is patterned during development is essential to the analyses of somite complexes or of the suspension of somite addition. Giesbrecht (1913) initially proposed that during each molt to a new copepodid stage, one new somite is added immediately anterior to, and adjacent to, the posterior abdominal somite, also known as the anal somite, which bears the caudal rami. That is, each new somite is added from a growth zone that is located in the anterior part of the posterior abdominal somite. Dudley (1966) recognized that changes in the notodelphyid body during copepodid development could be explained in this way, and Hulsemann (1991b) generalized the model for all copepods. A cellular basis for somite addition of copepods has yet to be proposed, although several cellular models exist for other crustaceans (Ooishi, 1959; Dohle & Scholtz, 1997). Alternative models, in which new somites are added either anteriorly or posteriorly from the posterior thoracic somite or anterior abdominal somite, have not been proposed and are not explored here.

Among gymnoplean copepods, the second thoracic somite usually articulates anteriorly with the cephalothorax during early stages of the copepodid phase of development. However, the arthrodial membrane separating the second thoracic somite from the cephalothorax may fail to form, so that the second thoracic somite becomes incorporated into the cephalothorax at CIII (fig. 23A, B; table III), e. g., *Euchaeta japonica* (see Campbell, 1934), or at CVI (fig. 23C, D; table IV), e. g., *Scopelatum vorax* (see Ferrari & Steinberg, 1993) and *Parkius karenwishnerae* (see Ferrari & Markhaseva, 1996).

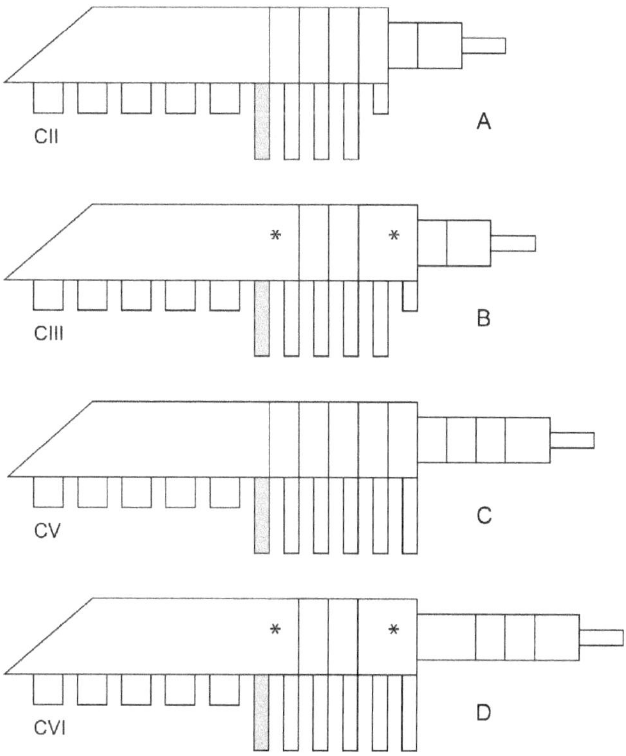

Fig. 23. *Euchaeta japonica*, schematic of the body and limbs (modified from Campbell, 1934); A, CII; B, CIII; asterisks on unarticulated thoracic somites 1 and 2, and unarticulated thoracic somites 5 and 6. *Parkius karenwishnerae*, schematic of the body and limbs (modified from Ferrari & Steinberg, 1993); C, CV; D, CVI; asterisks on unarticulated thoracic somites 1 and 2, and unarticulated thoracic somites 5 and 6. See fig. 21 for interpretation.

TABLE III

Somites of *Euchaeta japonica* at CI, CII and CIII. An initial capital letter indicates a somite or complex of the anterior part of the body; a somite of the posterior part of the body is in italics; an arthrodial membrane is indicated by a comma; absence of an arthrodial membrane is indicated by a dash; complexes of thoracic somites 1 + 2 and thoracic somites 5 + 6 are in **bold**; posterior is right. Cph, cephalon; th, thoracic somite; abd, abdominal somite

CI:	Cph – Th1, Th2, Th3, Th4, *th5*, *abd1*
CII:	Cph – Th1, Th2, Th3, Th4, Th5, *th6*, *abd1*
CIII:	Cph – **Th1-Th2**, Th3, Th4, **Th5-Th6**, *th7*, *abd1*

Among podopleans, failure of an arthrodial membrane to separate the second thoracic somite from the cephalothorax occurs only during the molt to CI, e. g., *Bryocamptus zschokkei alleganiensis* (see Carter & Bradford,

TABLE IV

Somites of the *Parkius karenwishnerae* female at CV and CVI; complexes of thoracic somites 1 + 2 and thoracic somites 5 + 6 are in **bold**; CI unknown. Explanations as in table III

CII:	Cph – Th1, Th2, Th3, Th4, Th5, *th6*, *abd1*
CIII:	Cph – Th1, Th2, Th3, Th4, Th5, Th6, *th7*, *abd1*
CIV:	Cph – Th1, Th2, Th3, Th4, Th5, Th6, *th7*, *abd2*, *abd1*
CV:	Cph – Th1, Th2, Th3, Th4, Th5, Th6, *th7*, *abd2*, *abd3*, *abd1*
CVI:	Cph – **Th1-Th2**, Th3, Th4, **Th5-Th6**, *th7-abd2*, *abd3*, *abd4*, *abd1*

TABLE V

Somites of *Pleuromamma xiphias* at CII and CIII; complex of thoracic somites 5 + 6 is in **bold**. Explanations as in table III

CI:	Cph – Th1, Th2, Th3, Th4, *th5*, *abd1*
CII:	Cph – Th1, Th2, Th3, Th4, Th5, *th6*, *abd1*
CIII:	Cph – Th1, Th2, Th3, Th4, **Th5-Th6**, *th7*, *abd1*

1972), *Hemicyclops japonicus* (see Itoh & Nishida, 1995), *Leptinogaster major* (see Humes, 1986), or *Taeniacanthus lagocephali* (see Izawa, 1986a).

The arthrodial membrane separating the fifth and sixth thoracic somites on the body of gymnopleans like *Euchaeta japonica* (see Campbell, 1934) and *Pleuromamma xiphias* (see Ferrari, 1985) may fail to form at CIII (fig. 23A, B; tables III, V).

CIII also is the stage at which the sixth thoracic somite is incorporated into the anterior section of the body. One stage earlier, at CII, the limbless sixth thoracic somite is a part of the posterior section of the body and articulates with the fifth thoracic somite that forms a part of the anterior section. The fifth thoracic somite at CII bears the bud of swimming leg 4. At CIII, the presence of the bud of limb 5 on the anterior section of the body supports the hypothesis that the sixth thoracic somite has become part of the anterior section of the body as it has formed a somite complex with the fifth thoracic somite; the arthrodial membrane separating the two somites at CII has failed to form at CIII. This arthrodial membrane remains suppressed throughout the remaining stages of copepodid development.

The arthrodial membrane between the fifth and sixth thoracic somites fails to form later in development, during the molt to CVI (fig. 23C, D; table IV), on other gymnopleans, e. g., *Scopelatum vorax* (see Ferrari & Steinberg, 1993) and *Parkius karenwishnerae* (see Ferrari & Markhaseva, 1996).

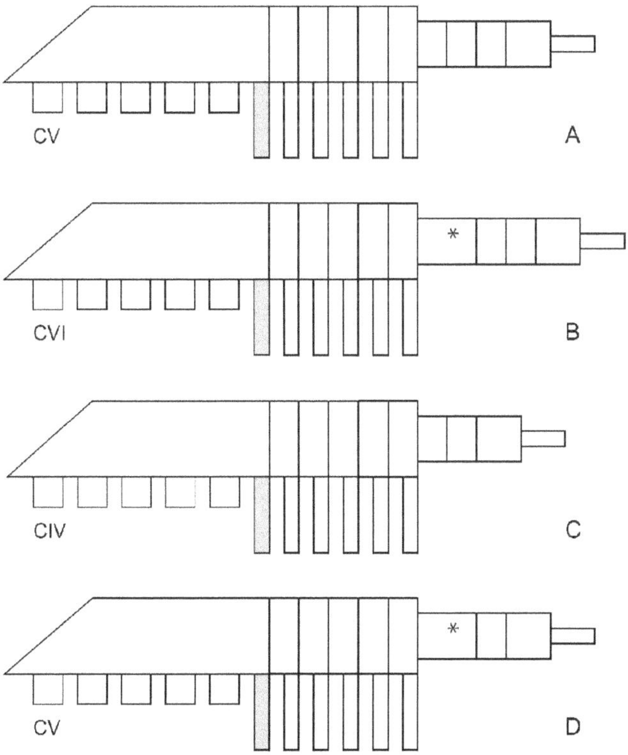

Fig. 24. *Ridgewayia klausruetzleri*, schematic of the body and limbs (modified from Ferrari, 1995); A, CV; B, CVI; asterisk on unarticulated posterior thoracic somite (thoracic somite 7) and anterior abdominal somite (abdominal somite 2). *Pontella chierchiae*, schematic of the body and limbs; C, CIV; D, CV; asterisk on unarticulated posterior thoracic somite (thoracic somite 7) and anterior abdominal somite (abdominal somite 2). See fig. 21 for interpretation of morphology.

TABLE VI

Somites of the *Ridgewayia klausruetzleri* female at CI-CVI; the genital complex of thoracic somite 7 and abdominal somite 2 is in **bold**. Explanations as in table III

CI:	Cph – Th1, Th2, Th3, Th4, *th5*, *abd1*
CII:	Cph – Th1, Th2, Th3, Th4, Th5, *th6*, *abd1*
CIII:	Cph – Th1, Th2, Th3, Th4, Th5, Th6, *th7*, *abd1*
CIV:	Cph – Th1, Th2, Th3, Th4, Th5, Th6, *th7*, *abd2*, *abd1*
CV:	Cph – Th1, Th2, Th3, Th4, Th5, Th6, th7, *abd2*, *abd3*, *abd1*
CVI:	Cph – Th1, Th2, Th3, Th4, Th5, Th6, ***th7-abd2***, *abd3*, *abd4*, *abd1*

TABLE VII

Somites of the *Pontella chierchiae* female, a centropagoidean calanoid, at CIV-CVI; the genital complex of thoracic somite 7 and abdominal somite 2 is in **bold**. Explanations as in table III

CIV:	Cph – Th1, Th2, Th3, Th4, Th5, Th6, *th7*, *abd2*, *abd1*
CV:	Cph – Th1-Th2, Th3, Th4, Th5-Th6, ***th7-abd2***, *abd3*, *abd1*
CVI:	Cph – Th1-Th2, Th3, Th4, Th5-Th6, ***th7-abd2***, *abd3*, *abd4*, *abd1*

TABLE VIII

Somites of the *Acartia tonsa* female at CIII-CVI. The genital complex of thoracic somite 7 and abdominal somite 2 is in **bold**. Explanations as in table III

CIII:	Cph – Th1, Th2, Th3, Th4, Th5, Th6, *th7*, *abd1*
CIV:	Cph – Th1, Th2, Th3, Th4, Th5, Th6, ***th7-abd2***, *abd1*
CV:	Cph – Th1, Th2, Th3, Th4, Th5, Th6, ***th7-abd2***, *abd3*, *abd1*
CVI:	Cph – Th1, Th2, Th3, Th4, Th5, Th6, ***th7-abd2***, *abd3*, *abd4*, *abd1*

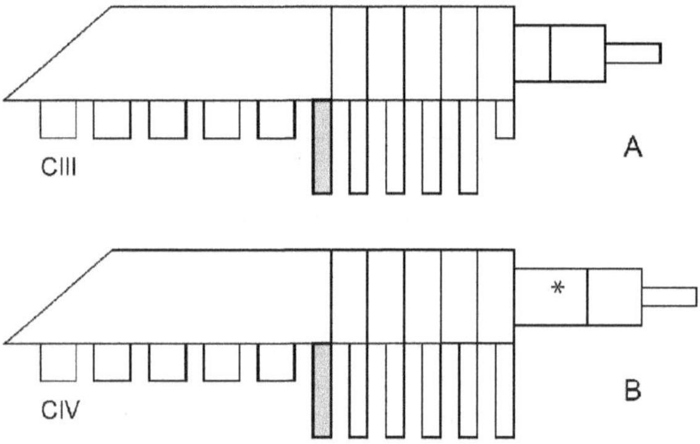

Fig. 25. *Acartia erythraea*, schematic of the body and limbs. A, CIII; B, CIV; asterisk on unarticulated posterior thoracic somite (thoracic somite 7) and anterior abdominal somite (abdominal somite 2). See fig. 21 for interpretation of morphology.

The genital complex of adult female copepods usually results from a failure of the arthrodial membrane to form between the posterior, or seventh, thoracic somite and the anterior, or second, abdominal somite. This failure is observed among gymnopleans, podopleans and thaumatopsylloids. Formation of the female genital complex usually occurs during the molt to CVI (fig. 24A, B; table VI). However, failure of the arthrodial membrane to form between the seventh thoracic and the second abdominal somites may take place during the molt to CV (fig. 24C, D; table VII) in many female centropagoidean calanoids (Ferrari & Ueda, 2005) and a few eucalanid calanoid females, e. g., *Eucalanus hyalinus* and *E. attenuatus* (see Geletin, 1976), as well as in

TABLE IX

Somites of the *Caligus elongatus* female at chalimus 3 and chalimus 4; the complex that includes thoracic somite 6 and thoracic somite 7 is in **bold**. Explanations as in table III

Chalimus 3:	Cph – Th1-Th2-Th3, Th4-Th5, *th6*, *th7-abd2-abd1*
Chalimus 4:	Cph – Th1-Th2-Th3, Th4-Th5, ***th6-th7-abd2-abd3-abd1***

the females of several poecilostomes, e. g., *Oncaea media* (see Malt, 1982) and *Neoergasilus japonicus* (see Urawa et al., 1980b) as pointed out by Izawa (1991).

Failure of an arthrodial membrane to form between the seventh thoracic and second abdominal somites also may occur earlier in development, during the molt to CIV (fig. 25A, B; table VIII) of females of all species of *Acartia*, e. g., *A. tonsa* (see Sabatini, 1990) or *A. californiensis* (see Trujillo-Ortiz, 1986), as well as of *Eucalanus subtenuis* (see Geletin, 1976).

A genital complex does not form in females of a few copepods, e. g., *Benthomisophria palliata* (see Boxshall & Roe, 1980), *Platycopia orientalis* (see Ohtsuka & Boxshall, 1994), *Notodelphys ascidicola* (see Allman, 1847), and *Scottomyzon gibberum* (see Ivanenko et al., 2001). In general, the female body architecture of these copepods is similar to that of the male.

Among caligid females, a different kind of complex of the genital somite is formed when the sixth thoracic somite fails to articulate posteriorly with the seventh thoracic (or genital) somite during the molt to chalimus 4 (table IX), e. g., *Caligus elongatus* (see Piasecki, 1996) or the comparable first pre-adult, e. g., *Caligus spinosus* (see Izawa, 1969) or *Lepeophtheirus dissimulatus* (see Lewis, 1963). Among these caligid siphonostomatoids, a number of segment complexes may form during chalimus development (table IX); chalimus 4 corresponds to CV of other copepods (see previous chapter).

During the terminal molt to the adult copepodid, CVI, suppression of the formation of the fourth abdominal somite has been reported for females of eucalanid calanoids, e. g., *Eucalanus attenuatus* and *E. subtenuis* (see Geletin, 1976), and for males of poecilostomes (Izawa, 1991), e. g., *Taeniastrotos pleuronichthydis* (see Izawa, 1986b, as *Anchistrotos pleuronichthydis*). One difficulty with confirming any hypothesis for suppression of formation of an abdominal somite is that there is no unambiguous way of determining whether an abdominal somite is absent or whether it has formed a somite complex with another somite. Abdominal somites do not bear limbs and so lack a clear marker for their position on the body, unlike thoracic somites whose limb is expressed in a predictable setose, bud-like configuration one

TABLE X

Somites of the *Pleuromamma xiphias* female at CV, and four different interpretations of CVI; 1: suspension of formation of the fourth abdominal somite; 2-4 masking one or more abdominal somites; different somite complexes on the posterior part of the body are in **bold**. Explanations as in table III

CV:	Cph – Th1, Th2, Th3, Th4, Th5-Th6, *th7*, *abd2*, *abd3*, *abd1*
1: CVI:	Cph – Th1, Th2, Th3, Th4, Th5-Th6, ***th7-abd2***, *abd3*, *abd1*
2: CVI:	Cph – Th1, Th2, Th3, Th4, Th5-Th6, ***th7-abd2-abd3***, *abd4*, *abd1*
3: CVI:	Cph – Th1, Th2, Th3, Th4, Th5-Th6, ***th7-abd2***, ***abd3-abd4***, *abd1*
4: CVI:	Cph – Th1, Th2, Th3, Th4, Th5-Th6, ***th7-abd2***, *abd3*, ***abd4-abd1***

stage after the somite has formed. In any hypothesis for suppression of an abdominal somite, an alternate explanation should be considered, that one or more arthrodial membranes separating abdominal somites may have failed to form, masking more than one abdominal somite within an abdominal somite complex. For example, the posterior section of the body of females of *Pleuromamma xiphias* at CV is composed of the seventh thoracic somite and the first three abdominal somites (table X). At CVI, the posterior section of the body is composed of a large genital complex anteriorly, a middle abdominal somite or an abdominal somite complex, and the posterior abdominal somite or a complex of the posterior abdominal somite and another abdominal somite.

One interpretation of the posterior section of the body of *P. xiphias* is that the formation of the fourth abdominal somite has been suppressed. In this case, the genital complex is of the usual architecture, formed by the suppression of the arthrodial membrane between the seventh thoracic somite and the second abdominal somite; the middle part of the posterior section is the third abdominal somite, which articulates with the posterior abdominal, or anal, somite (table X (1)). However, if the formation of the fourth abdominal somite has not been suppressed, that somite may be present but undetected if the genital complex is composed of the seventh thoracic, and the second and third abdominal somites. The middle part of the urosome would then be the fourth abdominal somite, which articulates with the posterior (= first) abdominal, or anal, somite (table X (2)). Alternatively, the fourth abdominal somite may comprise a complex with the third abdominal somite to make up the middle part of the urosome (table X (3)), and finally it may comprise a posterior somite complex with the anal somite (table X (4)). Evidence that a newly formed somite may remain unarticulated with the posterior abdominal somite also is observed in the copepodid stages of *Caribeopsyllus amphiodiae* (see previous chapter).

Among copepods, most somite complexes along the copepodid body result from failure to express one or more arthrodial membranes that separate somites. This failure of expression usually occurs after the arthrodial membranes initially have been expressed earlier in development. There are only a few examples in which failure to express an arthrodial membrane occurs at the same stage that the somite initially is added to the body; examples of these somite complexes will include the posterior abdominal, or anal, somite.

Patterning processes, like the one above for the addition of somites to the body, provide a way to infer conditions in which serially homologous structures may fail to form during development. Toward this end, the following Rule of Serial Homologs can be formulated:

"If serial homologs which are formed later during the normal course of patterning are present, then serial homologs which are formed earlier during the normal course of patterning also should be present."

An inaptly defined pseudosomite between the sixth and seventh thoracic somite of some harpacticoids and cyclopoids (see Klie, 1949; Huys & Boxshall, 1991; Martínez Arbizu, 1997, 1999) provides an example of the utility of the Rule of Serial Homologs. This area of sclerotization is described as located posterior to the developmentally older, sixth thoracic somite and anterior to the developmentally younger, seventh thoracic somite. Sixth and seventh thoracic somites without an intervening pseudosomite appear to be present in species that are both ancestral and descendent to those with the pseudosomite. If the pseudosomite were a part of the body comparable to but distinct from the sixth and seventh thoracic somites, then the normal anterioposterior patterning of the body would have to be suspended in order to accommodate its addition. A more likely explanation is that the pseudosomite results from an intermediate section of weak sclerotization of the anterior part of the seventh thoracic somite.

PATTERNING THE APPENDAGES OF COPEPODS

Much less is known about how copepod limbs are patterned during post-embryonic development than about how the copepod body is patterned. Several types of evidence have been used to infer limb patterning. **Alignment analysis** locates segments or setae with unique morphology that are conserved during two or more developmental stages. Aligning images of limbs by juxtaposing the unique segments or setae at successive stages can be used to generate an hypothesis about the particular segments and setae that have been added to each successive stage. In the method of **formation homology**, some organization of the exoskeleton of the following stage of development may be identified within the exoskeleton of a specimen, because the exoskeleton of this present stage appears to be used as a template for the exoskeleton of the following developmental stage. New setae and arthrodial membranes that will be expressed in the following developmental stage often can be located from the internal organization of the present stage. **Youngest element analysis** identifies the developmental age of segmental elements, like setae and arthrodial membranes, of a limb and a basic hypothesis of limb patterning can be deduced from these developmental ages. One or more segments of the limb are here designated as source segments. A source segment appears to be homologous to the formative zone (Fuller, 1920) or meriston (Henson, 1947) of the antenna of hemimetabolous insects, which is the homolog of antenna 1 of crustaceans. A source segment can be located on the limb in the following two ways: if the limb is patterned either proximally or distally to the source segment, the source segment is located between the youngest and the oldest element; if the limb is patterned both proximally and distally from the source segment, the source segment is located among the younger elements.

Antenna 1

Antenna 1 of copepods is a uniramous limb throughout its development; protopodal segmentation and the identity of the ramus are not clear, although evidence for considering the ramus an endopod (Ferrari & Benforado, 1998b) seems reasonable. Variability in the number of segments expressed both phylogenetically and during its ontogeny has attracted significant interest in the development of antenna 1. Studies of formation homology for antenna 1 have focused almost exclusively on the last naupliar stage of calanoids, e. g., *Temora longicornis* (see Oberg, 1906), *Eurytemora velox* (see Gurney, 1931), *Diaptomus siciloides* (see Comita & Tommerdahl, 1960), *Diaptomus oregonensis* (see Comita & McNett, 1967), and *Drepanopus forcipatus* (see Hulsemann, 1991a). The primary purpose of these studies was to determine the homologous setae on antenna 1 of the last nauplius and first copepodid stages, because some setae on antenna 1 of the last nauplius of calanoids have no successor on antenna 1 of the first copepodid.

A valuable descriptive study compared the development of antenna 1 among representative species from six copepod orders (Boxshall & Huys, 1998). Although no model was provided to pattern the addition of new segment elements to this limb, an alignment analysis was used to determine new segments at each stage; the production of new segments seemed to be located at several different locations along the limb. An alignment analysis using an unusual segment in *Tigriopus japonicus* (see Itô, 1970: 496, fig. 12) or in *Thermomesochra reducta* (see Itô & Burton, 1980: 21, fig. 14), or unusual setae in *Notodelphys affinis*, *Pygodelphys aquilonaris* and *Doropygus* spp. (see Dudley, 1966: 132, table VI) was used in proposing a segment-splitting model for the development of antenna 1 of these two harpacticoids and three notodelphyid cyclopoids. An increase in the number of segments was modeled through a process of splitting segments, so that one large segment at an early copepodid stage split to form two segments or more at a later copepodid stage. The patterning of antenna 1 for 39 species in 23 genera of Cyclopidae apparently includes segment splitting and the production of more than one segment distally from different points along antenna 1 (Schutze et al., 2000). Most new segments on antenna 1 of these cyclopoids were added during the molt to the sixth copepodid, and most variation in configuration of the limb resulted from suspension of the addition of different segments during that molt.

In the above studies, the origin of new segments was not confined to (a) particular location or locations along antenna 1; instead the presentation of new segment elements appeared to be a rather diffuse phenomenon along the proximodistal axis of the limb. As a result, any segment during any step of limb development might split to form two segments. Another general weakness of these segment-splitting models is that a segment that is identified as new at a particular stage could have been united in the previous stage with the segment either proximal to it or distal to it. However, no method was proposed to make this important determination.

A survey of 33 species from 27 genera of harpacticoid copepods in 17 families (Dahms, 1989a) identified either segment fusion, resulting from the loss of a pre-existing arthrodial membrane, or the proliferation of new segments from a source segment as the two primary ways the harpacticoid antenna 1 is configured during development. Most segments added to the antenna 1 of these animals appear to result from patterning either proximally or distally from the segment initially adjacent to the proximal segment of antenna 1 at CI. However, a few new segments of a few species appear either proximally or distally to the segment adjacent to the distal segment at CI.

For a study of the calanoid copepods, *Ridgewayia klausruetzleri*, *Pleuromamma xiphias* and *Pseudocalanus elongatus*, the number of places on antenna 1 that new segments were allowed to originate was minimized so that patterning of the limb was well-focused (Ferrari & Benforado, 1998b). Three source segments were identified for antenna 1 (fig. 26), and the distal and middle source segments were responsible for most of the patterning. Both of these source segments added segments or segment elements either proximally or distally, and both could add elements of more than one segment during a molt. No attempt was made by Ferrari & Benforado (1998b) to determine the correspondence between the source segments of these calanoids and the source segment of the harpacticoids (Dahms, 1989a).

Among other crustaceans, only the middle two or three segments of antenna 1 of podocopid ostracodes are patterned during post-embryonic development (Maddocks, 2000); whether these segments are patterned from a source segment is not clear. A diffuse model of segment splitting also has been proposed for podocopan ostracodes (Smith & Tsukagoshi, 2005). The flagellum of the asellid isopod, *Asellus aquaticus* adds segment elements during two developmental steps (discussed but not illustrated in Maruzzo et al., 2007). The proximal flagellomere, which is the fifth article of antenna 1, is a source segment of the new segment elements. The proximal flagellomere produces, immediately distal to itself, one primary flagellomere during each of a series of molts. Primary flagellomeres, in turn, produce doublets of new flagellomeres during the following molt.

The proximal flagellomere of the flagellum of the asellid isopod, *Lirceus macrourus* (see Zeleny, 1907, as *Mancasellus macrourus*) and of the amphipod, *Gammarus chevreuxi* (see Sexton, 1924) also serves as a source segment to produce one new flagellomere after every molt. A proximal article of the lateral flagellum serves as a source segment for flagellomeres for antenna 1 of the decapod, *Panulirus argus*; three new flagellomeres are produced

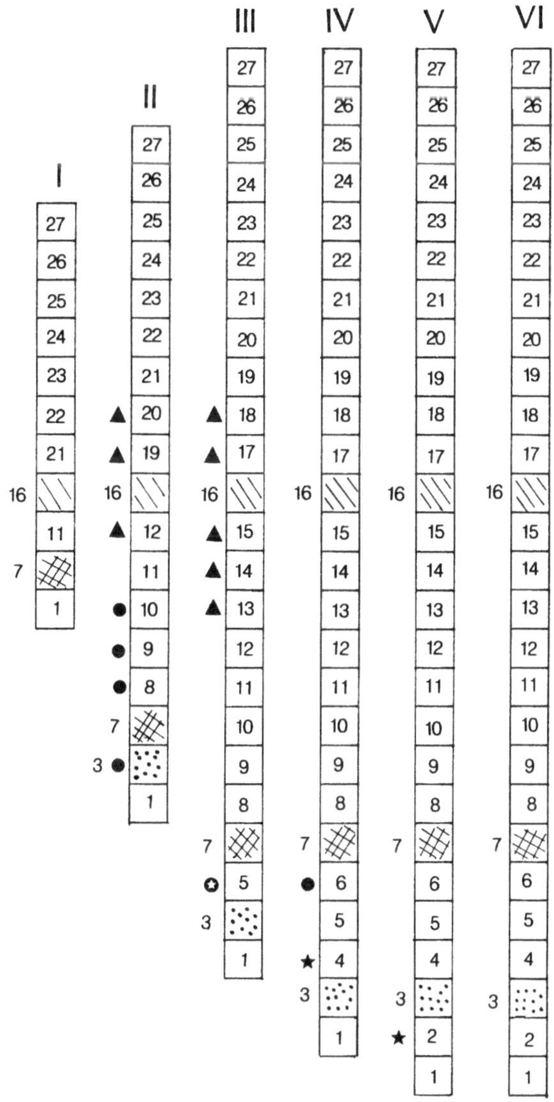

Fig. 26. Schematic of patterning of calanoid antenna 1 from CI = I to CVI = VI (modified from Ferrari & Benforado, 1998b). Horizontal lines delimit groups of setae that are segmental elements; horizontal lines may not correspond to arthrodial membranes. Distal source segment (16) is hatched, middle source segment (7) is cross-hatched, proximal source segment (3) is stippled; triangles to the left mark progeny of distal source segment, circles mark progeny of middle source segment, star marks progeny of proximal source segment; star-in-circle may be a progeny of middle or proximal source segment.

during each molt (Steullet et al., 2001). Observations of more derived pancrustaceans also are available (Fuller, 1920; Henson, 1947; Minelli, 2004).

Exopod of antenna 2

Nothing has been published about the patterning of antenna 2 of copepods. However, as noted in the chapter on "The nauplius and naupliar development", all of the segment elements of the exopod of antenna 2 are added during the naupliar phase of development or during the molt to the first copepodid. Development of antenna 2 of the calanoid copepod, *Calanus finmarchicus* is illustrated in fig. 1. Antenna 2 appears as a transformed limb at NI; the exopod is 6-segmented. The proximal segment of the exopod is unarmed; the adjacent segment is elongate and bears a ventral seta distally. The middle segment is short and bears a ventral seta. The antepenultimate segment and the penultimate segment also are short and have the same setal configuration as the middle segment. The distal segment bears a crown of three setae.

A short segment with a ventral seta is added at NII and a second short segment with a ventral seta is added at NIII. At NIV, a second ventral seta is added to the elongate segment; at NV and NVI, a third and then a fourth seta, respectively, are added to the elongate segment. At CI, the proximal segment is unarmed; the adjacent segment is short and bears a ventral seta distally. An elongate segment bears three ventral setae. Four short segments follow, each with a ventral seta. The penultimate segment is elongate and the terminal segment bears a crown of three setae. The setation of the exopodal segments of these stages is presented in table XI.

One way the data in table XI may be interpreted is by assuming that each exopodal segment, except the terminal segment, bears no more than one ventral seta; this ventral seta is a formation seta because it is present when the segment elements initially appear. In addition, the proximal, or finishing, arthrodial membrane of a segment is assumed to be more labile than its ventral seta; failure to form a finishing arthrodial membrane results

TABLE XI

Setation of the exopod of antenna 2 of *Calanus finmarchicus* with segments simply indicated by placement of arthrodial membranes (as commas); distal is right

NI:	0, 1, 1, 1, 1, 3
NII:	0, 1, 1, 1, 1, 1, 3
NIII:	0, 1, 1, 1, 1, 1, 1, 3
NIV:	0, 2, 1, 1, 1, 1, 1, 3
NV:	0, 3, 1, 1, 1, 1, 1, 3
NVI:	0, 4, 1, 1, 1, 1, 1, 3
CI:	0, 1, 3, 1, 1, 1, 1, 1, 3

in a segment complex. Here the elongate segment with more than one seta is a complex. The penultimate segment is a segment modified by slight elongation of its distal part so that the formation seta is found in the middle of the segment. The terminal segment is the only segment that may bear more than one seta. Segmentation and setation now are interpreted in table XII.

A simple way to explain the patterning of the segmental elements, such as setae and finishing arthrodial membranes, on this exopod (table XII) is to use the elongate segment as a marker and to assume that there is a source segment for new segment elements proximal to the distal ventral seta within the elongate segment, as indicated in table XIII. This source segment is located within a segment complex, does not bear a seta, and usually is not identified by a finishing arthrodial membrane.

TABLE XII

Setation of the exopod of antenna 2 of *Calanus finmarchicus* with one ventral seta per segment except for the distal segment (with a crown of three terminal setae); more than one seta on any other 'segment' indicates a segment complex; comma (,) is an arthrodial membrane, dash (-) indicates the location where an arthrodial membrane has failed to form, resulting in a segment complex; distal is right

NI:	0, 1, 1, 1, 1, 3
NII:	0, 1, 1, 1, 1, 1, 3
NIII:	0, 1, 1, 1, 1, 1, 1, 3
NIV:	0, 1-1, 1, 1, 1, 1, 1, 3
NV:	0, 1-1-1, 1, 1, 1, 1, 1, 3
NVI:	0, 1-1-1-1, 1, 1, 1, 1, 1, 3
CI:	0, 1, 1-1-1, 1, 1, 1, 1, 1, 3

TABLE XIII

Segments of the exopod of antenna 2 of *Calanus finmarchicus*. Roman numeral 'one' with asterisk (I*) is the source segment that patterns the ramus; comma (,) is an arthrodial membrane, dash (-) indicates the location where an arthrodial membrane has failed to form, resulting in a segment complex; distal is right. Lower case letters are segments distal to the source segment and 'b' is the oldest of these segments; Arabic numerals are segments proximal to the source segment and 2 is the oldest of these segments. Setae are not indicated in this table

NI:	2, I*-f , e, d, c, b
NII:	2, I*-g, f, e, d, c, b
NIII:	2, I*-h, g, f, e, d, c, b
NIV:	2, I* i-h, g, f, e, d, c, b
NV:	2, I*-j-i-h, g, f, e, d, c, b
NVI:	2, 3-I*-j-i-h, g, f, e, d, c, b
CI:	2, 3, I*-j-i, h, g, f, e, d, c, b

At NII-NIII, the new segments 'g', and 'h', respectively, are added distally and adjacent to the source segment (table XIII); each of these new segments has a finishing arthrodial membrane. Elements of the new segments 'i' and 'j' are added at NIV-NV; they also are distal to the source segment. These segments do not have a finishing arthrodial membrane, and so they make up part of a segment complex that includes the source segment. At NVI, the ventral seta of the new segment '3' is added adjacent, but proximally, to the source segment. Its distal finishing arthrodial membrane is added adjacent, but proximally, to the source segment during the molt to CI. This arthrodial membrane separates segment '3' from the adjacent elongate segment complex, which continues to include the source segment proximally.

Antenna 2 of polyarthran copepods also develops a multi-segmented exopod during the naupliar phase of development and retains it through the copepodid phase. For comparative purposes, development of antenna 2 of the polyarthran copepod, *Longipedia americana* is illustrated in fig. 2. The exopod is 6-segmented at NI, and differs from that of *Calanus finmarchicus* only in its distal segment, whose crown has two setae rather than three. NII of *L. americana* also differs from NII of *C. finmarchicus* only in the number of crown setae on the distal segment. NIII does not differ at all from that of *C. finmarchicus*, because a third seta has been added to the crown of *L. americana*. At NIV, a short segment with a ventral seta is added in *L. americana*; at the corresponding stage of *C. finmarchicus*, a second ventral seta has been added to the elongate segment. A second ventral seta is added to the elongate segment at NV of *L. americana*. There is no change in the exopod at NVI. At CI, the penultimate segment is short and an arthrodial membrane is added, which divides the elongate second segment with two setae into a long segment with one seta and distal to it a new short segment with one seta. The new short segment is the flexion point for the exopod. The setation of the exopod of these stages is presented in table XIV.

Again, several assumptions can be applied to interpret the addition of segments and setae to this exopod. Each exopodal segment except the terminal segment bears no more than one ventral seta, this ventral seta is a formation seta. The proximal, finishing arthrodial membrane of a segment is more labile than its ventral seta, so that the failure of an arthrodial membrane to form results in a segment complex. These assumptions yield the interpretation in table XV.

Applying an assumption for this exopod of one source segment that does not bear a seta, that may not be identified by arthrodial membranes, and that

TABLE XIV

Setation of the exopod of antenna 2 of *Longipedia americana* with segments simply indicated by placement of arthrodial membranes (as commas); distal is right

NI:	0, 1, 1, 1, 1, 2
NII:	0, 1, 1, 1, 1, 1, 2
NIII:	0, 1, 1, 1, 1, 1, 1, 3
NIV:	0, 1, 1, 1, 1, 1, 1, 1, 3
NV:	0, 2, 1, 1, 1, 1, 1, 1, 3
NVI:	0, 2, 1, 1, 1, 1, 1, 1, 3
CI:	0, 1, 1, 1, 1, 1, 1, 1, 1, 3

TABLE XV

Setation of the exopod of antenna 2 of *Longipedia americana* with one ventral seta per segment except for distal segment (with a crown of two or three terminal setae); more than one seta on any other 'segment' indicates a segment complex; comma (,) is an arthrodial membrane, dash (-) indicates the location where an arthrodial membrane has failed to form, resulting in a segment complex; distal is right

NI:	0, 1, 1, 1, 1, 2
NII:	0, 1, 1, 1, 1, 1, 2
NIII:	0, 1, 1, 1, 1, 1, 1, 3
NIV:	0, 1, 1, 1, 1, 1, 1, 1, 3
NV:	0, 1-1, 1, 1, 1, 1, 1, 1, 3
NVI:	0, 1-1, 1, 1, 1, 1, 1, 1, 3
CI:	0, 1, 1, 1, 1, 1, 1, 1, 1, 3

TABLE XVI

Segments of the exopod of antenna 2 of *Longipedia americana*. Roman numeral 'one' with asterisk (I*) is the source segment that patterns the ramus; comma (,) is an arthrodial membrane, dash (-) indicates the location where an arthrodial membrane has failed to form, resulting in a segment complex; distal is right. Lower case letters are segments distal to the source segment and 'b' is the oldest of these segments; Arabic numerals are segments proximal to the source and '2' is the oldest of these segments. Setae are not indicated in this table

NI:	2, I*-f, e, d, c, b
NII:	2, I*-g, f, e, d, c, b
NIII:	2, I*-h, g, f, e, d, c, b
NIV:	2, I*-i, h, g, f, e, d, c, b
NV:	2, I*-j-i, h, g, f, e, d, c, b
NVI:	2, I*-j-i, h, g, f, e, d, c, b
CI:	2, I*-j, i, h, g, f, e, d, c, b

is located proximal to the distal ventral seta on the elongate segment of NI, results in the identity of segmental elements expressed in table XVI.

Comparing tables XI and XII for *Calanus finmarchicus* with tables XIV and XV, respectively, for *Longipedia americana*, several differences can be identified in the two exopods. The ventral seta of segment '3' of *C. finmarchicus*, which is added at NVI, plus its finishing arthrodial membrane, which is added at CI, both fail to form in *L. americana*. The finishing arthrodial membrane of segment 'i' of *L. americana* initially appears at CI; this finishing arthrodial membrane never forms in *C. finmarchicus* so that segment 'i' remains a part of its elongate complex.

Among other crustaceans, the flagellum of antenna 2 of the asellid isopod, *Asellus aquaticus* results from the addition of a set of quartets of new flagellomeres; each set is produced by the activity of a primary flagellomere. One primary flagellomere separates distally from the proximal flagellomere at each molt and subsequently produces its quartet of flagellomeres by a stereotypic process (Maruzzo et al., 2007). The first flagellomere of the flagellum of valviferan isopods belonging to the genus *Idotea* serves as a source segment that patterns a new flagellomere after every molt (noted but not illustrated in Naylor, 1955: 482).

Exopod of the mandible

Little is known about how the copepod mandible is patterned. During the naupliar phase of development, one ventral seta is added to the proximal exopodal segment of copepods like *Longipedia americana* (see Onbé, 1984), *Scutellidium hippolytes* (see Dahms, 1993b), and *Hemicyclops japonicus* (see Itoh & Nishida, 1997). The addition of this seta to the mandibular exopod suggests that the seta is a ventral formation seta of a new segment, without its finishing arthrodial membrane, that has been added to the exopod. This situation is similar to the exopod of antenna 2, with its source segment near the proximal border of the limb.

Maxilla 1 and maxilla 2

Nothing has been published on the patterning of maxilla 1 and maxilla 2 of copepods. The protopod comprises a significant part of these limbs, and some ideas about how these limbs are patterned will be suggested later in this chapter when a model of protopodal patterning is presented.

Endopod of the maxilliped

Unlike the exopod of antenna 2, the copepod maxilliped develops almost exclusively during the copepodid phase, although the limb bud, bearing a crown of two setae, is present at NVI of most calanoids. Efforts to understand how the copepod maxilliped is patterned have focused on the endopod (Ferrari, 1995; Ferrari & Dahms, 1998; Ferrari & Ivanenko, 2001). The addition of setae and arthrodial membranes to the endopod of the maxilliped of the calanoid, *Ridgewayia klausruetzleri* is shown in fig. 15 and table XVII.

The distal segment of the endopod at CI includes a crown group of two terminal setae plus a subterminal dorsal seta and a subterminal ventral seta. This segment appears to be a complex of two segments, a distal one bearing the terminal crown of two setae, and an adjacent segment bearing a ventral and a dorsal formation seta. The setation of segments of the maxilliped of *Ridgewayia klausruetzleri* then can be revised as in table XVIII.

A source segment has been observed directly from formation homology for the endopod of the maxilliped of *R. klausruetzleri* (see Ferrari, 1995 and fig. 15), and for *Eurytemora affinis* (see Ferrari & Ivanenko, 2001). This source segment of the endopod is the proximal segment at CI, and the ante-

TABLE XVII

Setation of the endopod of the maxilliped of *Ridgewayia klausruetzleri*, except for NVI from *Eurytemora affinis*, with segments simply indicated by the placement of arthrodial membranes (as commas); distal is right

NVI:	2
CI:	1, 4
CII:	1, 1, 4
CIII:	1, 1, 2, 4
CIV:	2, 2, 1, 2, 4
CV:	3, 3, 2, 3, 4
CVI:	4, 4, 3, 4, 4

TABLE XVIII

Setation of the endopod of the maxilliped of *Ridgewayia klausruetzleri*, except for NVI from *Eurytemora affinis*; comma (,) is an arthrodial membrane, dash (-) indicates the location where an arthrodial membrane has failed to form, resulting in a segment complex; distal is right

NVI:	2
CI:	1, 2-2
CII:	1, 1, 2-2
CIII:	1, 1, 2, 2-2
CIV:	2, 2, 1, 2, 2-2
CV:	3, 3, 2, 3, 2-2
CVI:	4, 4, 3, 4, 2-2

TABLE XIX

Segments of the endopod of the maxilliped of *Ridgewayia klausruetzleri*. Roman numeral 'one' with asterisk (I*) is the source segment that patterns the ramus; comma (,) is an arthrodial membrane, dash (-) indicates the location where an arthrodial membrane has failed to form, resulting in a segment complex; distal is right. Lower case letters are segments distal to the source segment, 'b' is the oldest of these segments; Arabic numerals are segments proximal to the source, '2' is the oldest of these segments. Setae are not indicated in this table

NVI:	I*- b
CI:	I*, c-b
CII:	2, I*, c-b
CIII:	2, 3, I*, c-b
CIV:	2, 3, 4, I*, c-b
CV:	2, 3, 4, I*, c-b
CVI:	2, 3, 4, I*, c-b

penultimate segment of all copepodid stages. As noted above, the arthrodial membrane between the penultimate and distal segments fails to form in these copepods. Unlike the source segment for the exopod of antenna 2, the source segment for the endopod of the maxilliped bears a formation seta ventrally, and can be defined by arthrodial membranes proximally and distally.

A dorsal seta is added to the source segment or antepenultimate segment during the molt to CIII. Ventral setae added to the source segment and segments proximal to it during the molts to CIV, CV and CVI are post-formation setae, because they are added after the formation seta and the distal, finishing arthrodial membrane initially have been presented. Post-formation setae are added to the proximal segment and to the segment adjacent to it during the molt to CIV, and to all segments except the distal segment during the molts to CV and CVI, respectively. Post-formation setae do not represent segment complexes but instead are supernumerary setae. Applying these assumptions, the following interpretation about the identity of segments of the endopod of the maxilliped is presented in table XIX.

Post-formation setae are common on the endopod of the maxilliped of polyarthrans as well as calanoids (see Ferrari & Dahms, 1998), but are rare in copepods of most other orders. Table XX shows segments and setae of the endopod of the maxilliped of the cyclopoid, *Procyclopina feiticeira* (from Ferrari & Ivanenko, 2001), the harpacticoid, *Macrosetella gracilis* (from Ferrari & Dahms, 1998), and the siphonostomatoid, *Scottomyzon gibberum* (from Ivanenko et al., 2001); no new segments or setae are added to the cyclopoid after CIV, to the harpacticoid after CI, or to the siphonostomatoid after CIII.

TABLE XX

Setation of the endopod of the maxilliped of, A, the cyclopoid, *Procyclopina feiticeira* (from Ferrari & Ivanenko, 2001); B, the harpacticoid, *Macrosetella gracilis* (from Ferrari & Dahms, 1998); and, C, the siphonostomatoid, *Scottomyzon gibberum* (from Ivanenko et al., 2001); comma (,) is an arthrodial membrane, dash (-) indicates the location where an arthrodial membrane has failed to form, resulting in a segment complex; distal is right

	A	B	C
NVI:	not present	not present	not present
CI:	1, 2-2	1, 1	1, 2
CII:	0, 1, 2-2	1, 1	1-1, 2
CIII:	0, 1, 1, 2-2	1, 1	1-1-1, 2
CIV:	1, 1, 1-1, 2-2	1, 1	1-1-1, 2
CV:	1, 1, 1-1, 2-2	1, 1	1-1-1, 2
CVI:	1, 1, 1-1, 2-2	1, 1	1-1-1, 2

TABLE XXI

Segments of the endopod of the maxilliped of, A, the cyclopoid, *Procyclopina feiticeira* (from Ferrari & Ivanenko, 2001); B, the harpacticoid, *Macrosetella gracilis* (from Ferrari & Dahms, 1998); and C, the siphonostomatoid, *Scottomyzon gibberum* (from Ivanenko et al., 2001). Roman numeral 'one' with asterisk (I*) is the source segment that patterns the ramus; comma (,) is an arthrodial membrane, dash (-) indicates the location where an arthrodial membrane has failed to form, resulting in a segment complex. Lower case letters are segments distal to the source segment, 'b' is the oldest of these segments; Arabic numerals are segments proximal to the source, '2' is the oldest of these segments. Setae are not indicated in this table

	A	B	C
CI:	I*, c-b	I*, b	I*, b
CII:	2, I*, c-b	I*, b	2-I*, b
CIII:	2, 3, I*, c-b	I*, b	2-3-I*, b
CIV:	2, 3, 4-I*, c-b	I*, b	2-3-I*, b
CV:	2, 3, 4-I*, c-b	I*, b	2-3-I*, b
CVI:	2, 3, 4-I*, c-b	I*, b	2-3-I*, b

Applying the calanoid patterning model to the cyclopoid, *P. feiticeira* (from Ferrari & Ivanenko, 2001), the harpacticoid, *M. gracilis* (from Ferrari & Dahms, 1998), and the siphonostomatoid, *S. gibberum* (from Ivanenko et al., 2001) results in the following interpretations of the identity of segments (table XXI).

Rami of the remaining limbs

More studies have been made of development of the four copepod swimming legs than of the remaining thoracopods, including the maxilliped. Two types of data have been used to suggest how swimming legs are patterned. Formation homology includes information from the internal organization of the present stage about the future location of an arthrodial membrane that will separate segments in the following copepodid stage or the future location of a formation seta. The arthrodial membrane that will separate the proximal segment of the adult exopod of limb 5 of *Acartia clausi* and *A. longiremis* was illustrated within the limbs of CV (see Kraefft, 1910, pl. 1 figs. 48-49). The middle segment and the distal segment of the exopod of swimming legs 1 and 3 of *Megacyclops viridis* [given as *Cyclops viridis*] were observed and illustrated within those limbs at CV (see Lucks, 1926: 15, figs. 40-41), and the arthrodial membranes that will separate the middle and the distal segment of the exopod of swimming legs 2-3 were illustrated within the limbs of CIV of *Macrocyclops albidus* (see Defaye, 1984, pls. 8, 9). It should be noted that *Macrocyclops albidus* and *Megacyclops viridis* belong to two different monophyletic lineages of cyclopid copepods and the arthrodial membranes that separate the middle and distal exopodal segments of their swimming legs form at CV and CVI, respectively (Ferrari, 1998). An apparent arthrodial membrane within the limb of CIV, however, was not reported in the following stage of *Pontellina* sp. (see Hulsemann & Fleminger, 1975, figs. 3, 4) and its absence passed without comment.

Failure of an arthrodial membrane to form between segments results in a segment complex on the ramus of a swimming leg. Examples of such complexes have been hypothesized from adult morphology of calanoids like *Euchirella rostrata*, whose proximal and middle exopodal segment of swimming leg 1 form a complex (see Giesbrecht, 1893a, pl. 15 fig. 11) and *Temora stylifera*, whose proximal and middle endopodal segment of swimming leg 4 form a complex (see Giesbrecht, 1893a, pl. 17 fig. 13). Homologs of the finishing arthrodial membrane that fails to form in the above two species are present in related adult calanoids, so the interpretation is a straightforward deduction. Similar complexes also have been inferred indirectly from development of the exopod of swimming leg 2 of *Enhydrosomella* (see Fiers, 1996: 22, fig. 12).

However, if an homologous arthrodial membrane does not form in related species, as is the case for the distal complex of the exopod and endopod of swimming legs of copepods, inferences are not as easily deduced (see Ferrari & Benforado, 1998a). Examples of such complexes on calanoid swimming leg 1 are given below. The formation seta of the presumptive proximal segment and the presumptive middle segment of the rami of swimming legs

1-4 initially appears on the distal segment complex of both rami. The finishing arthrodial membrane, which separates the formation seta from the distal complex and allows visualization of the distal boundary of these segments, forms during the molt to a later copepodid stage. This finishing arthrodial membrane separates the proximal part of the distal segment, including the proximal dorsal and/or ventral formation seta, from the rest of the distal complex. The process initially was described by Illg (1949: 411) for the second (middle) segment of the swimming legs 1-4 of the poecilostome, *Paranthessius columbiae*. Later, observations for the proximal segment of the exopod of swimming leg 2 at CII of *Harpacticus uniremis* by Itô (1971: 252, fig. 14), for both proximal and middle segments of the swimming leg 2 of *Alebion lobatus* (see Benz, 1989), for both proximal and middle segments of the swimming legs of the calanoids, *Ridgewayia klausruetzleri*, *Pleuromamma xiphias* and *Temora longicornis* plus the cyclopoid *Dioithona oculata* (see Ferrari & Benforado, 1998a), and for limb 5 of several centropagoidean calanoids (Ferrari & Ueda, 2005) have confirmed the general applicability of this process during limb development. The process was termed "setal precedence" (Ferrari & Benforado, 1998a) because the dorsal and/or ventral formation seta(e), which will be found on the presumptive segment, precedes the formation of the arthrodial membrane that allows visualization of the boundary of that segment.

The copepodid stage at which specific setae first appear on swimming legs 1-4 was illustrated initially by Dudley (1966: 137, fig. 51) for the notodelphyid cyclopoids, *Notodelphys affinis*, *Pygodelphys aquilonaris*, *Scolecodes huntsmani* and a composite of species of *Doropygus* including *D. seclusus*, *D. bayeri*, *D. mohri* and *D. fernaldi*. Later Kô (1969d: 97, fig. 1) deduced similar information for the poecilostome *Ostrincola koe*. An example from Dudley (1966) is shown in fig. 27F. Terminal elements on the buds of the presumptive rami of swimming legs 1-2 of the fourth nauplius (corresponding to NVI of a copepod with six naupliar stages), were described by Dudley (1966) as pointed processes rather than as setae or spines, while terminal elements on the buds of the presumptive rami of swimming legs 3-4 on CI and CII, respectively were described as cuticularized sacs. As a result, the crown setae on the distal complex of the rami of the transformed limb are not differentiated by age from the remaining distodorsal or distoventral setae. Buds of swimming legs 1-4 were not described by Kô (1969d) for *Ostrincola koe*, so these crown setae also were not differentiated by age from the other setae of the distal complex.

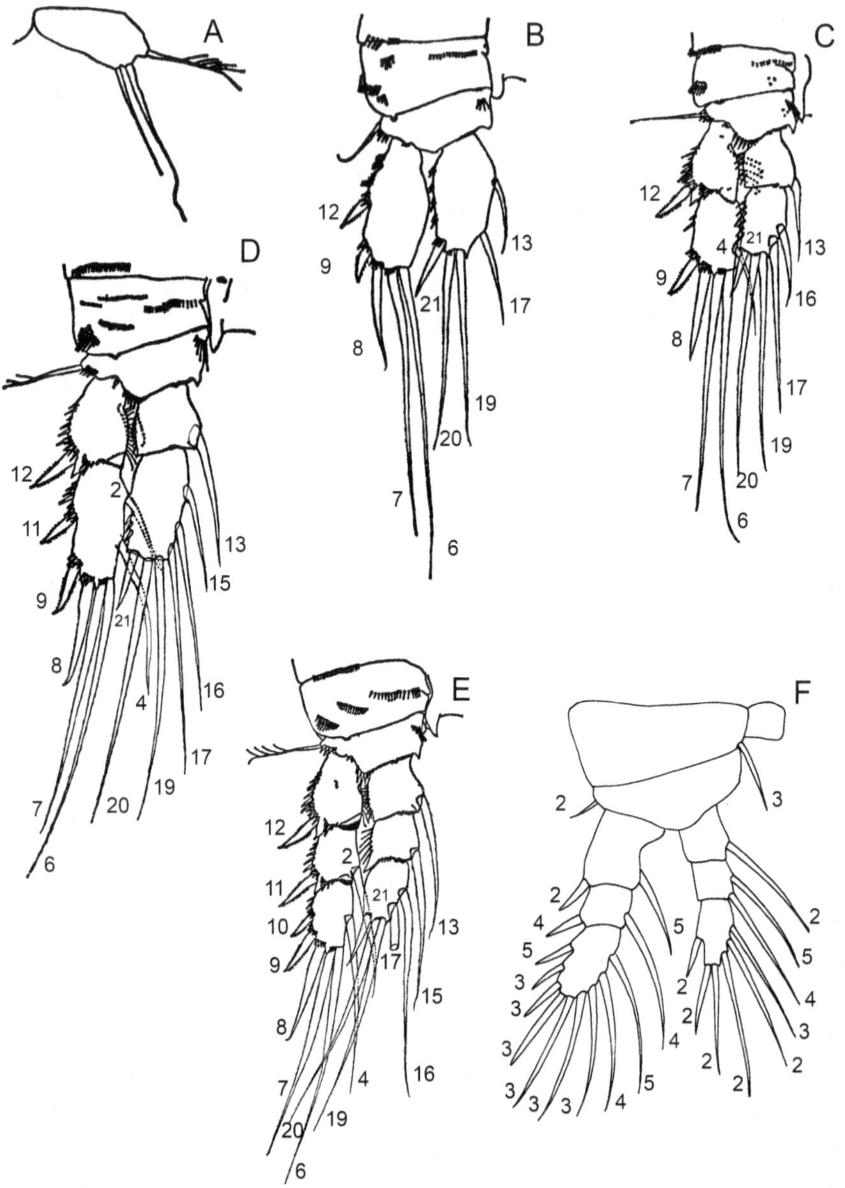

Fig. 27. *Paramphiascella fulvofasciata*, five steps in the development of swimming leg 3 (modified from Rosenfield & Coull, 1974, figs. 80-84); A, CI; B, CII; C, CIII; D, CIV; E, CV; numbers indicate the relative position of setae from dorsal to ventral of exopod [2-12], and dorsal to ventral of endopod [13-21]. F, *Notodelphys affinis*, swimming leg 3 (modified from Dudley, 1966, fig. 51c), numbers show specific copepodid stage at which setae first appear.

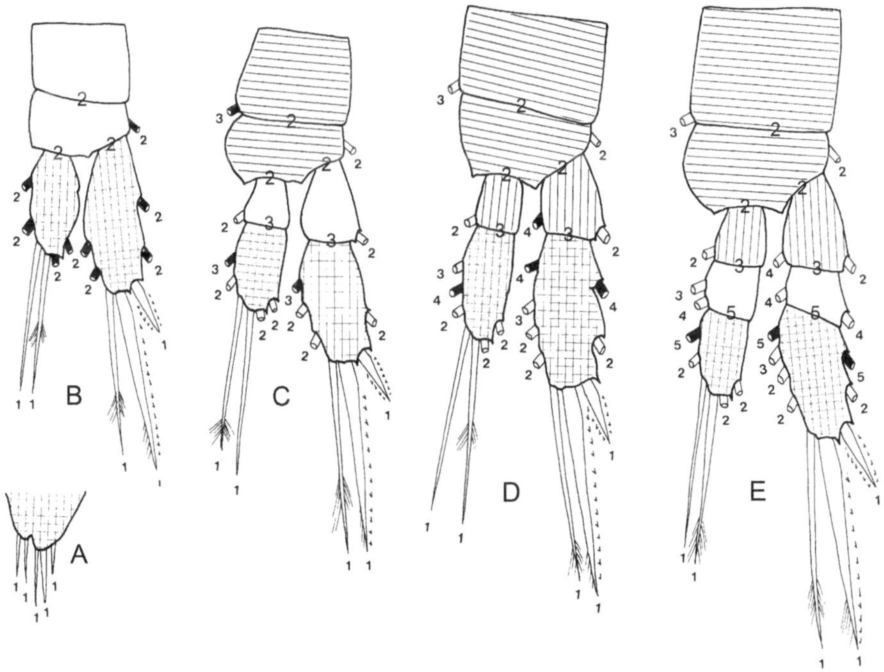

Fig. 28. *Dioithona oculata*, five steps in the development of swimming leg 3 (modified from Ferrari & Ambler, 1992). A, CI (limb bud); B, CII (transformed limb); C, CIII; D, CIV; E, CV. Numbers are stage at which setae and arthrodial membranes first appear; 1, CI; 2, CII; etc.; darkened setae are new.

A variation on this setal identification system was formulated for the harpacticoid, *Paramphiascella fulvofasciata* by Rosenfield & Coull (1974, figs. 68-88) who assigned a unique number to each seta of the swimming leg in order to determine homologous setae during development. However, the number was not linked to the stage at which the seta initially appeared. An example from Rosenfield & Coull (1974) is shown in fig. 27A-E.

The exact relationship between the appearance of specific setae and the appearance of arthrodial membranes was proposed for swimming leg 3 of *Dioithona oculata* (see Ferrari & Ambler, 1992, fig. 13). In this analysis, the age of the crown setae of the exopod and endopod are determined (fig. 28). However, new setae of the following stage were inferred to be those not enclosed within the exoskeleton of the present limb, and this assumption may not be correct in all cases. All setae on swimming legs 1-4 of the poecilostome, *Conchyliurus quintus*, including new setae added at each stage of development, were identified by Kim (1994). The location of new elements at each stage agrees in general with Dudley (1966), Kô (1969d) and Ferrari & Ambler (1992).

However, the data from Dudley (1966), Kô (1969d), and Ferrari & Ambler (1992) suggest that the proximal seta on the middle endopodal segment of swimming legs 1-4 is presented one stage later than the distal seta on that segment so that this proximal seta is younger than the distal seta. This configuration would imply that the endopod of the swimming leg has two source segments while the exopod is patterned from only one. The implication from the study of Kim (1994) is that the proximal seta on the middle endopodal segment is presented one stage earlier than the distal seta of that segment. As a result, only one source segment is required to pattern the endopod from Kim's (1994) observations. We agree and follow this inference of Kim (1994) in our fig. 28. The primary result from these studies is that the source segment of each ramus of swimming legs 1-4 is located within the distal segment complex and toward its proximal boundary.

Caudal ramus

This appendage is unsegmented in copepods, and its conformation differs from the protopod/ramus [or: rami] configuration of cephalic and thoracic limbs. There have been a number of attempts to understand how setae are added to the caudal ramus of harpacticoid copepods (Dahms, 1992a, 1993a; Huys & Böttger-Schnack, 1994; Fiers, 1996; George, 2001). These analyses include assumptions about setal homologies based on setal size or setal morphology, and hypotheses about the displacement of setae along the appendage. As yet no consensus has been reached on these issues.

Contrasting early development of swimming leg 1 with swimming legs 2-4

The number of setae on swimming legs during early steps of limb development has been documented for a number of copepods (Ferrari, 2000). An example of this early development can be seen in fig. 28A-C for swimming leg 3 of *Dioithona oculata*. The buds of swimming legs 2-4 are presented at NVI, CI and CII, respectively. These buds bear at most three setae on the presumptive exopod and at most two setae on the presumptive endopod. A greater percentage of the surveyed species bears this setal configuration than any other configuration. The transformed limb of swimming legs 2-4 forms at CI, CII and CIII, respectively. These limbs bear at most seven setae on the exopod and at most six setae on the endopod, and a greater percent of the surveyed species bears this setal configuration than any other. The next step in limb development is an apparently 2-segmented limb for swimming legs 2-4;

this step forms at CII, CIII and CIV, respectively. These limbs bear at most eight setae on the exopod, including a dorsal seta on the proximal exopodal segment, and at most seven setae on the endopod, including a ventral seta on the proximal segment; a greater percent of the surveyed species bears this setal configuration than any other configuration. Homologs for most of these setae, except the crown setae on the limb bud, as noted above, also have been determined for all swimming legs of several notodelphyid cyclopoids (see Dudley, 1966), and for the poecilostomes, *Ostrincola koe* and *Conchyliurus quintus* (see Kô, 1969d; Kim, 1994). A crown group of three setae (dorsal, terminal, ventral) on the exopod of the transformed limb of swimming legs 2-4 is homologous to the three setae on the presumptive exopod of the limb bud. The crown group of two setae, both terminal, on the endopod of the transformed limb is homologous to the two setae on the presumptive endopod of the limb bud (fig. 28A, B). All other setae on the transformed limb are new: two dorsal and two ventral setae on the exopod, and two dorsal and two ventral on the endopod. The proximal dorsal seta of the exopod will be allocated to the proximal segment of the early 2-segmented exopod after the next molt (fig. 28C). In like manner, the proximal ventral seta of the endopod will be allocated to the proximal segment of the early 2-segmented endopod after the next molt. The new seta on the early 2-segmented exopod is the proximal seta on the distal segment complex, and this will be allocated to the middle segment later in development (fig. 28D). In like manner, the new seta on the early 2-segmented endopod is the proximal seta on the distal segment complex, and this will be allocated to the middle segment complex later in development.

The corresponding configurations for swimming leg 1 differ from those of swimming legs 2-4 (Ferrari, 2000). The bud of swimming leg 1 is presented at NVI and may bear up to four setae on the presumptive exopod of many copepods. In addition, there may be up to three setae on the presumptive endopod of calanoids, e. g., *Temora longicornis* (fig. 29A), although no more than two setae are present on the presumptive endopod of podopleans. On the transformed limb of swimming leg 1, there are no more than eight setae on the exopod and no more than seven on the endopod. On the early 2-segmented limb there may be no more than nine setae on the exopod and no more than eight on the endopod. A greater percent of surveyed species has these configurations on the transformed limb and the early 2-segmented limb of swimming leg 1 (Ferrari, 2000).

The fourth seta on the presumptive exopod of the limb bud of swimming leg 1 is located proximally and dorsally (fig. 29A); it is homologous to the

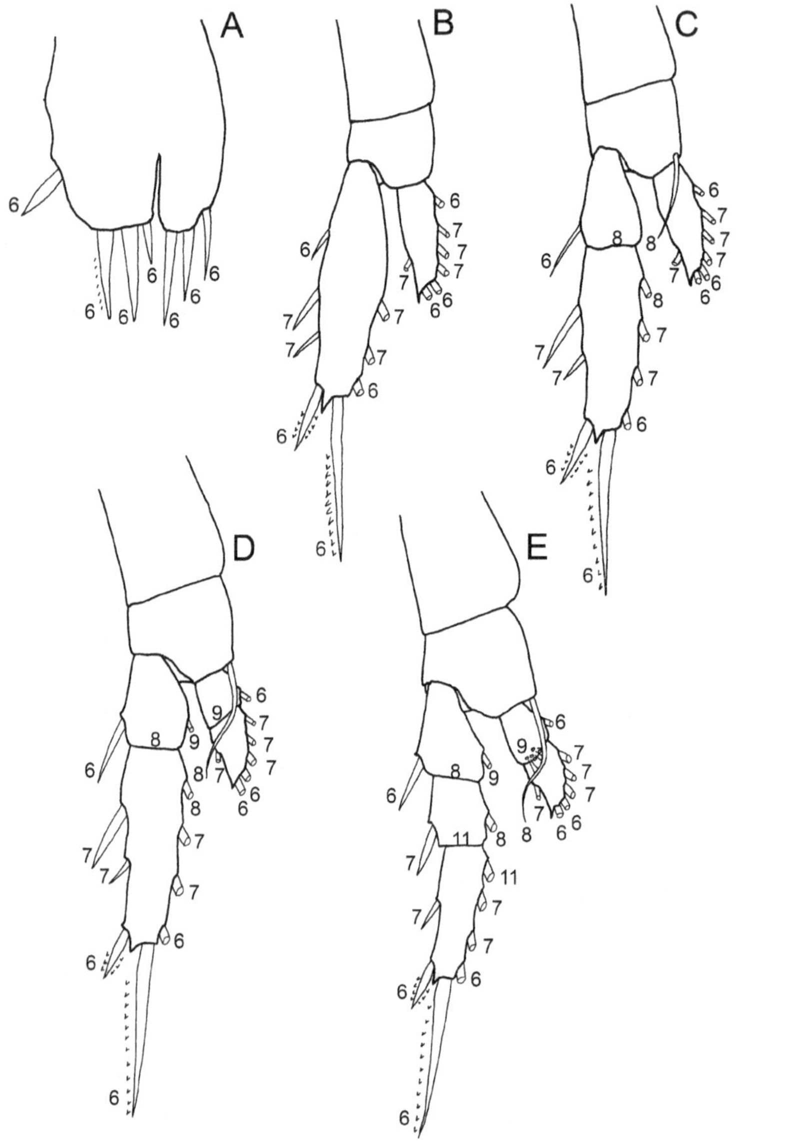

Fig. 29. *Temora longicornis*, five steps in the development of swimming leg 1 (modified from Ferrari & Benforado, 1998b). A, bud; B, transformed limb; C, early apparent 2-segmented step; D, late apparent 2-segmented step; E, apparent 3-segmented step. Numbers show stage at which setae and arthrodial membranes first appear; 6, NVI; 7, CI; etc.

proximodorsal seta of the transformed limb (fig. 29B). The third seta on the presumptive endopod of calanoids is located ventrally and somewhat proximally; it is homologous to the proximoventral seta of the transformed limb (fig. 29B). This proximodorsal seta on the exopod of the transformed limb will be allocated to the proximal segment of the early 2-segmented exopod during the molt to CII (fig. 29C), while the proximoventral seta of the endopod of the transformed limb will be allocated to the proximal segment of the early 2-segmented endopod during the molt to CIII (fig. 29D). The new, eighth seta on the exopod of the transformed limb is the middle dorsal seta and eventually it will be allocated to the middle segment of a 3-segmented exopod during the molt to CV (fig. 29E). The new, seventh seta of the endopod of the transformed limb is distal to the proximal seta. Its location remains on the distal segment of this 2-segmented endopod, because a distal finishing arthrodial membrane does not form to separate the presumptive middle segment from the distal segment. On swimming leg 1 of a species with a 3-segmented endopod, the seventh seta of the endopod of the transformed limb will be allocated to the middle segment of a 3-segmented endopod, usually during the molt to CV (fig. 30).

This interpretation is consistent with a hypothesis that for a group of limbs of similar configuration, like copepod swimming legs 1-4, one anterior limb or more begins development in a more advanced state, with more elements present, than the posterior limbs. Swimming leg 1 of copepods not only bears more elements at the bud stage, but the extra seta of the exopod and endopod of calanoids indicates that the rami are more advanced in their patterning than the buds of swimming legs 2-4. Furthermore, swimming leg 1 maintains this advanced state throughout its early development. This hypothesis, explaining the advanced configuration of one or more anterior limbs, is supported in a more general way by observations of the trunk limbs of the branchiopod, *Leptestheria kawachiensis*. Anterior limbs of this crustacean initially appear in an advanced stage of development relative to posterior trunk limbs (Ferrari & Grygier, 2003).

Von Vaupel Klein's Organ

Understanding how swimming legs are patterned during development is invaluable in determining the segmental homologies of copepod limbs. Von Vaupel Klein's Organ (VVKO; see Von Vaupel Klein, 1972) of calanoid copepods provides an excellent example of this kind of analysis. VVKO usually is made up of the ventral seta on the basis of swimming leg 1, which seta is curved and often recurved over the anterior face of the endopod of the limb. Many setules along the primary curve of this seta are directed toward a sensory area of pores and/or denticles on the anterior face of the endopod.

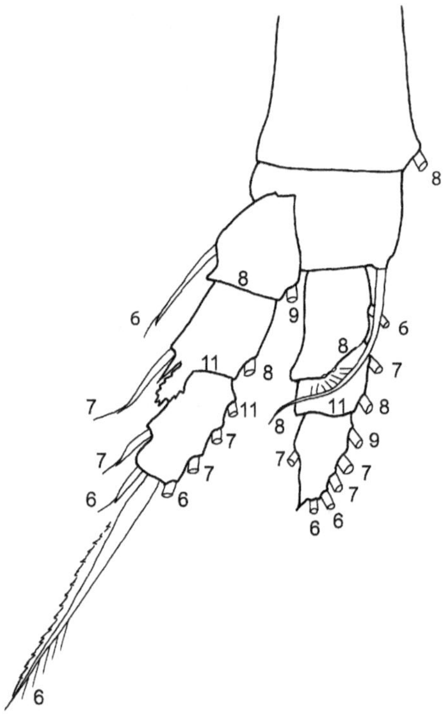

Fig. 30. *Ridgewayia klausruetzleri*, adult swimming leg 1 (proximal up). Numbers show stage at which setae and arthrodial membranes first appear; 6, NVI; 7, CI; 8, CII; etc. NVI unknown; setae for that stage inferred from configuration of *Temora longicornis*.

Often this sensory area of the endopod is found on or near a raised bump on the endopod. The association of the ventral seta of the basis with the anterior sensory area of the endopod is VVKO (Ferrari & Steinberg, 1993).

Among different calanoids, the sensory area of VVKO is found on the proximal segment of a 3-segmented endopod, e. g., *Ridgewayia klausruetzleri* (fig. 30), but also on the proximal segment of a 2-segmented endopod, e. g., *Acrocalanus gibber* (fig. 31), and at mid-length on a 1-segmented endopod, e. g., *Euchirella messinensis* (fig. 32). If it can be shown that this sensory area is homologous among all three kinds of limbs, then VVKO may be proposed as a synapomorphy for calanoid copepods, including the platycopiids.

In figs. 30-32, respectively, the naupliar or copepodid developmental stage at which a seta or arthrodial membrane first appears on swimming leg 1 of *R. klausruetzleri*, *A. gibber* and *E. messinensis* is indicated by Arabic numerals. The location of a source segment for each ramus can be identified for *R. klausruetzleri* as adjacent to the youngest elements: between the arthrodial membrane separating the middle segment from the distal complex and the

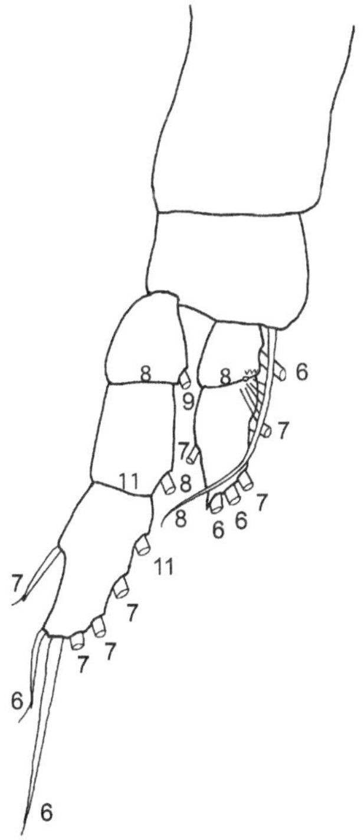

Fig. 31. *Acrocalanus gibber*, adult swimming leg 1 (proximal up). Numbers show stage at which setae and arthrodial membranes first appear; 6, NVI; 7, CI; 8, CII; etc.; see Ferrari (2000) for setation at NVI.

proximal seta of the distal complex; or between the proximal seta of the distal complex and its adjacent seta on the complex. For *A. gibber* and *E. messinensis* the location is comparable, as will be seen in the following tables, even though fewer elements are present to derive the inference for these two calanoids. In tables XXII-XXIV, respectively, the two rami of swimming leg 1 of *R. klausruetzleri*, *A. gibber* and *E. messinensis* at each copepodid stage are arrayed linearly so that non-terminal segments are represented by no more than one dorsal and/or ventral seta.

Only one source segment is required to specify the pattern of each ramus. The location of this source segment is between two of the youngest elements at CVI, the proximal seta of the distal complex and its adjacent seta on the distal complex. For CI-CV, the location can be derived through the systematic deletion of elements initially formed during the next older copepodid. Like the source segment of the exopod of antenna 2, which also is located within a segment complex, the source segment for rami of the swimming leg does not bear a seta and is not demarcated by two arthrodial membranes.

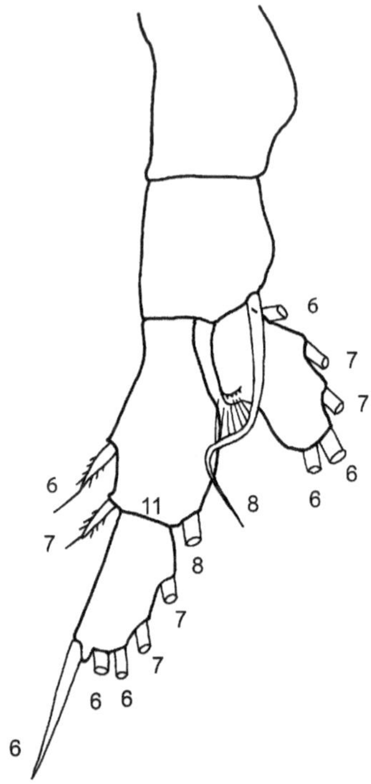

Fig. 32. *Euchirella messinensis*, adult swimming leg 1 (proximal up). Numbers show stage at which setae and arthrodial membranes first appear; 6, NVI; 7, CI; 8, CII; etc. NVI unknown; setae for that stage inferred from configuration of *Acrocalanus gibber*.

In tables XXV-XXVII, respectively, the segments of the rami of swimming leg 1 of *R. klausruetzleri*, *A. gibber* and *E. messinensis* proximal to the source segment are identified with Arabic numerals. The oldest segment proximal to the source segment has the smallest number '2'; the youngest proximal segment has the largest number, '4' for the exopod and '5' for the endopod. Distal to the source segment Latin letters identify each segment. The oldest segment distal to the source segment has the letter 'b'; the youngest segment has the letter 'd'. These limbs have rather different conformations, usually described as 3-segmented, 2-segmented, or 1-segmented, respectively, for the three species of calanoids. Despite these differences, specific limb patterning results in segment '2' of all three limbs as being the oldest segment proximal to the source segment of the endopod. Segment '2' always is adjacent to the basis and for these two reasons bears the sensory area of the endopod on the adult. Therefore, the sensory area on the anterior face of the endopod is homologous among the limbs of all three calanoids, and Von Vaupel Klein's

TABLE XXII

Setation of the rami of swimming leg 1 of copepodids of the calanoid, *Ridgewayia klausruetzleri*. Setation of the last nauplius is unknown but inferred from that of *Temora longicornis*. No more than one ventral seta (v) and one dorsal seta (d) per segment except for the distal segment of the exopod with one dorsal, one terminal and one ventral setae (dtv) and for the distal segment of the endopod with two terminal setae (2t). Comma (,) is an arthrodial membrane, dash (-) indicates the location where an arthrodial membrane has failed to form, resulting in a segment complex

NVI	Re:	d-dtv
	Ri:	v-2t
CI	Re:	d-d-v-dv-dtv
	Ri:	v-v-v-dv-2t
CII	Re:	d, dv-v-dv-dtv
	Ri:	v, v-v-v-dv-2t
CIII	Re:	dv, dv-v-dv-dtv
	Ri:	v, v-v-v-v-dv-2t
CIV	Re:	dv, dv-v-dv-dtv
	Ri:	v, v-v-v-v-dv-2t
CV	Re:	dv, dv, v-v-dv-dtv
	Ri:	v, v-v, v-v-dv-2t
CVI	Re:	dv, dv, v-v-dv-dtv
	Ri:	v, v-v, v-v-dv-2t

TABLE XXIII

Setation of the rami of swimming leg 1 of the calanoid, *Acrocalanus gibber*. No more than one ventral seta (v) and one dorsal seta (d) per segment except for the distal segment of the exopod with one dorsal, one terminal and one ventral setae (dtv) and for the distal segment of the endopod with two terminal setae (2t). Comma (,) is an arthrodial membrane, dash (-) indicates the location where an arthrodial membrane has failed to form, resulting in a segment complex

NVI	Re:	dt
	Ri:	v-2t
CI	Re:	v-dv-dtv
	Ri:	v-v-dv-2t
CII	Re:	0, v-v-dv-dtv
	Ri:	v, v-dv-2t
CIII	Re:	v, v-v-dv-dtv
	Ri:	v, v-dv-2t
CIV	Re:	v, v-v-dv-dtv
	Ri:	v, v-dv-2t
CV	Re:	v, v, v-v-dv-dtv
	Ri:	v, v-dv-2t
CVI	Re:	v, v, v-v-dv-dtv
	Ri:	v, v-dv-2t

TABLE XXIV

Setation of the rami of swimming leg 1 of copepodids of the calanoid, *Euchirella messinensis*. Setation of the last nauplius is unknown but inferred from that of *Acrocalanus gibber*. No more than one ventral seta (v) and one dorsal seta (d) per segment except for the distal segment of the exopod with one dorsal, one terminal and one ventral setae (dtv) and for the distal segment of the endopod with two terminal setae (2t). Comma (,) is an arthrodial membrane, dash (-) indicates the location where an arthrodial membrane has failed to form, resulting in a segment complex

NVI	Re:	dtv
	Ri:	v-2t
CI	Re:	d-dv-v-v-dtv
	Ri:	v-v-dv-2t
CII	Re:	d-dv-v-v-dtv
	Ri:	v-v-dv-2t
CIII	Re:	d-dv-v-v-dtv
	Ri:	v-v-dv-2t
CIV	Re:	d-dv-v-v-dtv
	Ri:	v-v-dv-2t
CV	Re:	d-dv, v-v-dtv
	Ri:	v-v-dv-2t
CVI	Re:	d-dv, v-v-dtv
	Ri:	v-v-dv-2t

TABLE XXV

Segments of the rami of the swimming leg 1 of the calanoid, *Ridgewayia klausruetzleri*. Segmentation of the last nauplius is unknown but inferred from that of *Temora longicornis*. Roman numeral 'one' with asterisk (I*) is the source segment for patterning the ramus; comma (,) is an arthrodial membrane, dash (-) indicates the location where an arthrodial membrane has failed to form, resulting in a segment complex. Lower case letters are segments distal to the source segment, 'b' is the oldest of these segments; Arabic numerals are segments proximal to the source, '2' is the oldest of these segments. Setae are not indicated in this table

NVI	Re:	2-I*-b
	Ri:	2-I*-b
CI	Re:	2-3-I*-d-c-b
	Ri:	2-3- I*-d-c-b
CII	Re:	2, 3-I*-d-c-b
	Ri:	2, 3-4-I*-d-c-b
CIII	Re:	2, 3-I*-d-c-b
	Ri:	2, 3-4-5-I*-d-c-b
CIV	Re:	2, 3-I*-d-c-b
	Ri:	2, 3-4-5-I*-d-c-b
CV	Re:	2, 3, 4-I*-d-c-b
	Ri:	2, 3-4, 5-I*-d-c-b
CVI	Re:	2, 3, 4-I*-d-c-b
	Ri:	2, 3-4, 5-I*-d-c-b

TABLE XXVI

Segments of the rami of the swimming leg 1 of the calanoid, *Acrocalanus gibber*. Roman numeral 'one' with asterisk (I*) is the source segment for patterning the ramus; comma (,) is an arthrodial membrane, dash (-) indicates the location where an arthrodial membrane has failed to form, resulting in a segment complex. Lower case letters are segments distal to the source segment, 'b' is the oldest of these segments; Arabic numerals are segments proximal to the source, '2' is the oldest of these segment. Setae are not indicated in this table

NVI	Re:	2-I*-b
	Ri:	2-I*-b
CI	Re:	2-I*-d-c-b
	Ri:	2-I*-c-b
CII	Re:	2, 3-I*-d-c-b
	Ri:	2, I*-c-b
CIII	Re:	2, 3-I*-d-c-b
	Ri:	2, I*-d-c-b
CIV	Re:	2, 3-I*-d-c-b
	Ri:	2, I*-d-c-b
CV	Re:	2, 3, 4-I*-d-c-b
	Ri:	2, I*-d-c-b
CVI	Re:	2, 3, 4-I*-d-c-b
	Ri:	2, I*-d-c-b

TABLE XXVII

Segments of the rami of the swimming leg 1 of the calanoid, *Euchirella messinensis*. Segmentation of the last nauplius is unknown but inferred from that of *Acrocalanus gibber*. Roman numeral 'one' with asterisk (I*) is the source segment for patterning the ramus; comma (,) is an arthrodial membrane, dash (-) indicates the location where an arthrodial membrane has failed to form, resulting in a segment complex. Lower case letters are segments distal to the source segment, 'b' is the oldest of these segments; Arabic numerals are segments proximal to the source, '2' is the oldest of these segments. Setae are not indicated in this table

NVI	Re:	2-I*-b
	Ri:	2-I*-b
CI	Re:	2-3-I*-d-c-b
	Ri:	2-I*-d-c-b
CII	Re:	2-3-I*-d-c-b
	Ri:	2-I*-d-c-b
CIII	Re:	2-3-I*-d-c-b
	Ri:	2-I*-d-c-b
CIV	Re:	2-3-I*-d-c-b
	Ri:	2-I*-d-c-b
CV	Re:	2-3, I*-d-c-b
	Ri:	2-I*-d-c-b
CVI	Re:	2-3, I*-d-c-b
	Ri:	2-I*-d-c-b

Organ can be proposed as a synapomorphy for calanoid copepods, including the platycopiids.

Thoracopods: maxilliped and swimming legs

Information obtained about the patterning of the rami of swimming legs also may be applied to the question of whether the copepod maxilliped is derived from a configuration like a swimming leg, or whether the maxilliped is derived from an older thoracopod morphology from which swimming legs also have been derived. Segments of the endopod of the maxilliped distal to the source segment are formed during the molt from a limb bud to the transformed limb; further patterning of the transformed maxilliped results in the addition of endopodal segments proximal to the source segment. This also is generally true for both rami of the swimming legs, although a dorsal seta without its arthrodial membrane may be added to the endopod during the copepodid phase. An example of the addition of a dorsal seta, mentioned above for the maxilliped, has been observed on the endopod of limb 5 of *Centropages abdominalis*. A new dorsal seta is added to the antepenultimate segment during the molt to CV, one stage after the limb bud has been re-configured (Ferrari & Ueda, 2005). Both the maxilliped and swimming legs of copepods may form segment complexes as a result of the failure of one or more arthrodial membranes to form; the distal segment complex is a hall-mark of these thoracopods. The maxilliped and swimming legs are similar in these two ways.

Copepod swimming legs, however, exhibit the process of setal precedence. The dorsal and/or ventral formation seta that will be found on segments proximal to the source segment initially are added to the distal segment complex. Subsequently, these formation setae will be allocated to the proximal or middle segment when the distal, finishing arthrodial membrane is formed one molt or more later. Setal precedence appears to be a derived process of swimming leg development, relative to the process of development of the maxilliped, whose formation seta and arthrodial membrane initially are added in register, during the same molt. The copepod maxilliped does not appear to be derived from a swimming leg. Swimming legs are significantly transformed thoracopods whose derived states are setal precedence and an interpodal bar between contralateral coxal segments.

A branchiopod is the only other crustacean for which the patterning of post-cephalic limbs has been hypothesized. Patterning the endopod of trunk limbs of *Leptestheria kawachiensis* is now understood by observations of for-

TABLE XXVIII

Representation of the segmentation of the endopod of an anterior trunk limb of *Leptestheria kawachiensis* from Ferrari & Grygier (2003); steps 1-3 from a posterior limb; step 4 from the anterior trunk limb of the male. Roman numeral 'one' with asterisk (I*) is the source segment that patterns the ramus; dash (-) indicates the location where an arthrodial membrane has failed to form, resulting in a segment complex. Lower case letters are segments distal to the source segment, 'b' is the oldest of these segments; Arabic numerals are segments proximal to the source, '2' is the oldest of these segments. Setae are not indicated in this table

Step 1:	I*
Step 2:	2-I*
Step 3:	2-I*-b
Step 4 [male 1st trunk limb]:	2-I*-c-b

mation homology (Ferrari & Grygier, 2003). The endopodal segment proximal to the source segment forms first, followed by the distal endopodal segment (table XXVIII). A fourth endopodal segment, distal and adjacent to the source segment, is added last to the first trunk limb of males. This fourth segment is elongate, bears a palp, and previously was misinterpreted as part of an articulating palp until the analysis of Ferrari & Grygier (2003). Patterning of the branchiopod trunk limb then is unlike the thoracopods of copepods, because proximal patterning occurs early in development and distal patterning occurs later.

The protopod

In the previous sections, patterning has referred to the addition of setae and arthrodial membranes to the rami of the limbs. Much less is known about how the protopod of copepod limbs is patterned, because the arthrodial membranes and all endites that will bear setae on the protopod of the adult copepod limb appear to be present in the transformed limb (Ferrari, 1995; Ferrari & Dahms, 1998). It should be noted that the copepodid stage at which setae are added to specific endites has some power to predict the identity of some protopodal endites (Ferrari & Ivanenko, 2001). Fiers (1991a: 41) suggested that the dorsal seta on the basis of swimming leg 4 and limb 5 of the harpacticoid, *Galapalaophonte biarticulata* is present on the limb bud; this would imply that protopodal patterning is initiated during the formation of the limb bud. Recent observations of steps in development of the trunk limbs of the branchiopod, *Leptestheria kawachiensis* (see Ferrari & Grygier, 2003) suggest that the protopod is patterned from a point where the limb articulates with the body (table XXIX).

TABLE XXIX

Representation of the segmentation of an anterior trunk limb of *Leptestheria kawachiensis* from Ferrari & Grygier (2003); steps 1-4 from a posterior limb; step 5 from the anterior trunk limb of the male. Roman numeral 'one' with asterisk (I*) is the source segment that patterns the endopod; dash (-) indicates the location of an arthrodial membrane that has failed to form, resulting in a segment complex. Lower case letters are segments distal to the source segment, 'b' is the oldest of these segments; Arabic numerals are segments proximal to the source, '2' is the oldest of these segments; plus (+) indicates location of protopodal patterning; 'bd' is body; upper case letters are B, basis; C, coxa; Pd, distal praecoxal endite; Pm, middle praecoxal endite; Pp, proximal praecoxal endite. Setae are not indicated in this table

Step 1:	bd+Pd-C-B-I*
Step 2:	bd+Pm-Pd-C-B-2-I*
Step 3:	bd+Pp-Pm-Pd-C-B-2-I*
Step 4:	bd+Pp-Pm-Pd-C-B-2-I*-b
Step 5 [male 1st trunk limb]:	bd+Pp-Pm-Pd-C-B-2-I*-c-b

The basis appears to be developmentally the oldest part of the protopod of the branchiopod; this supports the inference of Fiers (1991a) for copepod thoracopods. The coxa with a single endite is the next oldest segment on the protopod of the branchiopod. The youngest segment is the praecoxa; its distal endite is younger then the coxal endite, but the distal endite is developmentally older than its middle endite. The proximal endite of the praecoxa, which is closest to the body, is developmentally the youngest part of the protopod.

Indirect support for this patterning process can be deduced from the configuration of the maxilliped of several cyclopoid copepods. At most, only the middle and distal praecoxal endites of the calanoid maxilliped are present on the cyclopoid maxilliped. These two endites, bearing fewer setae than in calanoids, are expressed among phylogenetically older species of Cyclopidae, e. g., *Euryte longicauda* (see Ferrari & Ivanenko, 2001). However, only the distal praecoxal endite is present in derived species of Cyclopidae [e. g., *Speocyclops racovitzai*]. From a calanoid maxilliped with three praecoxal endites, it appears that there has been a step-wise truncation of development of the praecoxa, affecting the number of praecoxal endites as the evolution of cyclopoid copepods has proceeded. As a result, the number of praecoxal endites has been progressively reduced beginning with the proximal praecoxal endite, which is closest to the body when it is present, followed by the middle praecoxal endite, which is closest to the body when the proximal praecoxal endite is absent (Ferrari & Ivanenko, 2001). Recent observations of 'Orsten'-type fossil arthropods without a praecoxa and with a simple, lobe-like coxa (Walossek & Müller, 1997) imply that patterning of the protopod may not have been expressed in the ground plan of the crustacean limb, and that the segmented nature of the ancestral crustacean limb was the sole result of ramal patterning.

Hansen (1893) was the first carcinologist to recognize a 3-segmented protopod for those limbs of crustaceans posterior to antenna 1. He based his conclusions on a study of antenna 2 and the mandible of metanauplii and adults of calanoid copepods, and of antenna 2 of *Microsetella* (as *Setella*). Lang (1946) confirmed, but did not illustrate, a naupliar praecoxal joint on antenna 2 and on the mandible from NII onwards. Dahms (1990c) showed that a distinctly 3-segmented protopod of antenna 2 is present at NI of some species of harpacticoids.

Interpretations of the praecoxa of post-mandibular limbs have been more recent. The praecoxa of the maxilliped of some cyclopoid copepods is an articulating segment with two endites; these two endites are proximal to the single endite of the coxal segment. A syncoxa with four endites including three praecoxal endites is found on the maxilliped of most calanoid copepods (Ferrari & Ivanenko, 2001). An alternate hypothesis that the proximal segment of the maxilliped is a syncoxa with three coxal endites and one praecoxal endite (Huys & Boxshall, 1991) is not supported by the Rule of Serial Homologs. Patterning of the branchiopod protopod also has been used to infer for copepods that the proximal segment of maxilla 2 is a syncoxa with two endites (Ferrari & Ivanenko, 2005; Suárez-Morales et al., 2006). The distal endite of the syncoxa is the coxal endite; the proximal endite is a praecoxal endite that is homologous to the distal praecoxal endite of the maxilliped (Ferrari & Ivanenko, 2005; Suárez-Morales et al., 2006). An alternate hypothesis that the proximal segment of maxilla 2 is a coxa with two endites (Huys & Boxshall, 1991) is not supported by the Rule of Serial Homologs. On maxilla 1, an articulating praecoxal segment with one endite is proximal to the single endite of the coxal segment of copepods (Boxshall, 1985).

Generalities of limb patterning

As described above, the structure of copepod limbs is basically a bifurcate, linear topology in which each bifurcation is a ramus. Segment patterning of the limb is complicated, because segments can be added from at least three points: distally from the point at which the protopod, at one end of the line, meets the body; and proximally and/or distally from at least one source segment located along the length of each bifurcation. Although some aspects of the addition of segments may be obscured, patterning of this kind of linear topology should be understandable if not necessarily intuitive.

The earlier observations here suggest that limb patterning has been a persistent process during the evolutionary history of copepods, as well as of other crustaceans. For the ancestral crustacean, a set of reasonable assumptions is that: the serially segmented limbs were patterned identically; there was no patterning of the protopod; the source segment of each ramus remained adjacent to the basis so that elements of each new ramal segment were added only distally to the source segment; all of the elements of only one new segment were added at each molt; and, with the exception of the distal two segments, the formation seta of most new endopodal segments was ventral, while the formation seta of most new exopodal segments was dorsal.

Variations in the above initial conditions, expressed among limbs of copepods, include: both proximal and distal patterning from a source segment with proximal patterning preceding distal patterning or vice versa; elements of one segment or more than one segment added at each molt. An exopod for which most setae are ventral, as is the case for antenna 2 and mandible of copepods, is assumed to have been derived from a configuration like the exopod of antenna 2 of nauplii of *Derocheilocaris typica*, for which most setae are dorsal (fig. 3A); a mechanism that has been proposed for dorsal exopodal setae early in development (Ferrari & Grygier, 2003). In addition to these differences, variations in patterning expressed in the flagellum of antenna 1 and antenna 2 of malacostracans also should be registered.

Because the arthropod limb has been historically persistent, the process of its patterning should be understandable from a few principles, irrespective of the fact that evolution of the patterning process has resulted in the above variability of that process. There should be no need to derive the patterning of rami empirically for each limb on every species of crustacean, nor should determination of segment homologies of crustaceans be peculiar. Some of the principles that are applicable to patterning copepod rami are as follows:

- Ramal segments are serially repeated, homologous structures composed of similar exoskeletal elements, e. g., setae and arthrodial membranes.
- The addition of new segmental elements to a ramus takes place from one source segment (most rami studied) or occasionally from more than one source segment (calanoid antenna 1).
- New segmental elements may form either proximal or distal to a source segment.

- Initially, a ramal segment is comprised of no more than one dorsal seta and one ventral seta (formation setae) associated with a finishing arthrodial membrane.
- The dorsal and/or ventral formation seta initially may be presented out of register with the formation of the finishing arthrodial membrane that defines one boundary of the segment (setal precedence of swimming legs).
- The presentation of the dorsal formation seta also may be out of register with that of the ventral formation seta. If this is the case, the dorsal formation seta of an exopodal segment usually precedes the presentation of its ventral formation seta, while the ventral formation seta of an endopodal segment usually precedes the presentation of its dorsal formation seta.

THE MOLT FROM THE LAST NAUPLIUS TO THE FIRST COPEPODID AND THE NUMBER OF NAUPLIAR SOMITES

The molt from the last nauplius to the first copepodid represents a significant change in the architecture of the copepod body, and also includes a reconfiguration of most of the appendages. In contrast, changes during molts within the naupliar phase or within the copepodid phase of development are more modest. In this chapter, structural and functional changes during the molt from the last nauplius to the first copepodid are itemized for copepods in which the last nauplius is an NVI that bears the bud of swimming legs 1-2. Changes in morphology and a simple model for patterning the body and limb buds during the copepodid phase of development are used to extrapolate back through the naupliar phase of development in order to infer the number of somites present at various stages of naupliar development.

Specific changes in body architecture and limb configuration result in the major differences between the nauplius and the copepodid. Arthrodial membranes separate many somites of copepodids, including the first copepodid, but arthrodial membranes do not separate naupliar somites. There are two differences in limb configuration between nauplii and copepodids. During the molt to the first copepodid, up to five post-mandibular naupliar setose limb buds plus the setose bud of the caudal ramus of the last nauplius are reconfigured into transformed appendages. In addition, the naupliar arthrite of antenna 2 fails to form on antenna 2 of any copepodid, and its function is assumed by the gnathobase of the mandibular coxa.

During the naupliar phase of development, only antenna 1, antenna 2, and the mandible appear as transformed limbs. The more posterior limbs of the last nauplius, including maxilla 1, maxilla 2, the maxilliped and swimming legs 1-2, plus the caudal ramus, are expressed only as limb buds, although maxilla 2 and the maxilliped may not form on nauplii of some copepod orders (see table I). Limb buds do not appear to articulate with their somite. During the molt to the first copepodid, these naupliar limb buds and the caudal ramus are reconfigured and appear as transformed appendages along with the three previously transformed limbs of the last nauplius. Exceptions include the maxilliped of some poecilostomes like *Ergasilus hypomesi* (see Kim, 2004), which may not be expressed until a later copepodid stage. In most copepods, each of the transformed appendages clearly articulates with its somite. The only limb bud present on the first copepodid is that of swimming leg 3; it does not articulate with its somite.

The transformed limbs of the naupliar phase of development, antenna 1, antenna 2 and mandible, seldom lose segments or setae during the molt from the last nauplius to the first copepodid. Exceptions to this rule include the loss of setae to antenna 1 of most calanoids and some harpacticoids (see Oberg, 1906; Fahrenbach, 1962; Hulsemann, 1991b); reduction in the number of segments making up the exopod of antenna 2 of harpacticoids and siphonostomatoids; reduction of the exopod of antenna 2 to a poorly-sclerotized, sac-like structure in cyclopoids and poecilostomes; reduction of the mandibular palp to a poorly-sclerotized, sac-like structure in the Cyclopidae and poecilostomes; and loss of the naupliar arthrite of antenna 2 in almost all copepods (see Fahrenbach, 1962; and see also above).

Significant functional changes of the limbs also occur during the molt from the last nauplius to the first copepodid (Storch, 1928). Nauplii swim using antenna 1, and the well-developed exopods of antenna 2 and of the mandible. Antenna 2 and mandible also are used to create feeding currents (Paffenhöfer & Lewis, 1989), and the arthrite of antenna 2 is used to push food through the mouth. The first copepodid of calanoids uses the exopod of antenna 2 and the exopod of the mandible in glide-like swimming and calanoid copepodids also can use swimming legs 1-2 plus movement at the articulation between the anterior and posterior parts of the body to produce a jump-like swimming. This is a mode of movement restricted to the copepodid stages. Reduced segmentation of the exopod of antenna 2 of the first copepodid of harpacticoids and siphonostomatoids, as well as the absence of this ramus in copepodids of cyclopoids and poecilostomes, precludes a glide-like swimming during the copepodid phase of development of these copepods. Movements of these copepodids are restricted to the use of swimming legs 1-2. With the loss of the naupliar arthrite of antenna 2, the mandibular gnathobase pushes food into the mouth of the copepodid. The mandibular gnathobase is present only during the copepodid phase of development of most copepods; NIV-NVI of calanoids provide the only exception. On calanoid nauplii IV-VI, the function of the mandibular gnathobase, relative to the function of the naupliar arthrite of antenna 2, has not been investigated.

The duration of the last nauplius and the size of the first copepodid are dependent on the availability of food to calanoids like *Calanus finmarchicus* (see Irigoien et al., 2003). Susceptibility of some calanoids to predation may (Landry, 1978) or may not (Eiane & Ohman, 2004; Ohman et al., 2004) decrease from the last nauplius to the first copepodid. Calanoid copepods may exhibit a significant change in motion and swimming speed between the last nauplius and the first copepodid (Paffenhöfer et al., 1996).

The number of somites that make up the body of any naupliar stage cannot be observed directly because there are no unambiguous signs that arthrodial membranes separate somites of the body during the naupliar phase of development. Incomplete arthrodial membranes that appear to separate some body segments of some calanoid nauplii have been illustrated, e. g., for *Eucalanus elongatus* (see Johnson, 1937). In addition, the posterior border of the cephalic shield of calanoids has been interpreted mistakenly for an arthrodial membrane. In contrast, arthrodial membranes are prominent and separate many somites of the first copepodid. These somites include the third thoracic somite of the first copepodid, which bears the transformed swimming leg 2, the fourth thoracic somite bearing the bud of swimming leg 3, and the limbless fifth thoracic somite. All of these somites articulate both anteriorly and posteriorly. The second thoracic somite of the first copepodid, bearing the transformed swimming leg 1, also may articulate anteriorly, although in some species of poecilostomes, harpacticoids and calanoids, the anterior arthrodial membrane between the second thoracic somite and the cephalothorax fails to form (see chapter "Patterning the copepod body"). The first thoracic somite of the first copepodid, which bears the transformed maxilliped in all copepods, never articulates anteriorly with the cephalon and so forms the cephalothorax with it. The body architecture including limb configuration of the first copepodid is remarkably conserved among copepods, with the exception of the thaumatopsylloids. It consists of a cephalon with five appendages and five thoracic somites with the first being united to the cephalon. The limbs of the thoracic somites include the maxilliped, swimming legs 1-2 as transformed limbs with unarticulated rami, swimming leg 3 as a limb bud, and the posterior abdominal somite with its caudal ramus. This combination of body architecture and limb configuration has been identified as the phylotypic stage of copepods (Ferrari, 2003).

During the molt from the last nauplius to the first copepodid, the addition of an articulating limbless thoracic somite anterior to the anal somite and posterior to the articulating, bud-bearing fourth thoracic somite is the result of a process central to patterning of the body during copepodid development. With each molt to a new copepodid stage, one new, limbless somite is presented anterior to the anal somite. This new, limbless somite is more easily identified during the following molt. On the second through fourth copepodids, a setose limb bud is added to the thoracic somite that formed one stage earlier. This older thoracic somite with its setose limb bud is adjacent to the new, limbless somite.

The bud of swimming leg 4, and buds of limb 5 and limb 6 initially are formed on the thoracic somites 5-7 during the molts to the second through fourth copepodids, respectively. Thoracic somites 5-7 initially are presented at the first to third copepodids, respectively. During molts to the fourth through sixth copepodid, one abdominal somite is added at each molt. These three abdominal somites remain limbless throughout development. The absence of limbs on these abdominal somites may compromise identification of the pattern of somite addition (see above).

The process of patterning the body during the copepodid phase of development, i. e., one new somite added per molt with its setose limb bud added one stage later, can be extrapolated back through the naupliar phase. This model results in the first appearance of setose limb buds and an inference of the first appearance of somites for each naupliar stage according to table XXX.

Corollaries to this extrapolated model of naupliar development include: the initial appearance of a new setose limb bud on the most recently added pre-existing somite; pre-existing limb buds are not transformed during the naupliar phase; setae may be added to pre-existing limb buds as well as to pre-existing transformed limbs. This simple, extrapolated model does not accurately describe the naupliar development of copepods, as will be seen in the following paragraph. However, it may model the naupliar development of a more primitive crustacean from which the exceptions of copepods are derived attributes.

Evidence among copepods for the extrapolated model is equivocal. Certainly, the fifth thoracic somite is the new and most posterior thoracic somite on CI. Swimming leg 3, which is the limb on the fourth thoracic somite, is the new and most posterior thoracic limb. Swimming leg 3 is a bud and

TABLE XXX

Setose limb buds and somites added during naupliar molts according to the extrapolated model that assumes that one new somite is added per molt and that the setose limb bud of the new somite is added one stage later

CI:	swimming leg 3 and the fifth thoracic somite
NVI:	swimming leg 2 and the fourth thoracic somite
NV:	swimming leg 1 and the third thoracic somite
NIV:	maxilliped and the second thoracic somite
NIII:	maxilla 2 and the first thoracic somite
NII:	maxilla 1 and the fifth cephalic somite

usually has three terminal setae on the presumptive exopod and two terminal setae on the presumptive endopod. These setae will be allocated to the terminal crown of setae on the exopod and endopod, respectively, when the limb is reconfigured during the molt to CII (fig. 28). The most posterior limb on the last nauplius, NVI, is swimming leg 2; it is the limb of the third thoracic somite. Its configuration is a bilobe setose bud usually with three terminal setae on the presumptive exopod and two terminal setae on the presumptive endopod. Thus, swimming leg 2, like swimming leg 3, appears as predicted by the extrapolated model (table XXX).

Swimming leg 1 initially is presented at the sixth nauplius, not at the fifth nauplius as predicted by the model. However, this setose bud exhibits evidence of being one stage further along in development than that of swimming leg 2 (see previous chapter). Specifically, among many copepods, one dorsal and three terminal setae are found on the presumptive exopod of the bud of swimming leg 1, and on some calanoids there are one ventral and two terminal setae on the presumptive endopod of the bud (Ferrari, 2000). The dorsal seta on the presumptive exopod is the seta that will be found on the proximal segment of the exopod of the adult. The ventral seta on the presumptive endopod of calanoids is the seta that will be found on the proximal segment of the endopod of the adult. On other swimming legs, these two setae initially are present on the transformed limb, and are never found at the limb bud stage. Therefore, the bud of swimming leg 1 on the last nauplius seems to be advanced in its development and delayed in its initial appearance from the extrapolated model by one naupliar stage.

The maxilliped initially appears as a simple bud armed with 2 terminal, crown endopodal setae on the sixth nauplius of calanoids. This configuration is not advanced in its development although the initial appearance of this limb is delayed from the prediction of the extrapolated model by two naupliar stages.

Maxilla 2 initially is presented on the fifth nauplius of calanoids as a simple bud with 1 terminal seta. The initial appearance of this limb on the fifth nauplius is delayed by two naupliar stages, relative to the prediction of the extrapolated model. Maxilla 2 on the sixth nauplius of calanoids is a complex bud armed with terminal ramal setae and a series of enditic lobes with one or more setae; this configuration is advanced in its development at this stage.

Maxilla 1 initially appears on the second nauplius of cyclopoids and harpacticoids as predicted by the extrapolated model (table XXX). At this stage it is a simple bud with one terminal ramal seta. At the fourth naupliar stage, maxilla 1 is configured as a complex bud with setae on the presumptive rami and on a series of enditic lobes; more setae are added to the limb bud during the following two molts. However, maxilla 1 of calanoids initially appears as a simple bud on the third nauplius. In the fourth nauplius of calanoids it is expressed as a complex bud similar in general configuration to that of cyclopoids at the fourth nauplius, and to maxilla 2 of calanoids at the sixth nauplius. The initial appearance of maxilla 1, delayed until the third nauplius of calanoids from the prediction of the extrapolated model, explains why the second nauplius of calanoids has been referred to as a second orthonauplius, while the second nauplius of cyclopoids and harpacticoids is a metanauplius (Dietrich, 1915). Maxilla 1 initially appears as a simple bud on the sixth nauplius of siphonostomatoids, and is reconfigured to a transformed limb during the molt to the first copepodid.

In summary, maxilla 1 of cyclopoids and harpacticoids, and swimming leg 2 of most copepods initially appear at NII and NVI, respectively, as predicted by the extrapolated model of somite addition during the naupliar phase of development. The initial appearance of the swimming leg 1 is delayed one stage, while the initial appearance of maxilla 2 and the maxilliped is delayed two stages. Maxilla 1, maxilla 2, maxilliped and swimming leg 2 each first appear as a simple setose limb bud; the configuration of swimming leg 1 initially appears as a more complex setose limb bud.

If the extrapolated model is correct, the body of the first nauplius of copepods should be composed of four cephalic somites plus the posterior abdominal somite bearing the bud of the caudal ramus. The fifth cephalic somite and thoracic somites one through four then should be added progressively during the five consecutive naupliar molts, respectively.

Some poecilostomes, however, are missing one or more intermediate naupliar stages (Izawa, 1987), and some siphonostomatoids molt directly from an orthonauplius to the first copepodid (Kabata, 1972; Johnson & Albright, 1991). Their development appears to require the addition of more than one somite and limb bud during at least one molt. Furthermore, the extrapolated model does not agree with the hypothesis of Dudley (1966), who observed a series of superficial subexuvial structures (i. e., structures internal to the naupliar exoskeleton) in the first nauplius of notodelphyids. These superficial subexuvial structures were interpreted as the armed post-mandibular limbs of the first copepodid, and were identified as maxilla 1, maxilla 2, maxilliped, swimming leg 1, swimming leg 2 and swimming leg 3. According to Dudley's hypothesis, the body of the orthonauplius of copepods consists of all of the cephalic somites and at least the first four thoracic somites. During the copepodid phase of development, Dudley (1966) described the addition of somites in a manner consistent with a proliferation zone in front of the anal somite so that somites are added progressively during the copepodid phase, one somite at each molt. Dudley's hypothesis then would imply that body patterning during the copepodid phase of development is decoupled from that of the naupliar phase. An advantage of Dudley's hypothesis, however, is that the addition of more than one somite and limb at any molt are not required during the naupliar phase of development for those poecilostomes and siphonostomatoids that are missing naupliar stages, because all cephalic and at least the first four thoracic somites already are present in the body of the orthonauplius.

IMPLICATIONS OF DEVELOPMENT FOR PHYLOGENY

Information from the post-embryonic development of copepods has been applied to two kinds of phylogenetic studies. In the first, attributes of development are used to specify groups of presumably related species. Relationships among species within the groups and relationships among the groups are not specified. This kind of study is less rigorous than the second kind, in which attributes of development are used to specify some form of the ancestor-descendant relationships among all taxa considered.

Von Nordmann (1832) used the general architecture of the nauplius and the first copepodid stage to correctly group copepods that were quite different in their adult morphology. He studied parasites like *Achtheres percarum* and *Tracheliastes polycolpus*, whose relationships with other animals were not well understood. At the time, these parasites were known almost exclusively from their adult form and usually were placed among the molluscs. Free-living copepods were known as wingless insects but were not included with other crustaceans. Von Nordmann (1832) found that a nauplius hatched from the egg of the parasites, and that later in development the nauplius molted to a copepodid-like stage. Comparing these stages to the nauplii and the copepodids of free-living copepods, he concluded that the similarity among these two different stages of development indicated that the parasites and the free-living copepods are the same kind of animal, but that the parasites are significantly transformed later in their development from first copepodid to adult. However, comparisons of stages based on similarity alone also can be misleading. Claus (1876) incorrectly grouped copepods and decapods together based on the similarity of copepodid stages of copepods to the protozoeal stages of decapods.

Dudley (1966) was interested in determining whether parasitic copepods belonging to the Notodelphyidae should be grouped with gnathostome cyclopoids, like species of the Enterocolidae and Botryllophilidae (now Ascidicolidae), or with poecilostome cyclopoids, like species of Myicolidae and Mytilicolidae. She used attributes of the development of naupliar appendages like the lack of changes in segmentation to the exopod of antenna 2 during the naupliar phase, the proximal exopodal segment of the mandible fused or not to the basis, a 2-segmented mandibular endopod, and poor development of the post-mandibular appendages maxilla 1, maxilla 2 and maxilliped during later naupliar stages. She concluded that species of Notodelphyidae belong within the group of gnathostome cyclopoids.

Dahms (1990c) determined the following naupliar apomorphies for the oligoarthran harpacticoids: a mandibular exopod with long, terminal setae; one or two short, thick, terminal setae on the mandibular endopod; and buds of post-mandibular limbs located laterally on the body. Later, the following apomorphies for oligoarthran harpacticoids were added (Dahms, 2004a): the coxal gnathobase of antenna 2 is broad throughout naupliar development; the endopod of antenna 2 is elongate throughout naupliar development; the mandibular endopod is an elongate process. Polyarthran harpacticoid nauplii share several derived states (Dahms, 1990c) among themselves including: a 2-segmented mandibular endopod; a long and thick seta on the bud of the caudal ramus of the first nauplius; segments 3-5 of antenna 1 of the first nauplius are fused during the molt to the second nauplius, even for those species for which a new segment is added; and the terminal segment of antenna 1 is transformed from cylindrical to flat during the molt to the second nauplius. However, no naupliar apomorphies are shared between the oligoarthrans and polyarthrans, and Dahms (2004b) proposed removing the polyarthrans to a position as the sister-taxon of all remaining copepods.

Dahms (1990b) also discussed trends of reduction and specialization in naupliar morphology within the Thalestridae. This harpacticoid family includes free-living species like *Thalestris longimana* and *Parathalestris harpactoides*, species like *Diarthrodes cystoecus*, which is symbiontic with macroalgae, as well as *Thalestris rhodymeniae*, an obligate phytoparasite of macroalgae. Setae on antenna 1 of nauplii of the free-living species are long and have dense setulation; the parasites have fewer setae on antenna 1, and those setae have sparse setulation. The masticatory process of antenna 2 of the parasites is broader relative to that of the free-living species, and there is a delayed presentation and reduction in size of the limb buds of post-mandibular appendages of the parasites. Dahms (1990b) noted that several of the assumed derived states of thalestrid nauplii are shared with some species of the family Harpacticidae. Previously, the family Harpacticidae had not been placed in the same group of families as the family Thalestridae. Given the variation expressed among nauplii within the Thalestridae, Dahms (1990b) recommended a reassessment of the Thalestridae.

Ferrari (1991) abstracted segmentation patterns from the steps of development of swimming legs 1-4, and limbs 5-6 for species of the calanoid genus *Labidocera*, and for genera within the calanoid family Diaptomidae and within the cyclopoid family Cyclopidae. Segmentation patterns were derived simply by noting the addition of arthrodial membranes to the ramus of a limb, and as a result of this assumption, segments were not differentiated from segment complexes. The segmentation patterns then were used as character states to group ten species belonging to the calanoid genus *Labidocera*, 14 genera in the calanoid family Diaptomidae and 12 genera belonging to the cyclopoid family Cyclopidae. If states for different characters resulted in the possible placement of a taxon in more than one group, a simple method for determining the correct group was based on the number of times the different states had converged among all copepods whose development was known.

In describing the naupliar development of the diosaccid, *Stenhelia* (*Delavalia*) *palustris*, Dahms & Bresciani (1993) discovered several apomorphies such as the shape of the naupliar shield, the shape of the masticatory process of antenna 2, the segment number of the exopod of antenna 2, and the setation of the mandibular basis and endopod. Based on comparisons with other diosaccid species, they recommended removing this species from the family Diosaccidae.

The harpacticoid family Parastenheliidae includes only two genera. Naupliar development of *Parastenhelia megarostrum* was described by Dahms & Hicks (1996). They found that these nauplii are similar to nauplii of species of the family Tachidiidae in general shape of the body and location of setae on the basis of antenna 2, but that the nauplii do not exhibit a reduction in the size and setation of the limbs involved in feeding, a reduction notable among nauplii of the Tachidiidae. Based on naupliar morphology, the authors concluded that the Parastenheliidae are related to the Thalestridae. Ivanenko & Ferrari (2003) discussed the addition of the last abdominal somite and formation of a genital complex among three siphonostomatoid parasites, *Dermatomyzon nigripes*, *Asterocheres lilljeborgi* and *Scottomyzon gibberum*. *Scottomyzon gibberum* expresses more derived states, such as the absence of a genital complex and reduced setation of the maxilliped, than *Asterocheres lilljeborgi*. These two species, in turn, suppress the addition of a fourth abdominal somite, a derived state; this fourth abdominal somite is expressed in *Dermatomyzon nigripes*.

Schutze et al. (2000) placed 35 species from 29 genera of the cyclopoid family Cyclopidae into groups based on ten developmental patterns of antenna 1. New segments added at each step of limb development were determined by aligning six different marker setae plus another two pairs of marker setae. The groups were defined by differences in arthrodial membrane formation expressed prior to the terminal, adult molt. One resulting group contained most of the cyclopid species, and the terminal, adult molt of species in this group resulted in a limb of 11 to 17 segments, depending upon the species. The remaining groups contained fewer than four species, and six groups contained a single species. The largest group was composed of species from three different subfamilies; two species of *Apocyclops* were placed in two different groups. The groups that are based on the development of antenna 1 are not comparable to the lineages derived from the development of thoracopods (Ferrari, 1998; Ferrari & Ivanenko, 2005). Ferrari & Ueda (2005) identified a homologous ventral attenuation present on limb 5 of most females of calanoids of the superfamily Centropagoidea as a character state useful in grouping species into this superfamily. Formation of the genital complex during the molt to CV for most species or during the molt to CIV for *Acartia erythraea* was a second attribute used to group species into the superfamily. Both attributes were proposed as synapomorphies for the Centropagoidea because all other calanoids lack this ventral attenuation, and because the genital complex of most other copepods forms during the molt to CVI.

In providing a phylogenetic hypothesis relating derived oligoarthran harpacticoid copepods and calanoid copepods to thecostracan crustaceans, Dahms (2004a) proposed the following naupliar apomorphies for the ancestral copepod: a 3-segmented antenna 1; coxa of antenna 2 with two setae at NIII (one of the setae is the naupliar arthrite); antenna 2 with a 1-segmented endopod; thoracic limb buds medially juxtaposed; buds of second and third thoracic limbs present at NVI; six setae on the bud of the caudal ramus at NVI. In addition, the transformation of the last nauplius to the first copepodid suggested two more apomorphies of copepods: an anameric mode of somite addition; and the presence of biramous swimming legs 1-2 at the first copepodid stage.

Björnberg (1972) specified ancestor-descendant relationships of free-living cyclopoid, harpacticoid and calanoid copepods based on her study of the naupliar morphology of a large number of planktonic species. She assumed, in general, that the ancestral naupliar body would be simple or relatively undifferentiated, and that ancestral limbs would be composed of similarly repeated elements. Derived groups of copepods were expected to have a more complex body architecture, and the number of repeated elements on the limbs was assumed to be reduced in derived groups. She believed that the nauplii of many cyclopoid copepods had the simplest shape, and that the setae and muscles also were simple. Harpacticoids retained many of these simple attributes but the setation and musculature of calanoids was quite complex. Her analysis placed the cyclopoids at the base of a branching tree of copepods, with harpacticoids close to the base; calanoids were the most derived order. This conclusion challenged conventional wisdom then and now, that calanoids are close to the base of the Copepoda. Some species of oithonids and of cyclopids were placed close to the base of the Cyclopoida, while the polyarthrans (canuellids and longipediids) were placed close to the base of the Harpacticoida. Among the Calanoida, the nauplii of some centropagids and acartiids have the simplest shape and unmodified setae, and so these two families were placed at the base of that order.

Dahms et al. (1991) proposed phylogenetic relationships for six species of *Tisbe*, based on the morphology of the sixth naupliar stage. The types of character states considered included the shape and number of setae on particular limb segments, the relative length of setae and the presence of epicuticular extensions on setae, on limbs or on limb buds. The analysis was rigorous, and the resulting cladogram was compared to one in which both naupliar and adult characters were used. These two cladograms did not align particularly well because several naupliar apomorphies were allowed to converge so that a monophyletic lineage could be diagnosed by several adult characters. This was explained by a comment that more information was known about adult characters than about naupliar characters.

While describing the naupliar development of *Scutellidium hippolytes*, Dahms (1993a) presented a cladogram of three of the 25 genera in the Tisbidae. He used the naupliar autapomorphies of *Tisbe gracilis* and *Drescheriella glacialis* as synapomorphies of their respective genera. The types of character states considered included the stage at which specific setae are added, shape of the masticatory process (naupliar arthrite), shape and number of setae on particular limb segments, and presence of a presumptive endopod on the paired buds of swimming legs. Determining an ancestral state for these characters was difficult, because there was no information about the nauplii of other tisbid genera and no information about the nauplii of species related to the Tisbidae. *Tisbe* was found to be most closely related to *Drescheriella*, with *Scutellidium* sharing fewer derived states. In a second study, Dahms (1993c) investigated *Tegastes clausi* and *Alteutha interrupta* from the families Tegastidae and Peltidiidae, respectively, two families related to the Tisbidae within the Tisbidimorpha. Species within Tisbidae were found to share a derived branched setal complex on the caudal rami at the first copepodid stage; this setal complex was not present on *Tegastes clausi* and *Alteutha interrupta*. The species *Tegastes clausi* and *Alteutha interrupta* were found, in turn, to share the delayed development of setae on the swimming legs.

Ferrari (1998) abstracted information about how the maxilliped, swimming legs 1-4 and limbs 5-6 of species of Cyclopidae developed, and used this information to derive ancestor-descendant relationships among genera within the family. Development of rami of swimming legs 1-4 often is linked or coordinated. Cyclopid species either retain the ancestral condition for this coordinated limb development, or express one of two derived, coordinated states: development of swimming legs 1-4 truncated; or development of swimming legs 1-4 delayed. Individual rami of species expressing either ancestral, truncated, or delayed development may be further derived from the coordinated hierarchy. Later, Ferrari & Ivanenko (2005) used the development of the endopod of the maxilliped to further refine the evolution of those cyclopids that express the ancestral condition of development of the swimming legs. The results of these analyses did not support the traditional subfamilial groupings of cyclopid genera based on the setation of limb 5. Furthermore, species of *Diacyclops* in which the development of swimming leg rami is truncated are not predicted to be closely related to species of *Diacyclops* in which development of swimming leg rami is delayed.

These successes aside, the phylogenetic analyses utilizing changes during copepod development can be hampered in several ways: (1) there may be difficulties in obtaining a complete set of all developmental stages; (2) a greater effort in time may be needed to observe each species, relative to the time needed to examine the adult stage, because many developmental stages must be examined; (3) the small size of early developmental stages requires more sophisticated dissection and observation techniques than those needed for the adult stage; (4) fewer characters are available from earlier stages of development, particularly naupliar stages; (5) less comparative data are available from developmental stages of other species relative to the large database available for adults; (6) analysing characters that may be expressed as several different states during several different stages of development is difficult (Ferrari, 1998; Dahms, 2004a). In addition, when studies of developmental stages lead to conclusions that contradict results from comparative adult morphology, the standard of acceptance for developmental studies is raised appreciably. Nevertheless, comparative analysis of the structure of developmental stages and changes during development have provided valuable phylogenetic information, particularly about relationships of transformed parasites to free-living species (Von Nordmann, 1832), and more recently on the adequacy of the subfamilies of the Cyclopidae (cf. Ferrari, 1998; Schutze et al., 2000; Ferrari & Ivanenko, 2005) and the relationship of the polyarthrans to the other copepods (Dahms, 2004b).

SUMMARY AND RECOMMENDED STUDIES

The post-embryonic development of copepods is divided into two phases, a naupliar phase and a copepodid phase. Each phase is divided into a series of stages during which the exoskeleton does not change. Naupliar stages of copepods can be diagnosed by three features: somites of the body are not separated by arthrodial membranes; post-mandibular appendages maxilla 1, maxilla 2, the maxilliped, swimming legs 1-2 and the caudal ramus, if present, are expressed as unarticulated, setose buds; an elongate, ventral arthrite originates on the coxa of antenna 2.

The ventral arthrite of antenna 2 is an articulating element that is moved by a pair of muscles originating on the dorsal wall of the coxa and attaching anteriorly and posteriorly to the base of the arthrite. The arthrite is present on the nauplii of many species of copepods, as well as on the nauplii of other crustaceans. However, its presence is not universal among copepods. A naupliar arthrite is not present on nauplii of caligid species with a naupliar phase of one or two free-swimming stages that lack a mouth opening. For calanid or pseudodiaptomid calanoids, the arthrite also is absent from early naupliar stages that lack a mouth opening; however, the arthrite develops on antenna 2 of these calanoids during later naupliar stages in which a mouth opening is present. A relationship between the presence of a naupliar arthrite and the presence of a mouth is not direct; for example, a naupliar arthrite has been observed on a monstrillid species whose only free-living nauplius lacks a mouth, gut or anus. Furthermore, an early well-developed naupliar arthrite is reduced on the sixth nauplius of tachidiids and harpacticids in which a mouth is present throughout the naupliar phase.

No more than six naupliar stages have been reported for any copepod. Six stages are known for most free-living copepods studied to date and for many copepods associated with other invertebrates. Six naupliar stages are hypothesized to be the ancestral condition for calanoids, harpacticoids and cyclopoids including poecilostomes. Species of siphonostomatoids, misophrioids and thaumatopsylloids, and some species of poecilostomes have a naupliar phase of fewer than six stages. Molting between naupliar stages usually involves changes in the morphology of the exoskeleton, although consecutive but unchanged naupliar stages have been reported for some siphonostomatoids. A model of development extrapolated from the copepodid phase, in which one somite is added at each molt, results in a body architecture for the first nauplius of four cephalic somites and the posterior abdominal somite. Consistency in changes to the exoskeleton, and particularly to the setose limb buds, provides some predictive power in identifying the stages of a six-stage naupliar phase that are suppressed in species with fewer than six stages in the naupliar phase. Of special interest are poecilostomes whose consecutive loss of NII, NII-NIII, NII-NIV, and NII-NV explains naupliar phases of 5, 4, 3, and 2 stages, respectively.

Most copepods develop through a six-stage copepodid phase. No more than six copepodid stages are known to be separated by molts for any copepod. During the copepodid phase of development a naupliar arthrite on antenna 2 is never present, the thoracic and abdominal somites often articulate both anteriorly and posteriorly, and there are up to nine transformed appendages on the first copepodid stage: antenna 1, antenna 2, mandible, maxilla 1, maxilla 2, maxilliped, swimming legs 1-2 and the caudal ramus. Swimming leg 3 is a setose bud. The body increases in somite number and usually increases in size during the copepodid phase. Each remaining thoracic limb is added to its somite, as an unarticulated setose bud, one stage later than its somite is added to the body. Segment elements also are added to most limbs during the copepodid phase. Among caligid-like parasites, the second to fifth copepodid stages appear to be derived stages and are called chalimus stages 1-4. Reports of more than six stages for some caligid siphonostomes may be the result of an incorrect diagnosis of polymorphisms expressed in one or more stages. A copepodid phase of fewer stages has been reported for benthopelagic calanoids and a number of different parasitic copepods.

With the exception of copepods of the order Thaumatopsylloida, the body architecture of the first copepodid stage has been called the phylotypic stage of copepods because it is remarkably conserved in the following ways. The body of CI includes a cephalon with five appendages, five thoracic somites and the posterior abdominal somite. The first thoracic somite always is fused anteriorly to the cephalon; the second to fifth thoracic somites usually articulate anteriorly and posteriorly. Of the appendages, swimming legs 1-2 always are transformed limbs with unarticulated rami and interpodal bars, and swimming leg 3 is a setose bud; the posterior abdominal somite bears a transformed caudal ramus.

During development of gymnopleans, the anterior section of the copepodid body attains the architecture of the adult prosome at CIII. Podoplean copepods attain the architecture of the adult prosome at CII, while the adult prosome of thaumatopsylloid copepods is present at CI. These differences result in a different number of thoracic somites being incorporated into the posterior part of the adult body: one thoracic somite to the urosome of adult gymnopleans, two thoracic somites to the urosome of adult podopleans, and three thoracic somites to the urosome of adult thaumatopsylloids. Thaumatopsylloids also differ from gymnopleans and podopleans by the inclusion of the setose bud of swimming leg 4 at CI and the initial appearance of the setose buds of limbs 5-6 at CII-CIII, respectively, one stage earlier than in gymnopleans and podopleans. It also seems probable that the thaumatopsylloid body is comprised of one somite more than the body of gymnopleans and podopleans at comparable copepodid stages.

For many years, the gymnoplean architecture has been considered a synapomorphy for calanoids plus platycopiids, although there is no direct evidence to support this hypothesis from any analysis of the different extant copepod architectures and a possible ancestral architecture. A review of copepod development here, however, suggests the following gymnoplean synapomorphies: presentation of the bud of maxilla 1 delayed until NIII; mandibular gnathobase on NIV-NVI; presumptive endopod of the bud of swimming leg 1 at NVI with 3 setae, proximoventral seta to be allocated to the proximal endopodal segment; presence of Von Vaupel Klein's Organ on swimming leg 1; exopod of male limb 5 on the side of the genital opening with denticles or sensilla that aid in the transfer of the spermatophore to the female. Of these, the last two can be observed on the adult animals and they are quite widespread among Calanoida, including the Platycopiidae.

The copepodid body is patterned from a growth zone located in the anterior part of the anal (posterior, or first abdominal) somite; new somites are added only in the anterior direction and initially are presented adjacent to the anal somite. One new somite is added during the molt to each new copepodid stage. Complexes resulting from the failure to express an arthrodial membrane between two or more somites explain much of the variation in the body architecture of copepods within Gymnoplea and Podoplea. A limb is added as a setose bud to a thoracic somite one stage after that thoracic somite has formed. A model derived from this information simplifies the determination of homologous somites for many nauplii. Most somite complexes along the copepodid body result from failure of expression of the arthrodial membranes, which separates two somites; this failure of expression usually occurs after the arthrodial membrane initially has been expressed earlier in development.

Segmental patterning of copepod limbs is more complicated than the process of addition of somites to the body, although in general, variation in limb segmentation results from the truncation of patterning during development. The structure of copepod limbs is basically a bifurcate, linear topology in which the protopod is the base and each branch of the bifurcation is a ramus. Segment elements can be added from at least three points. The protopod appears to be patterned from the point at which the limb meets the body wall so that developmentally older segments and endites of the protopod are always distad, and the basis is the oldest protopodal segment. In contrast, each ramus is patterned from at least one source segment, and new segmental elements, including setae and arthrodial membranes, may be added either proximally or distally to the source segment. As a result of this patterning process, there is no direct correlation between the developmental age of a ramal segment, relative to other ramal segments, and the distance of that ramal segment from the basis of the protopod. For this reason, determination of homologous segments of a ramus usually is not straightforward.

Relatively much more is known about changes during development of the copepod exoskeleton than of any other organ. Enough information has been published about stage-specific changes of the exoskeleton so that several kinds of analyses could be discussed here. These analyses include the variability in the order of presentation of limb buds during the naupliar phase of development, the appearance of a conserved architecture for the first copepodid, and patterning of the rami of many limbs. However, many of the publications that provide information for these analyses describe the development of free-living species, particularly species of Calanoida, Harpacticoida and Cyclopoida. In contrast, much developmental information remains to be discovered about the less speciose orders Gelyelloida, Misophrioida, Monstrilloida, Mormonilloida, and Thaumatopsylloida. Furthermore, information about the exoskeleton of parasitic poecilostomes, as well as the number of naupliar stages of these parasites, only recently has begun to be analysed. Even less is known about development of the many species of Siphonostomatoida, particularly those associated with invertebrates.

Studies of the post-embryonic development of copepods began in the middle of the eighteenth century, and much has been learned in the intervening two and a half centuries. Nevertheless, more remains to be discovered, and these discoveries will suggest more sophisticated analyses. Furthermore, there are few genera for which any aspect of the development of more than two or three species has been published. As a result, specific predictions about unstudied species of even common, free-living genera, are difficult. And, of course, there are many genera and families for which no information about development is available.

A recent analysis of the nauplii of polyarthrans placed this group of copepods among the early branching events during the evolutionary history of copepods. The presence of the bud of swimming leg 4 on the first copepodid of thaumatopsylloids suggests an ordinal category for this lineage is appropriate, as well as its placement among the early branching events of copepods. The order Platycopioida often is assumed to have resulted from the earliest branching event of the copepods. However, questions remain about the comparability in number or the equivalency in degree of its few synapomorphies. An hypothesis in which the Platycopiidae result from a branching event within the Calanoida has not been explored in a systematic manner. It seems clear that a more nuanced analysis of the presumed basal taxa of the Copepoda would be a timely contribution, and that comparative development should provide critical information for this analysis.

Questions about the number of abdominal somites expressed during the copepodid development of different siphonostomatoid copepods points to a weakness in analyses of copepod body patterning. In the caligid-like parasites of fishes, the thoracic and abdominal somites, which make up the posterior part of the body, often are poorly sclerotized. This poor sclerotization may make difficult a determination of the number of abdominal somites present at any stage. Suppression of abdominal somite formation has been suggested for the siphonostomatoid parasites of invertebrates, for some calanoids and for some poecilostomes. Because an abdominal somite does not bear a limb, it is difficult to differentiate a single abdominal somite from complexes of two or more abdominal somites. The expression of antibodies raised to regulatory genes required for somite formation of crustaceans may be usefully applied in detecting the abdominal somite complexes among juvenile stages of these copepods.

A model of body patterning has been derived from the addition of somites and limb buds during the copepodid phase. This model then can be extrapolated back through the naupliar phase to predict somite number at any naupliar stage. The usefulness of this extrapolated model to predict somite number during the naupliar phase also would benefit from a cellular model based on the study of the expression of antibodies raised to regulatory genes required for somite formation.

The naupliar arthrite appears to function as an aid in moving food through the mouth. The mandibular gnathobase assumes this function during the copepodid phase of development. However, NIV-VI of many calanoids bear both a naupliar arthrite on antenna 2 and a gnathobase on the mandible. A study of how the naupliar arthrite and mandibular gnathobase act together during feeding of NIV-VI of calanoids would be a valuable addition to the functional morphology of nauplii.

Patterning the protopod of copepod limbs has not been available to direct observation because neither arthrodial membranes nor endites are added during development of the few limbs that have been studied. Inferences from a phylogenetic analysis of development of the maxilliped supports what generally is understood about patterning of the protopod from direct observations of branchiopod trunk limb development. As the development of more copepod limbs is studied, protopodal patterning may be available from the direct observations of the presence of setae on the different protopodal endites; this should provide a useful data set in a comparative analysis, including the existing information on the branchiopod trunk limbs.

More is known about how the rami of copepod limbs are patterned than on how the protopod is patterned, but ramal patterning is complicated because new segmental elements can be added either proximally or distally from at least one source segment. Currently, the degree to which proximal or distal patterning configures a particular limb cannot be anticipated. Whether proximal patterning or distal patterning is more likely to be expressed on a particular ramus, e. g., exopod vs. endopod, or on particular limbs, e. g., cephalic vs. thoracic limbs, or on rami of a particular configuration, e. g., rami with large numbers of segments vs. smaller numbers, remains to be determined.

There is a great deal of information about changes in antenna 1 during the naupliar and the copepodid phases of development. However, several incompatible models have been proposed to explain these changes. Given the number of observations, understanding how this important limb develops should be within reach once the analytical method for patterning is agreed upon.

Relative to stage-specific changes of the exoskeleton, very little is known about changes in the internal anatomy of copepods during development. Remarkably, much less seems to be known of changes in internal anatomy during the copepodid phase than during the naupliar phase. Development of some internal organ systems, like the digestive tract, are continuous and may not coordinate well with stage-specific changes in the exoskeleton. However, a stage-specific model of development should provide a useful comparative template for an analysis of changes in any continuously developing organ.

Other internal systems, like muscles or nerves, should show some correlation with stage-specific additions of somites because parts of these systems must be added as each new somite is added, and then become functional.

A better understanding is needed of the nature of the copepod urosome, which at present is defined almost exclusively by the architecture of the adult body. Analyses of the development of innervation of somites and the stage at which the longitudinal muscles that move particular somites become striated, may provide useful information for a diagnosis of the copepod urosome.

Little is known of the stage-specific differences in limb function during the naupliar or copepodid phase, and little is known about stage-specific differences of naupliar behavior or ecology. Studies of stage-specific differences in behavior and ecology among copepodid stages of species other than *Calanus finmarchicus*, and especially of non-planktonic copepods, will be a welcome addition to copepod post-embryonic development.

This volume celebrates some of the fascinating discoveries about post-embryonic development of copepod crustaceans. As can been seen from the above discussion, the door has opened only slightly. More and more is yet too little.

BIBLIOGRAPHY

(in three parts; parts II and III, respectively, starting on p. 141 and p. 144)

Differentiation of publications from the same year by the same author(s) has been done as usual, i. e., by adding a, b, ... etc. to the year, but readers should be aware of the fact that these letters may run across all three parts of this bibliography, here.

PAPERS ON THE POST-EMBRYONIC DEVELOPMENT OF COPEPODS CITED HEREIN

ALEKSEEV, V. R., 1982. Fotoperiodizm i sezonnoe razvitie populyatsii *Mesocyclops leuckarti* (Claus) (Crustacea, Copepoda). Photoperiodism and seasonal development in *Mesocyclops leuckarti* (Claus) (Crustacea, Copepoda) populations. Gidrobiologicheskii Zhurnal, Saratov, **23**: 22-28. [In Russian; English translation in Hydrobiological Journal, **23** (4)].

ALLAN, J. D., R. RICHMAN, D. R. HEINLE & R. HUFF, 1977. Grazing in juvenile stages of some estuarine calanoid copepods. Marine Biology, Berlin, **43**: 317-331.

ALLMAN, G. J., 1847. Biological contributions. No. III. Description of a new genus and species of Entomostraca (*Notodelphys*). Annals and Magazine of Natural History, **20**: 1-9, pls. 1-2.

ALVAREZ, V. & H. G. KEWALRAMANI, 1970. Naupliar development of *Pseudodiaptomus ardjuna* Brehm (Copepoda). Crustaceana, **18**: 269-276.

ALVAREZ-VALDERHAUG, V. & H. G. KEWALRAMANI, 1979. Larval development of *Apocyclops dengizicus* Lepeshkin (Copepoda). Crustaceana, **36**: 1-8.

AMBLER, J. W., F. D. FERRARI & J. FORNSHELL, 1991. Population structure and swarm formation of a cyclopoid copepod, *Dioithona oculata*, near mangrove cays. Journal of Plankton Research, **13**: 1257-1272.

ANDERSEN, H. R., L. WOLLENBERGER, B. HALLING-SORENSEN & K. O. KUSK, 2001. Development of copepod nauplii to copepodites: a parameter for chronic toxicity including endocrine disruption. Environmental Toxicology and Chemistry, **20**: 2821-2829.

AURIVILLIUS, P. O. C., 1879. On a new genus and species of Harpacticida. Kongliga Svenska Vetenskaps-Akademiens Handlingar, **5**: 3-16 + 4 pls.

AUVRAY, C. & B. DUSSART, 1966. Rôle de quelques facteurs du milieu sur le développement post-embryonnaire des Cyclopides (Crustacés Copépodes). I. - Généralitiés. Cas de *Eucyclops*. Bulletin de la Société Zoologique de France, **91**: 477-491.

BATCHELDER, H. P., 1985. Seasonal abundance, vertical distribution, and life history of *Metridia pacifica* (Copepoda: Calanoida) in the oceanic subarctic Pacific. Deep-Sea Research, (A, Oceanographic Research Papers) **32**: 949-964.

BENEDETTI, I., L. MOLA & M. A. SABATINI, 1992. Morphogenesis of the gut in the nauplius stages of the parasitic copepod *Lernaea cyprinacea*. Bollettino di Zoologia, **59**: 245-250.

BENNET, P. S., 1962. *Peroderma cylindricum* Heller, a copepod parasite of *Sardinella albella*. Journal of the Marine Biological Association of India, **3**: 70-74.

BENZ, G., 1989. Developmental stages of *Alebion lobatus* Cressey, 1970 (Copepoda: Euryphoridae) found parasitic on the sandbar shark (*Charcharhinus plumbeus* (Nardo, 1827)) in the western North Atlantic, and a phylogenetic analysis of the genus *Alebion* Krøyer, 1863. Canadian Journal of Zoology, **67**: 1578-1598.

BIRGE, E. A. & C. JUDAY, 1908. A summer resting stage in the development of *Cyclops bicuspidatus* Claus. Transactions of the Wisconsin Academy of Sciences, Arts, and Letters, **16**: 1-9.

BJORNBERG, M. H., 1984. Descrição de *Bryocyclops caroli* sp. n. estudo comparativo do desenvolvimento pósembrionário de *B. caroli* e *B. absalomi* Por, 1981 (Copepoda, Cyclopoida, Cyclopidae): 1-75. (M.Sc. Thesis, Universidade de São Paulo, São Paulo).

BJÖRNBERG, T. K. S., 1972. Developmental stages of some tropical and subtropical planktonic marine copepods. Studies on the Fauna of Curaçao and other Caribbean Islands, **40**(136): 1-185.

——, 1986a. Aspects of the appendages in development. Syllogeus, **58**: 51-66.

BOCQUET, C., 1948. Recherches sur les *Porcellidium* (Copépodes) de Roscoff. Archives de Zoologie Expérimentale et Générale, **85**: 237-259.

BOXSHALL, G. A., 1974a. The developmental stages of *Lepeophtheirus pectoralis* (Müller, 1776) (Copepoda: Caligidae). Journal of Natural History, London, **8**: 681-700.

BOXSHALL, G. A. & R. HUYS, 1998. The ontogeny and phylogeny of copepod antennules. Philosophical Transactions of the Royal Society of London, (B) **353**: 765-786.

BOXSHALL, G. A. & H. ROE, 1980. The life history and ecology of the aberrant bathypelagic genus *Benthomisophria* Sars, 1909 (Copepoda: Misophrioida). Bulletin of the British Museum (Natural History), (Zoology) **38**: 9-41.

BUNDY, M. H. & G. A. PAFFENHÖFER, 1997. The innervation of the setae on the appendages of the nauplius of *Eucalanus pileatus* Giesbrecht. Marine and Freshwater Behaviour and Physiology, **30**: 195-207.

BURMEISTER, H., 1835. Beschreibung einiger neuen oder weniger bekannten Schmarotzerkrebse, nebst allgemeinen Betrachtungen über die Gruppe, welcher sie angehören. Nova Acta Physico-Medica Academiae Caesareae Leopoldino-Carolinae Naturae Curiosorum (Acta der Kaiserlichen Leopoldinisch-Carolinischen Deutschen Akademie der Naturforscher), **17**: 269-336.

BUSKEY, E. J., 1994. Factors affecting feeding selectivity of visual predators on the copepod *Acartia tonsa*: locomotion, visibility and escape responses. Hydrobiologia, **292/293**: 447-453.

BUSKEY, E. J., C. COULTER & S. STROM, 1993. Locomotory patterns of microzooplankton: potential effects on food selectivity of larval fish. Bulletin of Marine Science, **53**: 29-43.

CALAFIORE, N. & G. COSTANZO, 1992. Larval stages of *Zygomolgus poucheti* (Canu, 1981) (Copepoda, Cyclopoida, Lichomolgidae) of the Lake Faro (Messina, Italy). Oebalia, (Supplement) **17**: 283-294.

CAMPBELL, M., 1934. The life history and post embryonic development of the copepods, *Calanus tonsus* Brady and *Euchaeta japonica* Marukawa. Journal of the Biological Board of Canada, **1**: 1-65.

CANDEIAS, A., 1966. Contribution to the knowledge of the development of *Halicyclops neglectus* Kiefer. Lisboa Revue de Biologie, **4**: 171-186.

CARTER, M. E. & J. M. BRADFORD, 1972. Postembryonic development of three species of freshwater harpacticoid Copepoda. Smithsonian Contributions to Zoology, **119**: 1-26.

CARTON, Y., 1964. Description de *Selioides bocqueti* n. sp., copépode cyclopoïde parasite de *Scalisetosus assimilis* MacIntosh (Aphroditidae) commensal d'*Echinus esculentus* L. Archives de Zoologie Expérimentale et Générale, (Notes et Revue) **104**: 84-103.

CARVALHO, M. DE, 1971. Desenvolvimento de *Ectocyclops rubescens* (Brady, 1904) (Copepoda, Crustacea). Boletim de Zoologia e Biologia Marinha, (n. ser.) **28**: 343-388.

CHANG, W. B. & C. H. LEI, 1993. Development and energy content of a brackish-water copepod, *Apocyclops royi* (Lindberg) reared in a laboratory. Bulletin of the Institute of Zoology, Academia Sinica, Taipei, **32**: 62-81.

CHANGEUX, J. P., 1961. Contribution à l'étude des animaux associés aux Holothurides. Actualités Scientifiques et Industrielles, **1284**: 1-124.

CICCHINO, G., 1975. Redescripcion y desarrollo postembrionario de *Pseudodiaptomus richardi inaequalis* (Brian) con notas acerca de sus afinidades con otras especies del genero (Copepoda, Pseudodiaptomidae). Physis, Buenos Aires, (B) **34**: 37-49.

CLAUS, C., 1858a. Ueber den Bau und die Entwicklung parasitischer Crustaceen: 1-32, 2 pls. (Druck und Verlag von Theodor Fischer, Marburg, Cassel).

——, 1858b. Zur Anatomie und Entwicklungsgeschichte der Copepoden. Archiv für Naturgeschichte, **24**: 1-76.

——, 1861. Ueber den Bau und die Entwicklung von *Achtheres percarum*. Zeitschrift für Wissenschaftliche Zoologie, **11**: 287-308.

——, 1863. Die frei lebenden Copepoden mit besonderer Berücksichtigung der Fauna Deutschlands, der Nordsee und des Mittelmeeres: 1-230, 37 pls. (Verlag von Wilhelm Engelmann, Leipzig).

——, 1876. Untersuchungen zur Erforschung der genealogischen Grundlage des Crustaceen-Systems. Ein Beitrag zur Descendenzlehre: 1-114, 19 pls. (Carl Gerhold's Son, Vienna).

COMITA, G. & S. MCNETT, 1976. The postembryonic developmental instars of *Diaptomus oregonensis* Lilljeborg, 1889 (Copepoda). Crustaceana, **30**: 123-163.

COMITA, G. & D. TOMMERDAHL, 1960. The postembryonic developmental instars of *Diaptomus siciloides* Lilljeborg. Journal of Morphology, **107**: 297-356.

COSTANZO, G., 1959. Sullo sviluppo de *Mytilicola intestinalis* Steuer (Crust. Cop.). Archivio Zoologico Italiano, **44**: 151-163.

——, 1969. Stadi naupliari e primo copepodite di *Lichomolgus canui* G.O. Sars (Copepoda, Cyclopoida) del Lago di Faro (Messina), allevata sperimentalmente. Bollettino di Zoologia, **36**: 143-153.

——, 1984. Gli stadi di sviluppo di *Modiolicola insignis* Aurivillius, 1882 (Copepoda, Cyclopoida, Sabelliphilidae), del Lago de Faro (Messina) da colture in laboratorio. Cahiers de Biologie Marine, **25**: 293-305.

COSTANZO, G. & N. CALAFIORE, 1985. Larval development of *Herrmannella rostrata* Canu, 1891 (Copepoda, Poecilostomatoida, Sabelliphilidae) of the Lake Faro (Messina) reared in the laboratory. Memorie de Biologia Marina e di Oceanografia, **15**: 141-154.

COSTANZO, G., N. CALAFIORE & N. CRESCENTI, 1994. Copepodids of *Doridicola longicauda* (Claus, 1860) (Copepoda, Poecilostomatoida, Lichomolgidae) associated with *Sepia officinalis* L. Journal of Crustacean Biology, **14**: 601-608.

CRAIN, J. A. & C. B. MILLER, 2000. Detection of sex and sex ratio in *Calanus finmarchicus* early stage fifth copepodites. ICES Journal of Marine Science, **57**: 1773-1779.

DAHMS, H.-U., 1987a. Postembryonic development of *Drescheriella glacialis* Dahms & Dieckmann (Copepoda, Harpacticoida) reared in the laboratory. Polar Biology, **8**: 81-93.

——, 1989a. Antennule development during copepodite phase of some representatives of Harpacticoida (Crustacea, Copepoda). Bijdragen tot de Dierkunde, Amsterdam, **59**: 159-189.

——, 1990a. Naupliar development of *Paraleptastacus brevicaudatus* Wilson, 1932 (Copepoda, Harpacticoida, Cylindropsyllidae). Journal of Crustacean Biology, **10**: 330-339.

——, 1990b. The first nauplius and the copepodite stages of *Thalestris longimana* Claus, 1863 (Copepoda, Harpacticoida, Thalestridae) and their bearing on the reconstruction of phylogenetic relationships. Hydrobiologia, **202**: 33-60.

——, 1990c. Naupliar development of Harpacticoida (Crustacea, Copepoda) and its significance for phylogenetic systematics. Mikrofauna Marina, **6**: 169-272.

——, 1992a. Metamorphosis between naupliar and copepodid phase in the Harpacticoida (Copepoda). Philosophical Transactions of the Royal Society of London, (B) **335**: 221-236.

——, 1993a. Naupliar development of *Scutellidium hippolytes* (Copepoda, Harpacticoida) and a comparison of nauplii within Tisbidae. Hydrobiologia, **250**: 1-14.

——, 1993b. Copepodid development in Harpacticoida (Crustacea, Copepoda). Microfauna Marina, **8**: 195-245.

——, 1993c. Comparative copepodid development in Tisbidimorpha sensu Lang, 1948 (Copepoda, Harpacticoida) and its bearing on phylogenetic considerations. Hydrobiologia, **250**: 15-37.

——, 2004a. Postembryonic apomorphies proving the monophyletic Copepoda. Zoological Studies, **43**: 446-453.

——, 2004b. Exclusion of the Polyarthra from the Harpacticoida and its reallocation as an underived branch of the Copepoda (Arthropoda, Crustacea). Invertebrate Zoology, **1**: 29-51.

DAHMS, H.-U. & M. BERGMANS, 1988. Postembryonic development of *Tisbe gracilis* (T. Scott, 1895) (Copepoda, Harpacticoida). Zoologica Scripta, **17**: 357-369.

DAHMS, H.-U. & J. BRESCIANI, 1993. Naupliar development of *Stenhelia* (*D.*) *palustris* from the North Sea. Ophelia, **37**: 101-116.

DAHMS, H.-U. & C. H. FERNANDO, 1992. Naupliar development of *Mesocyclops aequatorialis similis* and *Thermocyclops consimilis* (Copepoda: Cyclopoida) from Lake Awasa, a tropical Rift Valley lake in Ethiopia. Canadian Journal of Zoology, **70**: 2283-2297.

—— & ——, 1993a. Naupliar development of *Mesocyclops* cf. *thermocyclopoides* Harada, 1931 and *Thermocyclops decipiens* (Kiefer, 1929) (Copepoda: Cyclopoida) from Beira Lake, Sri Lanka. Journal of Plankton Research, **15**: 9-26.

—— & ——, 1993b. Naupliar development of *Phyllodiaptomus annae* Apstein, 1907 (Copepoda, Calanoida) from Sri Lanka. Zoological Journal of the Linnean Society, **108**: 197-208.

—— & ——, 1993c. Redescription of *Mesocyclops leuckarti* (Copepoda, Cyclopoida) including a study of its naupliar development. Internationale Revue der Gesamten Hydrobiologie, **78**: 589-609.

—— & ——, 1994. Redescription of female *Macrocyclops fuscus* (Jurine, 1820) (Copepoda, Cyclopoida) from Ontario with a description of naupliar stages. Journal of Plankton Research, **16**: 9-21.

—— & ——, 1995. Naupliar development of *Mesocyclops edax* (Copepoda, Cyclopoida). Journal of Crustacean Biology, **15**: 329-340.

DAHMS, H.-U. & G. R. F. HICKS, 1996. Naupliar development of *Parastenhelia megarostrum* (Copepoda, Harpacticoida) and its bearing on phylogenetic relationships. Journal of Natural History, **30**: 11-22.

DAHMS, H.-U., S. LORENZEN & H.-K. SCHMINKE, 1991. Phylogenetic relationships within the taxon *Tisbe* (Copepoda, Harpacticoida) as evidenced by naupliar characters. Zeitschrift für Zoologische Systematik und Evolutions-Forschung, **29**: 433-449.

DALE, T. & S. KAARTVEDT, 2000. Diel patterns in stage-specific vertical migration of *Calanus finmarchicus* in habitats with midnight sun. ICES Journal of Marine Science, **57**: 1800-1818.

DAMKAER, D. M., 2002. The copepodologist's cabinet: a biographical and bibliographical history. Memoirs of the American Philosophical Society, **240**: i-xix, 1-300.

DEFAYE, D., 1984. Développement et nutrition chez *Macrocyclops albidus* (Crustacé, Copépode): 1-107, pls. 1-19. (Ph.D. Thesis, Université Bordeaux I, Bordeaux).

DIETRICH, W., 1915. Die Metamorphose der freilebenden Süsswasser-Copepoden. I. Die Nauplien und das erste Copepoditstadium. Zeitschrift für Wissenschaftliche Zoologie, **113**: 252-324.

DO, T., T. KAJIHARA & J. HO, 1984. The life history of *Pseudomyicola spinosus* (Raffaele & Monticelli, 1885) from the blue mussel, *Mytilus edulis galloprovincialis* in Tokyo Bay, Japan, with notes on the production of atypical male. Bulletin of the Ocean Research Institute, University of Tokyo, **17**: 1-65.

DUDLEY, P., 1964. Some gastrodelphyid copepods from the Pacific coast of North America. American Museum Novitates, **2194**: 1-51.

——, 1966. Development and systematics of some Pacific marine symbiotic copepods. A study of the biology of the Notodelphyidae, associates of ascidians. University of Washington Publications in Biology, **21**: 1-282.

DURBIN, E. G., P. R. GARRAHAN & M. C. CASAS, 2000a. Depth distribution of *Calanus finmarchicus* nauplii on the Georges Bank during 1995 and 1996. ICES Journal of Marine Science, **57**: 1686-1693.

EIANE, K., D. L. AKSNES, M. D. OHMAN, S. WOOD & M. B. MARTINUSSEN, 2002. Stage-specific mortality of *Calanus* spp. under different predation regimes. Limnology and Oceanography, **47**: 636-645.

EIANE, K. & M. D. OHMAN, 2004. Stage-specific mortality of *Calanus finmarchicus*, *Pseudocalanus elongatus* and *Oithona similis* on Fladen Ground, North Sea, during a spring bloom. Marine Ecology Progress Series, **268**: 183-193.

ELGMORK, K., 1959. Seasonal occurrence of *Cyclops strenuus strenuus* in relation to environment in small water bodies in southern Norway. Folia Limnologica Scandinavica **11**: 1-196.

ELGMORK, K. & A. L. LANGELAND, 1970. The number of naupliar instars in Cyclopoida (Copepoda). Crustaceana, **18**: 277-282.

ESCRIBANO, R., L. RODRIGUEZ & C. IRRIBARREN, 1998. Temperature-dependent development and growth of *Calanus chilensis* Brodsky from northern Chile. Journal of Experimental Marine Biology and Ecology, **229**: 19-34.

EVSTIGNEEV, P. V., 1982. Izmenenie kharakteristik biolyuminestsentnogo signala v ontogeneze kopepod roda *Pleuromamma*. [Changes in characteristics of bioluminescent signals during ontogenesis of copepods of the genus *Pleuromamma*.] Biologiya Morya, **1982**: 55-59. [In Russian, with English summary.]

——, 1984. Stanovlenie reaktsii svetoizlucheniya v khode individual'nogo razvitiya kopepod roda *Pleuromamma*. Mekhanizmy Povedeniya, **1983**: 96-97.

FAHRENBACH, W., 1962. The biology of a harpacticoid copepod. La Cellule, **62**: 303-376 + 9 pls.

FANTA, E. S., 1973. Anatomy of the nauplii of *Euterpina acutifrons* (Dana) (Copepoda, Harpacticoida). Crustaceana, **23**: 165-181.

——, 1976. Anatomy of the nauplii of *Oithona ovalis* Herbst (Copepoda, Cyclopoida). Boletim do Museu Nacional do Rio de Janeiro, (Zoologia) **1**: 205-238.

——, 1982. Anatomy of the nauplii of *Pseudodiaptomus acutus* (Dahl) (Copepoda, Calanoida). Arquivos de Biologia e Tecnologia, Curitiba, **25**: 341-353.

FERNÁNDEZ, F., 1979. Nutrition studies on the nauplius larva of *Calanus pacificus* (Copepoda, Calanoida). Marine Biology, Berlin, **53**: 131-147.

FERRARI, F. D., 1985. Postnaupliar development of a looking-glass copepod, *Pleuromamma xiphias* (Giesbrecht, 1889), with analyses of distributions of sex and asymmetry. Smithsonian Contributions to Zoology, **420**: 1-55.

——, 1988. Developmental patterns in numbers of ramal segments of copepod postmaxillipedal legs. Crustaceana, **54**: 256-294.

——, 1991. Using patterns of appendage development to group taxa of *Labidocera*, Diaptomidae and Cyclopidae (Copepoda). Bulletin of the Plankton Society of Japan, (Special Volume): 115-127.

——, 1995. Six copepodid stages of *Ridgewayia klausruetzleri*, a new species of calanoid copepod (Ridgewayiidae) from the barrier reef in Belize, with comments on appendage development. Proceedings of the Biological Society of Washington, **108**: 180-200.

——, 1998. Setal developmental patterns of the thoracopods of cyclopid copepods (Cyclopoida) and their use in phylogenetic inference. Journal of Crustacean Biology, **18**: 471-489.

——, 2000. Patterns of setal numbers conserved during early development of swimming legs of copepods (Crustacea). Hydrobiologia, **417**: 81-90.

FERRARI, F. D. & J. AMBLER, 1992. Nauplii and copepodids of the cyclopoid copepod *Dioithona oculata* (Oithonidae) from mangrove cays in Belize. Proceedings of the Biological Society of Washington, **105**: 275-298.

FERRARI, F. D. & A. BENFORADO, 1998a. Relationships between arthrodial membrane formation and addition of setae to swimming legs 1-4 of *Dioithona oculata*, *Ridgewayia klausruetzleri*, *Pleuromamma xiphias*, and *Temora longicornis* (Crustacea, Copepoda). Crustaceana, **71**: 545-564.

—— & ——, 1998b. Setation and setal groups on antenna 1 of *Ridgewayia klausruetzleri*, *Pleuromamma xiphias*, and *Pseudocalanus elongatus* (Crustacea: Copepoda: Calanoida) during the copepodid phase of their development. Proceedings of the Biological Society of Washington, **111**: 209-221.

FERRARI, F. D. & H.-U. DAHMS, 1998. Segmental homologies of the maxilliped of some copepods as inferred by comparing setal numbers during copepodid development. Journal of Crustacean Biology, **18**: 298-307.

FERRARI, F. D., J. FORNSHELL, L. ONG & J. W. AMBLER, 2003. Diel distribution of copepods across a channel of an overwash mangrove island. Hydrobiologia, **499**: 147-159.

FERRARI, F. D. & V. N. IVANENKO, 2001. Interpreting segment homologies of the maxilliped of cyclopoid copepods by comparing stage-specific additions of setae during development. Organisms, Diversity and Evolution, **1**: 113-131.

—— & ——, 2005. Copepodid stages of *Euryte longicauda* (Cyclopoida, Cyclopidae, Euryteinae) from the White Sea associated with the bryozoan *Flustra foliacea*. Journal of Crustacean Biology, **25**: 353-374.

FERRARI, F. D. & E. L. MARKHASEVA, 1996. *Parkius karenwishnerae*, a new genus and species of calanoid copepod (Parkiidae, new family) from benthopelagic waters of the eastern tropical Pacific Ocean. Proceedings of the Biological Society of Washington, **109**: 264-285.

FERRARI, F. D. & D. STEINBERG, 1993. *Scopalatum vorax* (Esterly, 1911) and *Scolecithricella lobophora* Park, 1970 calanoid copepods (Scolecitrichidae) associated with a pelagic tunicate in Monterey Bay. Proceedings of the Biological Society of Washington, **106**: 467-489.

FERRARI, F. D. & H. UEDA, 2005. Development of the fifth leg of copepods belonging to the calanoid superfamily Centropagoidea (Crustacea). Journal of Crustacean Biology, **25**: 333-352.

FIERS, F., 1991a. Three new harpacticoid copepods from the Santa Maria Basin off the Californian Pacific Coast (Copepoda, Harpacticoida). Beaufortia, **42**: 13-47.

——, 1996. Redescription of *Enhydrosoma lacunae* Jakubisiak, 1933 (Copepoda, Harpacticoida), with comments on the *Enhydrosoma* species reported from West Atlantic localities, and a discussion of the cletodid development. Sarsia, **81**: 1-27.

FLEDDUM, A., S. KAARTVEDT & B. ELLERTSEN, 2001. Distribution and feeding of the carnivorous copepod *Paraeuchaeta norvegica* in habitats of shallow prey assemblages and midnight sun. Marine Biology, Berlin, **139**: 719-726.

FRIEDMAN, M. M. & J. R. STRICKLER, 1975. Chemoreceptors and feeding in calanoid copepods (Arthropoda: Crustacea). Proceedings of the National Academy of Sciences of the United States of America, **72**: 4815-4188.

GAARD, E., 2000. Seasonal abundance and development of *Calanus finmarchicus* in relation to phytoplankton and hydrography on the Faroe shelf. ICES Journal of Marine Science, **57**: 1605-1611.

GAULD, D. T., 1959. Swimming and feeding in crustacean larvae: the nauplius larva. Proceedings of the Zoological Society of London, **132**: 31-50.

GEER, C. DE, 1778. Mémoires pour servir à l'histoire des Insectes, **7**: i-xiv, 1-950. (Stockholm).

GELETIN, Y. V., 1976. Formirovanie abdomena v ontogeneze u kopepod rodov *Eucalanus* i *Rhincalanus* (Calanoida, Eucalanidae) i movaya siste. [Ontogenetic abdomen formation in copepods of genera *Eucalanus* and *Rhincalanus* (Calanoida, Eucalanidae) and new system of these copepods.] Issledovaniya Fauny Morei, **18**: 75-93.

GEORGE, K. H., 2001. First record of the "genus" *Ancorabolus* Norman, 1903 from the Southern Hemisphere, including analyses of copepodid development (Crustacea, Copepoda, Harpacticoida, Ancorabolidae). Senckenbergiana Biologica, **81**: 23-36.

GERRITSEN, J., 1978. Instar-specific swimming patterns and predation of planktonic copepods. Verhandlungen der Internationale Vereinigung für Theoretische und Angewandte Limnologie, **20**: 2531-2536.

GIESBRECHT, W., 1913. Crustacea. In: A. Lang (ed.), Handbuch der Morphologie der wirbellosen Tiere, **4**, Arthropoda: 9-252.

GISLASON, A. & O. S. ASTTHORSSON, 1996. Seasonal development of *Calanus finmarchicus* along an inshore-offshore gradient southwest of Iceland. Ophelia, **44**: 71-84.

GISLASON, A., O. S. ASTTHORSSON, H. PETURSDOTTIR, H. GUDFINNSSON & A. R. BODVARSDOTTIR, 2000. Life cycle of *Calanus finmarchicus* south of Iceland in relation to hydrography and chlorophyll a. ICES Journal of Marine Science, **57**: 1619-1627.

GOSWAMI, S. C., 1977a. Developmental stages of a cyclopoid copepod *Cyclopina longifera* (Sewell) reared in a laboratory. Mahasagar, **10**: 21-28.

——, 1978a. Developmental stages, growth and sex ratio in *Pseudodiaptomus binghami* Sewell (Copepoda, Calanoida). Indian Journal of Marine Science, **7**: 103-109.

GRABDA, J., 1963. Life cycle and morphogenesis of *Lernaea cyprinacea* L. Acta Parasitologica Polonica, **11**: 169-198.

GRAINGER, E. H. & A. A. MOHAMMED, 1991. Some diagnostic characters of copepodid stages of the cyclopoid copepod *Cyclopina schneideri* T. Scott and adults of the Arctic marine Cyclopinidae. Canadian Journal of Zoology, **69**: 2365-2373.

GREEN, E. P., R. P. HARRIS & A. DUNCAN, 1991. The naupliar development of marine calanoid copepods under high and low food conditions. Bulletin of Plankton Society of Japan, (Special Volume): 347-362.

GREEN, J., 1958. *Dactylopusioides macrolabris* (Claus) (Copepoda: Harpacticoida) and its frond mining nauplius. Proceedings of the Zoological Society of London, **131**: 49-54.

GREEN, S., A. W. VISSER, J. TITELMAN & T. KIØRBOE, 2003. Escape responses of copepod nauplii in the flow field of the blue mussel, *Mytilus edulis*. Marine Biology, Berlin, **142**: 727-733.

GRICE, G. D., 1969. The developmental stages of *Pseudodiaptomus coronatus* Williams (Copepoda, Calanoida). Crustaceana, **16**: 291-301.

GROBBEN, C., 1881. Die Entwicklungsgeschichte von *Cetochilus septentrionalis* Goodsir. Arbeiten aus dem Zoologischen Institut der Universität Wien, **3**: 243-282.

GRYGIER, M. J. & S. OHTSUKA, 1995. SEM observation of the nauplius of *Monstrilla hamatapex*, new species, from Japan and an example of upgraded descriptive standards for monstrilloid copepods. Journal of Crustacean Biololgy, **15**: 703-719.

GURNEY, R., 1930a. The larva of *Nicothoe astaci* and its systematic position. Journal of the Marine Biological Association of the United Kingdom, **16**: 453-459.

——, 1931. British freshwater Copepoda. I. General; Calanoida: 1-228. (The Ray Society, London).

——, 1932. British freshwater Copepoda. II. Harpacticoida: 1-336. (The Ray Society, London).

——, 1933a. Notes on some Copepoda from Plymouth. Journal of the Marine Biological Association of the United Kingdom, **19**: 299-304.

——, 1934a. The development of certain parasitic Copepoda of the families Caligidae and Clavellidae. Proceedings of the Zoological Society of London, **1934**: 177-217.

HAMBURGER, K. & F. BOETIUS, 1987. Ontogeny of growth, respiration and feeding rate of the freshwater calanoid copepod *Eudiaptomus graciloides*. Journal of Plankton Research, **9**: 589-606.

HARDING, J., 1954. The copepod *Thalestris rhodymeniae* (Brady) and its nauplius, parasitic in the seaweed *Rhodymenia palmata* (L.) Grev. Proceedings of the Zoological Society of London, **124**: 153-161.

HART, R. C., 1990. Copepod postembryonic durations: pattern, conformity, and predictability. The realities of isochronal and equiproportional development, and trends in the copepodid, naupliar duration ratio. Hydrobiologia, **206**: 175-206.

HERON, G. & T. BOWMAN, 1971. Postnaupliar developmental stages of the copepod crustaceans, *Clausocalanus laticeps*, *C. brevipes* and *Ctenocalanus citer* (Calanoida: Pseudocalanidae). Antarctic Research Series, **17**: 141-165.

HERON, G. & D. DAMKAER, 1986. A new nicothoid copepod parasitic on mysids from northwestern North America. Journal of Crustacean Biology, **6**: 652-665.

HIDALGO, P., R. ESCRIBANO & C. E. MORALES, 2005. Annual life cycle of the copepod *Eucalanus inermis* at a coastal upwelling site off Mejillones (23°S), northern Chile. Marine Biology, Berlin, **146**: 995-1003.

HIPEAU-JACQUOTTE, R., 1978. Développement post-embryonnaire du Copépode ascidicole Notodelphyidae *Pachypygus gibber* (Thorell, 1859). Crustaceana, **34**: 155-194.

HIRST, A. G., J. C. ROFF & R. S. LAMPITT, 2003. A synthesis of growth rates in marine epipelagic invertebrate zooplankton. Advances in Marine Biology, **44**: 1-142.

HUGGETT, J. A. & A. J. RICHARDSON, 2000. A review of the biology and ecology of *Calanus agulhensis* off South Africa. ICES Journal of Marine Science, **57**: 1834-1849.

HULSEMANN, K., 1991a. The copepodid stages of *Drepanopus forcipatus* Giesbrecht, with notes on the genus and a comparison to other members of the family Clausocalanidae (Copepoda Calanoida). Helgoländer Meeresuntersuchungen, **45**: 199-224.

——, 1991b. Tracing homologies in appendages during ontogenetic development of calanoid copepods. Bulletin of the Plankton Society of Japan, (Special Volume): 105-114.

HULSEMANN, K. & A. FLEMINGER, 1975. Some aspects of copepodid development in the genus *Pontellina* Dana (Copepoda: Calanoida). Bulletin of Marine Science, **25**: 174-185.

HUMES, A. G., 1986. Copepodids and adults of *Leptinogaster major* (Williams, 1907), a poecilostomatoid copepod living in *Mya arenaria* L. and other marine bivalve mollusks. Fishery Bulletin, United States, **84**: 227-245.

HUYS, R. & R. BÖTTGER-SCHNACK, 1994. Taxonomy, biology and phylogeny of Miracidae (Copepoda: Harpacticoida). Sarsia, **79**: 207-283.

HYGUM, B. H., C. REY & B. W. HANSEN, 2000. Growth and development rates of *Calanus finmarchicus* nauplii during a diatom spring bloom. Marine Biology, Berlin, **136**: 1075-1085.

ILLG, P., 1949. A review of the copepod genus *Paranthessius* Claus. Proceedings of the United States National Museum, **99**: 391-428.

INCZE, L. S., P. AAS & T. AINARE, 1996. Distributions of copepod nauplii and turbulence on the southern flank of Georges Bank: implications for feeding by larval cod (*Gadus morhua*). Deep-Sea Research, (II) **43**: 1855-1874.

IRIGOIEN, X., J. TITELMAN, R. P. HARRIS, D. HARBOUR & C. CASTELLANI, 2003. Feeding of *Calanus finmarchicus* nauplii in the Irminger Sea. Marine Ecology Progress Series, **262**: 193-200.

ITÔ, T., 1970. The biology of a harpacticoid copepod, *Tigriopus japonicus* Mori. Journal of the Faculty of Science Hokkaido University, (6, Zoology) **17**: 474-500, pls. 7-8.

——, 1971. The biology of a harpacticoid copepod, *Harpacticus uniremis* Kröyer. Journal of the Faculty of Science Hokkaido University, (6, Zoology) **18**: 235-255, 1 pl.

——, 1976. Morphology of the copepodid stages of *Zaus robustus* Itô and *Paratigriopus hoshidei* Itô from Japan, with reference to some biological observations (Harpacticoida: Harpacticidae). Journal of the Faculty of Science Hokkaido University, (6, Zoology) **20**: 211-229.

ITÔ, T. & J. BURTON, 1980. A new genus and species of the family Canthocamptidae (Copepoda, Harpacticoida) from a hot spring at Dusun Tua, Selangor, Malaysia. Zoologische Jahrbücher, (Systematik) **107**: 1-31.

ITÔ, T. & T. TAKASHIRO, 1981. The larval development of *Canthocamptus mirabilis* Stêrba (Copepoda, Harpacticoida), with reference to bionomics. Journal of the Faculty of Science Hokkaido University, (6, Zoology) **22**: 279-300.

ITOH, H., 2003. Ecology of copepods of the genus *Hemicyclops* (Poecilostomatoida, Clausidiidae) in Tokyo Bay and the Tama-River estuary: 1-94. (Ph.D. Thesis, The University of Tokyo, Tokyo). [In Japanese.]

ITOH, H. & S. NISHIDA, 1995. Copepodid stages of *Hemicyclops japonicus* Itoh and Nishida (Poecilostomatoida: Clausidiidae) reared in the laboratory. Journal of Crustacean Biology, **15**: 134-155.

—— & ——, 1997. Naupliar stages of *Hemicyclops japonicus* (Copepoda: Poecilostomatoida) reared in the laboratory. Journal of Crustacean Biology, **17**: 162-173.

IVANENKO, V. N. & F. D. FERRARI, 2003. Redescription of adults and description of copepodid development of *Dermatomyzon nigripes* (Brady & Robertson, 1876) and of *Asterocheres lilljeborgi* Boeck, 1859 (Copepoda: Siphonostomatoida: Asterocheridae). Proceedings of the Biological Society of Washington, **116**: 661-691.

IVANENKO, V. N., F. D. FERRARI & A. V. SMUROV, 2001. Nauplii and copepodids of *Scottomyzon gibberum* (Copepoda: Siphonostomatoida: Scottomyzontidae, new family), a symbiont of *Asterias rubens* (Asteroidea). Proceedings of the Biological Society of Washington, **114**: 237-261.

IZAWA, K., 1969. Life history of *Caligus spinosus* Yamaguti, 1939 obtained from cultured yellow tail, *Seriola quinqueradiata* T. & S. (Crustacea: Caligoida). Report of Faculty of Fisheries, Prefectural University of Mie, **6**: 127-157.

——, 1973. On the development of parasitic Copepoda. I. *Sarcotaces pacificus* Komai (Cyclopoida: Philichthyidae). Publications of the Seto Marine Biological Laboratory, **21**: 77-86.

——, 1975a. On the development of parasitic Copepoda. II. *Colobomatus pupa* Izawa (Cyclopoida: Philichthyidae). Publications of the Seto Marine Biological Laboratory, **22**: 147-155.

——, 1986a. On the development of parasitic Copepoda. III. *Taeniacanthus lagocephali* Pearse (Cyclopoida: Taeniacanthidae). Publications of the Seto Marine Biological Laboratory, **31**: 37-54.

——, 1986b. On the development of parasitic Copepoda. IV. Ten species of poecilostome cyclopoids, belonging to Taeniacanthidae, Tegobomolochidae, Lichomolgidae, Philoblennidae, Myicolidae, and Chondracanthidae. Publications of the Seto Marine Biological Laboratory, **31**: 81-162.

——, 1987. Studies on the phylogenetic implications of ontogenetic features in the poecilostome nauplii (Copepoda: Cyclopoida). Publications of the Seto Marine Biological Laboratory, **32**: 151-217.

——, 1991. Evolutionary reduction of body segments in the poecilostome Cyclopoida (Crustacea: Copepoda). Bulletin of Plankton Society of Japan, (Special Volume): 71-86.

——, 1997. The copepodid of *Peniculisa shiinoi* Izawa, 1965 (Copepoda, Siphonostomatoida, Pennellidae), a single free-swimming larval stage of the species. Crustaceana, **70**: 911-919.

JACOBS, J., 1961. Laboratory cultivation of the marine copepod *Pseudodiaptomus coronatus* Williams. Limnology and Oceanography, **6**: 443-446.

JOHNSON, M. W., 1937. The developmental stages of the copepod *Eucalanus elongatus* Dana var. *bungii* Giesbrecht. Transactions of the American Microscopical Society, **55**: 79-98.

——, 1948. The postembryonic development of the copepod *Pseudodiaptomus euryhalinus* Johnson, and its phylogenetic significance. Transactions of the American Microscopical Society, **67**: 319-330.

JOHNSON, S. C. & L. J. ALBRIGHT, 1991. Development, growth, and survival of *Lepeophtheirus salmonis* (Copepoda: Caligidae) under laboratory conditions. Journal of the Marine Biological Association of the United Kingdom, **71**: 425-436.

JURINE, L., 1820. Histoire des Monocles, qui se trouvent aux environs de Geneve & Paris: 1-260, pls. 1-22. (J. J. Pachoud, Geneva).

KABATA, Z., 1972. Developmental stages of *Caligus clemensi* (Copepoda: Caligidae). Journal of the Fisheries Research Board of Canada, **29**: 1571-1593.

——, 1986. The developmental stages of *Neobrachiella robusta* (Wilson, 1912), a parasitic copepod of *Sebastes* (Teleostei: Scorpaeniformes). Canadian Journal of Zoology, **65**: 1331-1336.

KABATA, Z. & B. COUSENS, 1973. Life cycle of *Salmincola californiensis* (Dana, 1852) (Copepoda: Lernaeopodidae). Journal of the Fisheries Research Board of Canada, **30**: 881-903.

KARAYTUG, S. & G. A. BOXSHALL, 1996. The life cycle of *Paracyclops fimbriatus* (Fischer, 1853) (Copepoda, Cyclopoida). Bulletin of the Natural History Museum of London, **62**: 41-70.

KATTNER, G., M. GRAEVE & W. HAGEN, 1994. Ontogenetic and seasonal changes in lipid and fatty acid/alcohol compositions of the dominant Antarctic copepods *Calanus propinquus*, *Calanoides acutus* and *Rhincalanus gigas*. Marine Biology, Berlin, **118**: 637-644.

KATTNER, G. & M. KRAUSE, 1987. Changes in lipids during the development of *Calanus finmarchicus* s.l. from copepodid I to adult. Marine Biology, Berlin, **96**: 511-518.

KIM, I.-H., 1994. Copepodid stages of *Conchyliurus quintus* Tanaka (Copepoda, Poecilostomatoida, Clausidiidae). Hydrobiologia, **292/293**: 616-169.

——, 2003. Copepodid stages of *Critomolgus anthopleurus* Kim (Copepoda, Poecilostomatoida, Rhynchomolgidae). Journal of Crustacean Biology, **23**: 558-567.

——, 2004. Copepodid stages of *Ergasilus hypomesi* Yamaguti (Copepoda, Poecilostomatoida, Ergasilidae) from a brackish lake in Korea. Korean Journal of Biological Science, **8**: 1-12.

KIM, I.-H. & J.-S. HO, 1992. Copepodid stages of *Hemicyclops ctenidis* Ho and Kim, 1990 (Clausidiidae), a poecilostomatoid copepod associated with a polychaete. Journal of Crustacean Biology, **12**: 631-646.

KÔ, Y., 1969d. External anatomy of a commensal copepod *Ostrincola koe* - I. Copepodid stage. Bulletin of the Faculty of Fisheries, Nagasaki University, **28**: 93-109. [In Japanese, with English summary.]

KRAEFFT, F., 1910. Über das Plankton in Ost- und Nordsee und den Verbindungsgebieten, mit besonderer Berücksichtigung der Copepoden. Wissenschafliche Meeresuntersuchungen, Herausgegeben von der Kommission zur Untersuchung der Deutschen Meere in Kiel und der Biologischen Anstalt auf Helgoland, Abteilung Kiel, (Neue Folge) **11**: 29-99.

KRØYER, H., 1863. Bidrag til Kundskab om Snyltekrebsene. Naturhistorisk Tidsskrift, (3) **2**: 75-426.

KUANG, P., 1962. Studies on the life history of *Lamproglena chinensis* Yu. Acta Hydrobiologica Sinica, **1**: 114-123, 2 pls. [In Chinese.]

——, 1980. A new genus of Lernaeidae (parasitic copepods) and its relation to affined genera. Acta Zootaxonomica Sinica, **5**: 124-128. [In Chinese.]

KUMAR, R. & T. RAMAMOHAMA RAO, 1998. Post-embryonic developmental rates as a function of food type in the cyclopoid copepod *Mesocyclops thermocyclopoides* Harada. Journal of Plankton Research, **20**: 272-287.

LANDRY, M. R., 1978a. Population dynamics and production of a planktonic marine copepod, *Acartia clausii*, in a small temperate lagoon on San Juan Island, Washington. Internationale Revue der Gesamten Hydrobiologie, **63**: 77-119.

LANGE, J. C., 1756. Laere om de Naturlige Vande, samt Grundig Undersogning af de Vande, som findes udi Kiobenhavn og dens Egne, og til allehaande Nytte af dens Indbyggere meest blive anvendte. Naest Oplosning paa det Sporsmaal: Hvilket Vand der er det bedste?: 1-134, 2 pls. (L. H. Lillie, Copenhagen).

LAPOTA, D., T. E. BOWMAN & J. R. LOSEE, 1988. Observations on bioluminescence in the nauplius of *Metridia longa* (Copepoda, Calanoida) in the Norwegian Sea. Crustaceana, **54**: 314-320.

LAPOTA, D. & J. R. LOSEE, 1984. Observations of bioluminescence in marine plankton from the Sea of Cortez. Journal of Experimental Marine Biology and Ecology, **77**: 209-240.

LEE, R. F., J. C. NEVENZEL & A. G. LEWIS, 1974. Lipid changes during life cycle of marine copepod, *Euchaeta japonica* Marukawa. Lipids, **9**: 891-898.

LEE, R. F., J. C. NEVENZEL & G.-A. PAFFENHÖFER, 1972. The presence of wax esters in marine planktonic copepods. Naturwissenschaften, **59**: 406-411.

LESCHER-MOUTOUÉ, F., 1966. Note sur la reproduction et le développement post-embryonnaire des *Speocyclops*. Annales de Spéléologie, **21**: 673-687.

LEWIS, A. G., 1963. Life history of the caligid copepod *Lepeophtheirus dissimulatus* Wilson, 1905 (Crustacea: Caligoida). Pacific Science, **17**: 195-242.

LIN, C. L., J. S. HO & S. N. CHEN, 1996. Developmental stages of *Caligus epidemicus* Hewitt, a copepod parasite of *Tilapia* cultured in brackish water. Journal of Natural History, London, **30**: 661-684.

——, —— & ——, 1997. Development of *Caligus multispinosus* Shen, a caligid copepod parasitic on the black sea bream (*Acanthopagrus schlegeli*) cultured in Taiwan. Journal of Natural History, London, **31**: 1483-1500.

LOWE, E., 1935. On the anatomy of a marine copepod, *Calanus finmarchicus* (Gunnerus). Transactions of the Royal Society of Edinburgh, (Earth Sciences) **58**: 561-603.

LUCKS, R., 1926. Zur Entwicklungsgeschichte des *Cyclops viridis* Jurine und seiner Stellung zum *Cyclops clausii* Heller. Schriften der Naturforschenden Gesellschaft in Danzig, **17**: 128-169.

LÜTZEN, J., 1968a. On the biology of the family Herpyllobiidae (parasitic copepods). Ophelia, **5**: 175-187.

MALT, S., 1982. Developmental stages of *Oncaea media* Giesbrecht, 1891 and *Oncaea subtilis* Giesbrecht, 1892. Bulletin of the British Museum (Natural History), (Zoology) **43**: 129-151.

MATTHEWS, J. B. L., 1964. On the biology of some bottom living copepods (Aetideidae and Phaennidae) from western Norway. Sarsia, **16**: 1-46.

MCMILLAN, G., 1991. Description of the life-stages of *Amphiascus undosus* Lang (Copepoda, Harpacticoida). Canadian Journal of Zoology, **69**: 512-527.

MILLER, C. B., J. A. CRAIN & C. A. MORGAN, 2000. Oil storage variability in *Calanus finmarchicus*. ICES Journal of Marine Science, **57**: 1786-1799.

MULLIN, M. M., 1995. Nauplii of the copepod, *Calanus pacificus*, off southern California in the El Niño winter-spring of 1992, and implications for larval fish. Journal of Plankton Research, **17**: 183-189.

MUNRO, I. G., 1974. The effect of temperature on the development of egg, naupliar and copepodite stages of two species of copepods, *Cyclops vicinus* Uljanin and *Eudiaptomus gracilis* Sars. Oecologia, **16**: 355-367.

NAKAMURA, K., T. KAJIHARA & M. OKA, 1979. Development of a cyclopoid copepod, *Pseudomyicola ostreae* Yamaguti, associated with a marine mussel, *Mytilus edulis galloprovincialis* Lamarck: external structure of nauplius in six stages. Proceedings of the Japan Society of Systematic Zoology, **17**: 38-44.

NORDMANN, A. VON, 1832. Mikrographische Beiträge zur Naturgeschichte der Wirbellosen Thiere. Part **2**: 1-150, 10 pls. (G. Reimer, Berlin).

——, 1864. Neue Beiträge zur Kenntnis parasitischer Copepoden. Bulletin de la Société des Naturalistes de Moscou, **37**: 461-520.

OBERG, M., 1906. Die Metamorphose der Plankton-Copepoden der Kieler Bucht. Wissenschaftliche Reihe der Deutschen Meeresuntersuchungen in Kiel, (Neue Folge) (Abteilung Kiel) **9**: 39-103.

OHMAN, M. D., K. EIANE, E. G. DURBIN, J. A. RUNGE & J. A. HIRCHE, 2004. A comparative study of *Calanus finmarchicus* mortality patterns in five localities in the North Atlantic. ICES Journal of Marine Science, **61**: 687-697.

OHMAN, M. D., J. A. RUNGE, E. G. DURBIN, D. B. FIELD & B. NIEHOFF, 2002. On birth and death in the sea. Hydrobiologia, **480**: 55-68.

OHMAN, M. D. & S. N. WOOD, 1996. Mortality estimation for planktonic copepods: *Pseudocalanus newmani* in a temperate fjord. Limnology and Oceanography, **41**: 126-135.

OHTSUKA, S. & G. [A.] BOXSHALL, 1994. *Platycopia orientalis* sp. nov. (Copepoda: Platycopioida) from the North Pacific with descriptions of copepodid stages. Journal of Crustacean Biology, **14**: 151-167.

ONBÉ, T., 1984. The developmental stages of *Longipedia americana* (Copepoda: Harpacticoida) reared in the laboratory. Journal of Crustacean Biology, **4**: 615-631.

OSGOOD, K. E. & B. W. FROST, 1994a. Ontogenetic diel vertical migration behaviors of the marine planktonic copepods *Calanus pacificus* and *Metridia lucens*. Marine Ecology Progress Series, **104**: 13-25.

—— & ——, 1994b. Comparative life histories of three species of planktonic calanoid copepods in Dabob Bay, Washington. Marine Biology, Berlin, **118**: 627-636.

PAFFENHÖFER, G. A. & K. D. LEWIS, 1989. Feeding behavior of nauplii of the genus *Eucalanus* (Copepoda, Calanoida). Marine Ecology Progress Series, **57**: 129-136.

PAFFENHÖFER, G. A. & P. A. LOYD, 1999. Ultrastructure of setae of the maxilliped of the marine planktonic copepod *Temora stylifera*. Marine Ecology Progress Series, **178**: 101-107.

PAFFENHÖFER, G. A., J. R. STRICKLER, K. D. LEWIS & S. RICHMAN, 1996. Motion behavior of nauplii and early copepodid stages of marine planktonic copepods. Journal of Plankton Research, **18**: 1699-1715.

PELSENEER, P., 1914. Éthologie de quelques *Odostomia* et d'un Monstrillide parasite de l'un d'eux. Bulletin Scientifique de la France et la Belgique, **48**: 1-14, 3pls.

PERKINS, P., 1983. The life history of *Cardiodectes medusaeus* (Wilson), a copepod parasite of lanternfishes (Myctophidae). Journal of Crustacean Biology, **3**: 70-87.

PETERSON, W. T., 2001. Patterns in stage duration and development among marine and freshwater calanoid and cyclopoid copepods: a review of rules, physiological constraints, and evolutionary significance. Hydrobiologia, **453/454**: 91-105.

PETERSON, W. T. & S. J. PAINTING, 1990. Developmental rates of the copepods *Calanus australis* and *Calanoides carinatus* in the laboratory, with discussion of methods used for calculation of development times. Journal of Plankton Research, **12**: 283-293.

PIASECKI, W., 1996. The developmental stages of *Caligus elongatus* Von Nordmann, 1832 (Copepoda: Caligidae). Canadian Journal of Zoology, **74**: 1459-1478.

PINCHUK, A. I. & A. J. PAUL, 1998. Growth of *Metridia pacifica* (Copepoda: Calanoida) nauplii in the laboratory. Plankton Biology and Ecology, **45**: 195-201.

RAMDOHR, K. A., 1805. Beiträge zur Naturgeschichte einiger deutschen Monoculusarten: 1-36, pls. 1-7. (Mikrographische Beiträge zur Entomologie und Helminthologie, Halle).

RANGA REDDY, Y. & C. RAMA DEVI, 1985. The complete postembryonic development of *Megadiaptomus hebes* Kiefer, 1936 (Copepoda, Calanoida) reared in the laboratory. Crustaceana, **48**: 40-63.

REDFIELD, G. W. & C. R. GOLDMAN, 1980. Diel vertical migration by males, females, copepodids and nauplii in a limnetic population of *Diaptomus* (Copepoda). Hydrobiologia, **74**: 241-248.

ROSENFIELD, D. & B. COULL, 1974. Adult morphology and larval development of *Paramphiascella fulvofasciata* n. sp. (Copepoda, Harpacticoida). Cahiers de Biologie Marine, **15**: 295-317.

RÖTTGER, R., 1969. Ökologie und Postlarvalentwicklung von *Scottomyzon gibberum*, eines auf *Asterias rubens* parasitisch lebenden Copepoden (Cyclopoida Siphonostoma). Marine Biology, Berlin, **2**: 145-202.

SABATINI, M. E., 1990. The developmental stages (copepodids I to VI) of *Acartia tonsa* Dana, 1849 (Copepoda, Calanoida). Crustaceana, **59**: 53-61.

SABATINI, M. E. & T. KIØRBOE, 1994. Egg production, growth and development of the cyclopoid copepod *Oithona similis*. Journal of Plankton Research, **16**: 1329-1351.

SABATINI, M. E., M. MARINI, B. FRATELLO, C. GELOSINI, L. MOLA & I. BENEDETTI, 1987. The digestive apparatus of the parasitic copepod *Lernaea cyprinacea* L. during development and in the adults. Acta Embryologiae et Morphologiae Experimentalis, **8**: 389-409.

SAZHINA, L. I., 1985. Naupliusy massovykh vidov pelagickeskikh kopepod Mirovogo okeana. [Nauplii of mass species of the world's oceans.]: 1-238. (Naukova Dumka, Kiev).

SCHRAM, T., 1979. The life history of the eye-maggot of the sprat *Lernaeenicus sprattae* (Sowerby) (Copepoda, Lernaeoceridae). Sarsia, **64**: 279-316.

SCHUTZE, M. L. M., C. E. F. ROCHA & G. A. BOXSHALL, 2000. Antennulary development during the copepodid phase in the family Cyclopidae (Copepoda, Cyclopoida). Zoosystema, **22**: 749-806.

SPROSTON, N. G., 1942. The developmental stages of *Lernaeocera branchialis* (Linn.). Journal of the Marine Biological Association of the United Kingdom, **25**: 441-466.

STORCH, O., 1928. Der Nahrungserwerb zweier Copepodennauplien (*Diaptomus gracilis* und *Cyclops strenuus*). Zoologische Jahrbücher, Jena, (Allgemeine Zoologie) **45**: 385-436.

SURIRAY, J. S. A., 1819. Sur les oeufs de *Calyge*. Journal de Physique, de Chimie, d'Histoire Naturelle et des Arts, **89**: 400.

TANDE, K. S., S. DROBYSHEVA, V. NESTEROVA, E. M. NILSSEN, A. EDVARDSEN & V. TERESCHENKO, 2000. Patterns in the variations of copepod spring and summer abundance in the northeastern Norwegian Sea and the Barents Sea in cold and warm years during the 1980s and 1990s. ICES Journal of Marine Science, **57**: 1581-1591.

TITELMAN, J. & T. KIØRBOE, 2003a. Motility of copepod nauplii and implications for food encounter. Marine Ecology Progress Series, **247**: 123-135.

—— & ——, 2003b. Predator avoidance by nauplii. Marine Ecology Progress Series, **247**: 137-149.

TORRES, C. G. & R. ESCRIBANO, 2003. Growth and development of *Calanus chilensis* nauplii reared under laboratory conditions: testing the effects of temperature and food resources. Journal of Experimental Marine Biology and Ecology, **294**: 81-99.

TRUJILLO-ORTIZ, A., 1986. Life cycle of the marine calanoid copepod *Acartia californiensis* Trinast reared under laboratory conditions. California Cooperative Fisheries Investigations, **27**: 188-204.

UCHIMA, M. & R. HIRANO, 1986. Food of *Oithona davisae* (Copepoda: Cyclopoida) and the effect of food concentration at first feeding of the larval growth. Bulletin of Plankton Society of Japan, **33**: 21-28.

UMMERKUTTY, A. N., 1964. Studies on Indian copepods. 6. The postembryonic development of two calanoid copepods *Pseudodiaptomus aurivilli* Cleve and *Labidocera bengalensis* Krishnaswamy. Journal of the Marine Biological Association of India, **6**: 48-60.

URAWA, S., K. MUROGA & S. KASAHARA, 1980a. Naupliar development of *Neoergasilus japonicus* (Copepoda: Ergasilidae). Bulletin of the Japanese Society of Scientific Fisheries, **46**: 941-987.

——, —— & ——, 1980b. Studies on *Neoergasilus japonicus* (Copepoda: Ergasilidae), a parasite of freshwater fishes - II. Development in copepodid stage. Journal of the Faculty of Applied Science, Hiroshima University, **19**: 21-58.

UNSTAD, K. H. & K. S. TANDE, 1991. Depth distribution of *Calanus finmarchicus* and *C. glacialis* in relation to environmental conditions in the Barents Sea. Polar Research, **10**: 409-420.

UYE, S.-I., I. AOTO & T. ONBÉ, 2002. Seasonal population dynamics and production of *Microsetella norvegica*, a widely distributed but little-studied marine planktonic harpacticoid copepod. Journal of Plankton Research, **24**: 143-153.

UYE, S.-I., Y. IWAI & S. KASAHARA, 1983. Growth and production of the inshore marine copepod *Pseudodiaptomus marinus* in the central part of the Inland Sea of Japan. Marine Biology, Berlin, **73**: 91-98.

UYE, S.-I. & T. ONBÉ, 1975. The developmental stages of *Pseudodiaptomus marinus* Sato (Copepoda, Calanoida) reared in the laboratory. Bulletin of the Plankton Society of Japan, **21**: 65-76.

WILLIAMS, R. & D. V. P. CONWAY, 1980. Vertical distribution of *Calanus finmarchicus* and *C. helgolandicus* (Crustacea: Copepoda). Marine Biology, Berlin, **60**: 57-61.

WILSON, C. B., 1905. North American parasitic copepods belonging to the family Caligidae. Part 1. The Caliginae. Proceedings of the United States National Museum, **28**: 479-672.

YAMAGUCHI, A., T. IKEDA, Y. WATANABE & J. ISHIZAKA, 2004. Vertical distribution patterns of pelagic copepods as viewed from the predation pressure hypothesis. Zoological Studies, **43**: 475-485.

ZÁNKAI, N. P., 1987. Post-embryonic development of cyclopoid copepods in various seasons at Lake Balaton (Hungary). Journal of Plankton Research, **9**: 1057-1068.

OTHER CITED LITERATURE

BLADES, P. I. & M. J. YOUNGBLUTH, 1980. Morphological, physiological, and behavioral aspects of mating in calanoid copepods. In: W. C. KERFOOT (ed.), Evolution and ecology of zooplankton communities, Special symposium, **3**: 39-51. (American Society of Limnology and Oceanography, Hanover, New Hampshire and London).

BOXSHALL, G. A., 1985. The comparative anatomy of two copepods, a predatory calanoid and a particle-feeding mormonilloid. Philosophical Transactions of the Royal Society of London, (B, Biological Sciences) **311**: 303-377.

BOXSHALL, G. A. & S. H. HALSEY, 2004. An introduction to copepod diversity, **1-2**: i-xv, 1-421, i-vii, 422-966. (The Ray Society, London, **166**).

DOHLE, W. & G. SCHOLTZ, 1997. How far does cell lineage influence cell fate specification in crustacean embryos? Seminars in Cell & Developmental Biology, **8**: 379-390.

FERRARI, F. D., 2003. [Review of] "The Copepodologist's Cabinet: A Biographical and Bibliographical History, volume one, Aristotle to Alexander von Nordmann (330 B.C. to A.D. 1832)". Journal of Crustacean Biology, **23**: 747-748.

FERRARI, F. D. & M. J. GRYGIER, 2003. Comparative morphology among trunk limbs of *Caenestheriella gifuensis* and of *Leptestheria kawachiensis* (Crustacea: Branchiopoda: Spinicaudata). Zoological Journal of the Linnean Society, London, **139**: 547-564.

FRYER, G., 1998. The role of copepods in freshwater ecosystems. Journal of Marine Systems, (Special Volume) **15**: 71-73.

FULLER, C., 1920. Studies on the post-embryonic development of the antennae of termites. Annals of the Natal Museum, **4**: 235-295.

GIESBRECHT, W., 1893a. Systematik und Faunistik der pelagischen Copepoden des Golfes von Neapel und der angrenzenden Meeres-Abschnitte. Fauna und Flora des Golfes von Neapel und der Angrenzenden Meeres-Abschnitte, Herausgegeben von der Zoologischen Station zu Neapel, **19**: 1-831, pls. 1-54.

HANSEN, H.-J., 1893. Zur Morphologie der Gliedmassen und Mundtheile bei Crustaceen und Insecten. Zoologischer Anzeiger, **16**: 193-198, 201-212.

HENSON, H., 1947. The growth and form of the head and antennae in the earwig (*Forficula auricularia* Linn.). Proceedings of the Leeds Philosophical and Literary Society, (Scientific Section) **5**: 21-32.

HO, J. S., M. DOJIRI, G. HENDLER & G. B. DEETS, 2003. A new species of Copepoda (Thaumatopsyllidae) symbiotic with a brittle star from California, U.S.A., and designation of a new order Thaumatopsylloida. Journal of Crustacean Biology, **23**: 582-594.

HUMES, A. G., 1994. How many copepods? Hydrobiologia, **292/293**: 1-7.

HUYS, R. & G. A. BOXSHALL, 1991. Copepod evolution: 1-468. (The Ray Society, London, **159**).

KLIE, W., 1949. Harpacticoida (Copepoda) aus dem Bereich von Helgoland und der Kieler Bucht. 1. Kieler Meeresforschungen, **6**: 90-128.

LANG, K., 1946. A contribition to the question of the mouth parts of the Copepoda. Arkiv för Zoologi, (A) **38**: 1-24.

MADDOCKS, R. F., 2000. The antennule in podocopid Ostracoda: chaetotaxy, ontogeny and morphometrics. Part 2. Ontogeny of the antennule of podocopid Ostracoda. Micropaleontology, **46** (Supplement 2): 38-62.

MARTÍNEZ ARBIZU, P., 1997. The monophyly of Erebonasteridae, with the description of *Centobnaster severnicus* sp. n. (Copepoda: Poecilostomatoida) from the Laptev Sea (Arctic Ocean). Zoologischer Anzeiger, **235**: 263-270.

——, 1999. New Erebonasteridae (Copepoda) from Vilkitzky Strait in the Arctic and from a Pacific hydrothermal vent site (northern Fiji Basin). Journal of Crustacean Biology, **19**: 93-105.

MARUZZO, D., A. MINELLI, M. RONCO & G. FUSCO, 2007. Growth and regeneration of the second antennae of *Asellus aquaticus* (Crustacea, Isopoda) in the context of arthropod antennal segmentation. Journal of Crustacean Biology, **27**: 184-196.

MINELLI, A., 2004. A segmental analysis of the beetle antenna. Studi Museo Tridentino di Scienze Naturali, (Acta Biologica) **81**: 91-101.

NAYLOR, E., 1955. The comparative external morphology and revised taxonomy of the British species of *Idotea*. Journal of the Marine Biological Association of the United Kingdom, **34**: 467-493.

OOISHI, S., 1959. Studies on the teloblasts in the decapod embryo: I. Origin of teloblasts in *Heptacarpus rectirostris* Stimpson. Embryologia, **4**: 283-309.

PARK, T., 1995. Taxonomy and distribution of the marine calanoid copepod family Euchaetidae. Bulletin of the Scripps Institution of Oceanography, **29**: 1-203, figs.

PARK, T., 2000. Taxonomy and distribution of the calanoid copepod family Heterorhabdidae. Bulletin of the Scripps Institution of Oceanography, **31**: 1-269.

QUEIROZ, K. DE & P. D. CONTINO, 2001. Phylogenetic nomenclature and the PhyloCode. Bulletin of Zoological Nomenclature, **58**: 254-271.

RUSHTON-MELLOR, S. K. & G. A. BOXSHALL, 1994. The developmental sequence of *Argulus foliaceus* (Crustacea: Branchiura). Journal of Natural History, **28**: 763-785.

SEIFRIED, S., 2003. Phylogeny of Harpacticoida (Copepoda): revision of 'Maxillipedasphalea' and Exanechentera: 1-259. (Cuvillier Verlag, Göttingen).

SEXTON, E. W., 1924. The moulting and growth-stages of *Gammarus*, with description of the normals and intersexes of *G. chevreuxi*. Journal of the Marine Biological Association of the United Kingdom, **13**: 340-401.

SMITH, R. J. & A. TSUKAGOSHI, 2005. The chaetotaxy, ontogeny and musculature of the antennule of podocopan ostracods (Crustacea). Journal of Zoology, London, **265**: 157-177.

STEULLET, P., H. S. CATE & C. D. DERBY, 2000. A spatiotemporal wave of turnover and functional maturation of olfactory receptor neurons in the spiny lobster *Panulirus argus*. Journal of Neuroscience, **20**: 3282-3294.

SUÁREZ-MORALES, E., F. D. FERRARI & T. M. ILIFFE, 2006. A new calanoid copepod (Crustacea: Copepoda: Epacteriscidae) from anchialine environments of the Yucatan Peninsula, Mexico, with comments on the biogeography of the family. Proceedings of the Biological Society of Washington, **119**: 222-238.

TIEMANN, H., 1984. Is the taxon Harpacticoida a monophyletic one? Crustaceana, (Supplement) **7**: 47-59.

VAUPEL KLEIN, J. C. VON, 1972. A new character with systematic value in *Euchirella* (Copepoda, Calanoida). Zoologische Mededelingen, Leiden, **47**: 497-512.

WALOSSEK, D. & K. MÜLLER, 1997. Cambrian 'Orsten'-type arthropods and the phylogeny of Crustacea. Systematics Association, (Special Volume) **55**: 139-153.

WALTER, T. C., S. OHTSUKA, S. PUTCHAKARN, K. PINKAEW & S. CHULLASORN, 2002. Redescription of two species of *Pseudodiaptomus* from Asia and Australia (Crustacea: Copepoda: Calanoida: Pseudodiaptomidae) with discussion of the female genital structure and zoogeography of Indo-West Pacific species. Proceedings of the Biological Society of Washington, **115**: 650-669.

WILLEN, E., 2000. Phylogeny of the Thalestridimorpha Lang, 1944 (Crustacea, Copepoda): 1-233. (Cuvillier Verlag, Göttingen).

WILLIAMS-HOWZE, J., 1997. Dormancy in the free-living copepod orders Cyclopoida, Calanoida, and Harpacticoida. Oceanography and Marine Biology, an Annual Review, **35**: 257-321.

ZELENY, C., 1907. The direction of differentiation in development. I. The antennule of *Mancasellus macrourus*. Archiv für Entwicklungsmechanik der Organismen, **23**: 324-343.

PAPERS ABOUT THE POST-EMBRYONIC DEVELOPMENT OF COPEPODS NOT CITED

ABDELHALIM, A., J. LEWIS & G. A. BOXSHALL, 1991. The life-cycle of *Ergasilus sieboldi* Nordmann (Copepoda: Poecilostomatoida), parasitic on British freshwater fish. Journal of Natural History, London, **25**: 559-582.

ABRAHAM, S. & U. K. GOPALAN, 1975. Growth of an estuarine copepod *Nitocra spinipes* Boeck cultured in the laboratory. Bulletin of the Department of Marine Science, University of Cochin, **7**: 309-318.

AKSNES, D. L., 1996. Natural mortality, fecundity and development time in marine planktonic copepods - implications of behaviour. Marine Ecology Progress Series, **131**: 315-316.

AKSNES, D. L. & J. BLINDHEIM, 1996. Circulation patterns in the North Atlantic and possible impact on population dynamics of *Calanus finmarchicus*. Ophelia, **44**: 7-28.

AKSNES, D. L. & T. MAGNESEN, 1983. Distribution, development, and production of *Calanus finmarchicus* (Gunnerus) in Lindaspollene, western Norway, 1979. Sarsia, **68**: 195-208.

—— & ——, 1988. A population dynamics approach to the estimation of production of four calanoid copepods in Lindalspollene, western Norway. Marine Ecology Progress Series, **45**: 57-68.

AKSNES, D. L. & M. D. OHMAN, 1996. A vertical life table approach to zooplankton mortality estimates. Limnology and Oceanography, **41**: 1461-1469.

ALEKSEEV, V. R., 1983. Growth, development and production of *Mesocyclops leuckarti* (Claus) (Copepoda, Crustacea) in water bodies of different type. Sbornik Nauchnykh Trudov Gosudarstvennogo Nauchno-issledovatel'skogo Instituta Ozernogo i Rechnogo Rybnogo Khozaistva, **204**: 48-56. [In Russian, with English summary.]

——, 2000. Taxonomic analysis of species characters for copepodid instars 4 and 5 of the subfamily Eucyclopinae of European Russia. Hydrobiologia, **417**: 57-79.

ALSTON, S., G. A. BOXSHALL & J. W. LEWIS, 1996. The life-cycle of *Ergasilus briani* Markewitsch, 1993 (Copepoda: Poecilostomatoida). Systematic Parasitology, **35**: 79-110.

AMBLER, J., F. D. FERRARI, J. FORNSHELL & E. BUSKEY, 1999. Diel cycles of molting, mating, egg production and hatching in the swarm forming cyclopoid copepod *Dioithona oculata*. Plankton Biology and Ecology, **46**: 120-127.

AMBLER, J. W. & C. B. MILLER, 1987. Vertical habitat-partitioning by copepodites and adults of subtropical oceanic copepods. Marine Biology, Berlin, **94**: 561-577.

AMELINA, L., 1927. Lichinki presnovodnykh Cyclopidae (Copepoda). [Die Süsswasser Cyclopidenlarven.]. Trudy Kosinskoi Biologicheskoy Stantsii, **5**: 31-39. [In Russian, with German summary.]

AMORES-SERRANO, R., 1978. Life histories and seasonal population dynamics of two cyclopoid copepods in Beaver Reservoir, Arkansas, including some observations on their post-embryonic development: i-xii, 1-89. (Thesis and Dissertation Series, report **17**, University of Arkansas).

ANDERSON, D. T. & G. T. ROSSITER, 1969. Hatching and larval development of *Haplostomella australiensis* Gotto (Copepoda, fam. Ascidicolidae), a parasite of the ascidian *Styela etheridgii* Herdman. Proceedings of the Linnean Society of New South Wales, **93**: 464-475.

—— & ——, 1969. Hatching and larval development of *Dissonus nudiventris* Kabata (Copepoda, fam. Dissonidae), a gill parasite of the Port Jackson shark. Proceedings of the Linnean Society of New South Wales, **93**: 476-481.

ANDERSON, J. T., 1994. Feeding ecology and condition of larval and pelagic juvenile redfish *Sebastes* spp. Marine Ecology Progress Series, **104**: 211-226.

ANDREAE, J. C., 1955. Some observations on the larval development of *Cyclops leuckarti* Claus and *Cyclops hyalinus* Rehberg. Zoologische Mededelingen, Leiden, **33**: 283-298.

ANDREWS, K. J. H., 1966. The distribution and life-history of *Calanoides acutus* (Giesbrecht). Discovery Reports, **34**: 117-162.

ANDRONOV, V. N., 1970. Nekotorye voprosy sistematiki semeistva Paracalanidae (Copepoda). [Some problems of taxonomy of the family Paracalanidae (Copepoda).]. Zoologicheskii Zhurnal, **49**: 980-985.

ANSTENSRUD, M. & T. A. SCHRAM, 1988. Host and site selection by larval stages and adults of the parasitic copepod *Lernaeenicus sprattae* (Sowerby) (Copepoda, Pennellidae) in the Oslofjord. Hydrobiologia, **167/168**: 587-595.

APSTEIN, C., 1908. *Nauplius pagurus*. In: Die Beteiligung Deutschlands an der Internationalen Meeresforschung, iv/v Jahresbericht, Berlin: 55-58.

ARASHKEVICH, E. G. & A. V. DRITS, 1997. Biochemical composition, respiration rate and gonad development in late-stage copepodites of *Calanoides carinatus* (Krøyer) from the Benguela upwelling region. Okeanologiya, **37**: 571-577. [In Russian.]

ARASHKEVICH, E. G., K. S. TANDE, A. F. PASTERNAK & B. ELLERTSEN, 2004. Seasonal moulting patterns and the generation cycle of *Calanus finmarchicus* in the NE Norwegian Sea, as inferred from gnathobase structures, and the size of gonads and oil sacs. Marine Biology, Berlin, **146**: 119-132.

ARAUJO-CASTRO, C. M. V. & L. P. SOUZA-SANTOS, 2005. Are the diatoms *Navicula* sp. and *Thalassiosira fluviatilis* suitable to be fed to the benthic harpacticoid copepod *Tisbe biminiensis*? Journal of Experimental Marine Biology and Ecology, **327**: 58-69.

ARCOS, F. & A. FLEMINGER, 1991. Morphological criteria for the separation of the late copepodid stages of *Temora stylifera* and *T. discaudata* (Copepoda, Calanoida). Journal of Plankton Research, **13**: 1177-1186.

ARGILLAS, A., 1931. Un Copépode parasite de *Mugil auratus* Risso, nouveau pour l'Algérie: *Caligus pageti* Russel. Bulletin de la Station d'Aquiculture et de Pêche Castiglione, **1930**: 93-106.

ARROYO, N. L., P. ABAUNZA & I. PRECIADO, 2002. The first naupliar stage of *Pennella balaenopterae* Koren and Danielssen, 1877 (Copepoda: Siphonostomatoida, Pennellidae). Sarsia, **87**: 333-337.

ASENCIO, G., E. CLASING, C. HERRERA, R. STEAD & J. NAVARRO, 1995. Larval development of *Amphiascopsis cinctus* Claus, 1866 (Harpacticoida: Diosaccidae) and *Laophonte parvula* Sars, 1908 (Harpacticoida: Laophontidae). Revista Chilena de Historia Natural, Santiago de Chile, **68**: 283-301. [In Spanish.]

ASTTHORSSON, O. S. & A. GISLASON, 1999. Inter-annual variation in abundance and development of *Calanus finmarchicus* in Faxafloi, West-Iceland. Rit Fiskideildar, **16**: 131-140.

—— & ——, 2003. Seasonal variations in abundance, development and vertical distribution of *Calanus finmarchicus*, *C. hyperboreus* and *C. glacialis* in the East Icelandic Current. Journal of Plankton Research, **25**: 843-854.

ATKINSON, A., 1989. Distribution of six major copepod species around South Georgia in early summer. Polar Biology, **9**: 353-363.

ATKINSON, A., S. B. SCHNACK-SCHIEL, P. WARD & V. MARIN, 1997. Regional differences in the life cycle of *Calanoides acutus* (Copepoda: Calanoida) within the Atlantic sector of the Southern Ocean. Marine Ecology Progress Series, **150**: 99-111.

ATKINSON, A. & J. D. SINCLAIR, 2000. Zonal distribution and seasonal vertical migration of copepod assemblages in the Scotia Sea. Polar Biology, **23**: 46-58.

AUEL, H. & W. HAGEN, 2005. Body mass and lipid dynamics of Arctic and Antarctic deep-sea copepods (Calanoida, *Paraeuchaeta*): ontogenetic and seasonal trends. Deep-Sea Research, (I) **52**: 1272-1283.

AURIVILLIUS, C. W. S., 1894. Redogörelse för de Svenska hydrografiska undersökningarne åren 1893-1894. III. Planktonundersökningar: animalisk plankton. Bihang till Kongliga Svenska Vetenskaps-Akademiens Handlingar, (4) **20**(3): 1-30, 7 tabs.

AUVRAY, C. & B. DUSSART, 1967. Rôle de quelques facteurs du milieu sur le développement post-embryonnaire des Cyclopides (Crustacés Copépodes). II. Cas des *Cyclops* et influences des facteurs extérieurs. Bulletin de la Société Zoologique de France, **92**: 11-22.

AVDEEV, G. V., 1978. Sistematicheskoe polozhenie roda *Tegobomolochus* Izawa, 1976 (Copepoda, Cyclopoida). Izvestiya Tikhookeanskogo nauchno-issledovatel'skogo Instituta Rybnogo Khozyaistva i Okeanografii, **102**: 119-122.

BACKHAUS, J. O., I. H. HARMS, M. KRAUSE & M. R. HEATH, 1994. An hypothesis concerning the space-time succession of *Calanus finmarchicus* in the northern North Sea. ICES Journal of Marine Science, **51**: 169-180.

BAIRD, W., 1850. The natural history of the British Entomostraca: 1-344, pls. 1-36. (Ray Society, London).

BAKKER, C. & P. VAN RIJSWIJK, 1987. Development time and growth rate of the marine calanoid copepod *Temora longicornis* as related to food conditions in the Oosterschelde estuary (southern North Sea). Netherlands Journal of Sea Research, **21**: 125-141.

—— & ——, 1990. Development of the marine copepod *Temora longicornis* in Lake Grevelingen: 11-13. (Progress Report 1989. Delta Institute for Hydrobiological Research).

BAN, S., 1994. Effect of temperature and food concentration on post-embryonic development, egg production and adult body size of calanoid copepod *Eurytemora affinis*. Journal of Plankton Research, **16**: 721-735.

BARNARD, J. L. & D. J. REISH, 1960. Wood-browsing habits of the harpacticoid copepod *Tisbe gracilis* (T. Scott) in southern California. Pacific Naturalist, **1**: 9-12.

BARNES, H. & M. BARNES, 1953. Biometry of the copepod *Calanus finmarchicus* (Gunn.) in stages V and VI. Journal of the Marine Biological Association of the United Kingdom, **32**: 305-313.

BARNETT, R. P. O., 1966. The comparative development of two species of *Platychelipus* Brady (Harpacticoida). In: H. BARNES (ed.), Some contemporary studies in marine science: 113-127. (George Allen and Unwin Ltd, London).

BARTSCH, I., 1996. Parasites of the Antarctic brittle star *Ophiacantha disjuncta* (Ophiacanthidae, Ophiuroidea): redescription of the copepod *Lernaeosaccus ophiacanthae* Heegaard, 1951. Mitteilungen aus dem Hamburgischen Zoologischen Museum und Institut, **93**: 63-72.

BATCHELDER, H. P. & C. B. MILLER, 1989. Life history and population dynamics of *Metridia pacifica*; results from simulation modelling. Ecological Modelling, **48**: 113-136.

BATTAGLIA, B., 1957. Ricerche sul ciclo biologico di *Tisbe gracilis* (T. Scott) (Copepoda, Harpacticoida), studiato in condizioni di laboratorio. Archivio di Oceanografia e Limnologia, **11**: 29-46.

BATTAGLIA, B. & P. TALAMINI, 1957. Osservazioni sullo sviluppo larvale di *Tisbe reticulata* Bocquet (Copepoda, Harpacticoida). Archivio di Oceanografia e Limnologia, **11**: 63-68.

BATTEN, S. D., D. W. WELCH, W. DAVID & T. JONAS, 2003. Latitudinal differences in the duration of development of *Neocalanus plumchrus* copepodites. Fisheries Oceanography, **12**: 201-208.

BECHEIKH, S., M. MICHAUD, F. THOMAS, A. RAIBAUT & F. RENAUD, 1998. Roles of resource and partner availability in sex determination in a parasitic copepod. Proceedings of the Royal Society of London, (B, Biological Sciences) **265**: 1153-1156.

BECHMANN, R. K., 1994. Use of life tables and LC50 tests to evaluate chronic and acute toxicity effects of copper on the marine copepod *Tisbe furcata* (Baird). Environmental Toxicology and Chemistry, **13**: 1509-1517.

——, 1999. Effect of the endocrine disrupter nonylphenol on the marine copepod *Tisbe battagliai*. Science of the Total Environment, **233**: 33-46.

BEJARANO, A. C. & G. T. CHANDLER, 2003. Reproductive and developmental effects of atrazine on the estuarine meiobenthic copepod *Amphiascus tenuiremis*. Environmental Toxicology and Chemistry, **22**: 3009-3016.

BEJARANO, A. C., G. T. CHANDLER & A. W. DECHO, 2005. Influence of natural dissolved organic matter (DOM) on acute and chronic toxicity of the pesticides chlorothalonil, chlorpyrifos and fipronil on the meiobenthic estuarine copepod *Amphiascus tenuiremis*. Journal of Experimental Marine Biology and Ecology, **321**: 43-57.

BELCIK, F., 1981. The male of *Ismaila monstrosa* Bergh, 1867 (Copepoda, Splanchnotrophidae). Crustaceana, **40**: 16-25.

BENEDEN, E. VAN, 1870. Recherches sur l'embryologie des Crustacés. IV. Développement des genres *Anchorella*, *Lerneopoda*, *Brachiella* et *Hessia*. Bulletin de l'Académie Royale de Belgique, (2) **29**: 223-254.

BENEDEN, P. J. VAN, 1848. Mémoire sur le développement et organisation des Nicothoés. Mémoires de l'Académie Royale de Belgique, **24**: 1-28.

BENEDETTI, I., L. MOLA, E. MONARI, M. A. SABATINI, M. MARINI & B. FRATELLO, 1989. Preliminary findings on the development of the parasitic copepod *Lernaea cyprinacea* L. from cleavage to hatching. Bollettino di Zoologia, **56**: 7-11.

——, ——, ——, ——, —— & ——, 1990. Development of the crustacean copepod *Lernaea cyprinacea* L. Acta Embryologiae et Morphologiae Experimentalis, **10**: 37.

BEN HASSINE, O. K., 1983. Les Copépodes parasites de poissons Mugilidae en Méditerranée occidentale (côtes Françaises et Tunisiennes) morphologie, bio-écologie, cycles évolutifs: i-vi, 1-452 (+ 19). (Ph.D. Thesis, Université des Sciences et Techniques du Languedoc).

BEN HASSINE, O. K. & A. RAIBAUT, 1981a. Développement larvaire expérimental de *Caligus pageti* Russel, 1926. Copépode parasite de poissons Mugilidae en Méditerranée. Bulletin de l'Office National des Pêches de Tunisie, **5**: 175-201.

—— & ——, 1981b. Réalisation expérimentale du cycle évolutif de *Ergasilus lizae* (Krøyer, 1863), copépode parasite de poissons Mugilidés. Premiers résultats de l'infestation. Archives de l'Institut Pasteur de Tunis, **58**: 423-430.

BENNETT, S. M., 1999. Egg development, hatching and early life-history of *Dissonus manteri* (Copepoda), parasitic on the gills of coral trout, *Plectropomus leopardus*. International Journal for Parasitology, **29**: 267-274.

BENNETT, S. M. & M. B. BENNETT, 1994. Pathology of attachment and vascular damage associated with larval stages of *Dissonus manteri* Kabata, 1966 (Copepoda: Dissonidae) on the gills of coral trout, *Plectropomus leopardus* (Lacépède) (Serranidae). Journal of Fish Diseases, **17**: 447-460.

BENZ, G., 1991. Description of some larval stages and augmented description of adult stages of *Albionella etmopteri* (Copepoda: Lernaeopodidae), a parasite of deep-water lanternsharks (*Etmopterus*: Squalidae). Journal of Parasitology **77**: 666-674.

BENZ, G. W. & S. A. M. ADAMSON, 1990. Disease caused by *Nemesis robusta* (Van Beneden, 1851) (Eudactylinidae: Siphonostomatoidea: Copepoda) infections on gill filaments of thresher sharks (*Alopias vulpinus* (Bonnaterre, 1758)), with notes on parasite ecology and life history. Canadian Journal of Zoology, **68**: 1180-1187.

BENZ, G. W., J. D. BORUCINSKA & S. A. GREENWALD, 2002. First descriptions of early- and middle-stage copepodids of *Anthosoma crassum* (Dichelesthiidae: Siphonostomatoida) and lesions on shortfin makos (*Isurus oxyrinchus*) infected with *A. crassum*. Journal of Parasitology, **88**: 19-26.

BENZ, G., P. MOHAN & G. CROW, 1992. Developmental stages of *Paralbion elongatus* from aquarium-held reef whitetip sharks (*Triaenodon obesus*) captured in Hawaiian waters. Journal of Parasitology, **78**: 1027-1035.

BENZ, G. W., B. E. SMITH & S. A. BULLARD, 2001. *Kroeyerina deetsorum* n. sp. (Copepoda: Kroyeriidae) from the olfactory sacs of Atlantic sharpnose sharks (*Rhizoprionodon terraenovae*) captured in the Gulf of Mexico and northwestern Atlantic and first report of copepodids representing Kroyeriidae. Journal of Parasitology, **87**: 1279-1290.

BERGGREEN, U., B. HANSEN & T. KIØRBOE, 1988. Food size spectra, ingestion and growth of the copepod *Acartia tonsa* during development: implications for determination of copepod production. Marine Biology, Berlin, **99**: 341-352.

BERGMANS, M., 1981. A demographic study of the life cycle of *Tisbe furcata* (Baird, 1837) (Copepoda: Harpacticoida). Journal of the Marine Biological Association of the United Kingdom, **61**: 691-705.

——, 1984. Life history adaptation to demographic regime in laboratory-cultured *Tisbe furcata* (Copepoda, Harpacticoida). Evolution, **38**: 292-299.

BERNARD, M., 1963a. Observations sur la biologie en aquarium de *Euterpina acutifrons*, Copépode pélagique. Rapports et Procès-verbaux des Réunions de la Commission Internationale pour l'Exploration Scientifique de la Mer Méditerranée, **17**: 545.

——, 1963b. Le cycle vital en laboratoire d'un copépode pélagique de Méditerranée *Euterpina acutifrons* Claus. Pelagos, **1**: 35-48.

——, 1964a. Le développement nauplien de deux Copépodes carnivores: *Euchaeta marina* (Prestandr.) et *Candacia armata* (Boeck) et observations sur le cycle de l'astaxanthine au cours de l'ontogenèse. Pelagos, **2**: 51-70.

——, 1964b. Observations sur la ponte et développement larvaire en aquarium d'un Copépode pélagique predateur: *Candacia armata* Boeck. Rapports et Procès-verbaux des Réunions de la Commission Internationale pour l'Exploration Scientifique de la Mer Méditerranée, **18**: 345-348.

——, 1965. Description du male et du premier stade nauplien de *Corina granulosa* (Giesbrecht), Copépode pélagique (Sapphirinidae). Pelagos, **2**: 45-50.

——, 1968. Premières observations sur la ponte, les oeufs, les stades naupliens et l'alimentation de quatre Pontellides (Copepoda). Rapports et Procès-verbaux des Réunions de la Commission Internationale pour l'Exploration Scientifique de la Mer Méditerranée, **19:** 525.

——, 1970. Quelques aspects de la biologie du Copépode pélagique *Temora stylifera* en Méditerranée. Essai d'écologie expérimentale. Pelagos, **11**: 3-196.

BIRD, N. T., 1968. Effects of mating on subsequent development of a parasitic copepod. Journal of Parasitology, **54**: 1194-1196.

BIRGE, E. A., 1898. Plankton studies on Lake Mendota. II. The Crustacea of the plankton from July, 1894, to December, 1896. Transactions of the Wisconsin Academy of Sciences, Arts, and Letters, **11**: 274-448.

BJÖRNBERG, T. K. S., 1965. Observations on the development and biology of the Miracidae Dana (Copepoda: Crustacea). Bulletin of Marine Science, **15**: 512-520.

——, 1966. The developmental stages of *Undinula vulgaris* (Dana) (Copepoda). Crustaceana, **11**: 65-76.

——, 1967. The larvae and young forms of *Eucalanus* Dana (Copepoda) from tropical Atlantic waters. Crustaceana, **12**: 59-73.

——, 1968. Estagios de desenvolvimento de alguns copepodos marinhos planctonicos tropicais e subtropicais: 1-175. (Ph.D. Thesis, Faculdade de Filosofia de Ciencias e Letras, Universidade de São Paulo, São Paulo).

——, 1986b. The rejected nauplius: a commentary. Syllogeus, **58**: 232-236.

——, 1999. Description of canuellid nauplii of São Sebastião Channel (southeastern Brazil). Nauplius, **6**: 155-160.

——, 2001. The naupliar stages of *Cyclopina yutimaete* Lotufo (Cyclopinidae, Cyclopoida). Hydrobiologia, **453/454**: 497-501.

BJÖRNBERG, T. K. S., R. M. LOPES & M. H. G. C. BJÖRNBERG, 1994. Charve para a identificacão de nauplios de copepodos planctonicos marinhos do Atlantico sul-ocidental. Nauplius, **2**: 1-16.

BLAINVILLE, H. M. D. DE, 1822. Mémoire sur les Lernées (*Lernaea*, Lin.). Journal de Physique, de Chimie, d'Histoire Naturelle et des Arts, Paris, **95**: 372-380, 437-447, pl. 1.

BLOCK, D. S., A. C. BEJARANO & G. T. CHANDLER, 2003. Ecdysteroid concentrations through various life-stages of the meiobenthic harpacticoid copepod, *Amphiascus tenuiremis* and the benthic estuarine amphipod, *Leptocheirus plumulosus*. General and Comparative Endocrinology, **132**: 151-160.

BOCQUET, C., 1953. Sur un Copépode Harpacticoïde mineur, *Diarthrodes feldmani* n. sp. Bulletin de la Société Zoologique de France, **78**: 101-105.

BOCQUET, C., J. BOCQUET-VÉDRINE & J. P. L'HARDY, 1970. Contribution à l'étude du développement des organes génitaux chez *Xenocoeloma alleni* (Brumpt), Copépode parasite de *Polycirrus caliendrum* Claparède. Cahiers de Biologie Marine, **11**: 195-208.

BOCQUET, C., J. GUILLET & J. H. STOCK, 1958. Sur le développement des expansions aliformes du Copépode *Nicothoe astaci* Audouin et Milne-Edwards. Comptes Rendus des Séances de l'Académie des Sciences, Paris, **246**: 836-839.

BOCQUET, C. & J. H. STOCK, 1958. Copépodes parasites d'invertébrés des côtes de la Manche. IV. Sur les trois genres synonymes de Copépodes Cyclopoïdes, *Leptinogaster* Pelseneer, *Strongylopleura* Pelseneer et *Myocheres* Wilson (Clausidiidae). Archives de Zoologie Expérimentale et Générale, **96** (Notes et Revue) (2): 71-89.

BOCQUET, C. & J. [H.] STOCK, 1959. Copépodes parasites d'invertébrés des côtes de France X. Sur les espèces de *Paranthessius* (Cyclopoida, Lichomolgidae) du groupe de *Herrmannella*, associées à des Pélécypodes. Proceedings Koninklijke Nederlandse Akademie van Wetenschappen, (C) **62**: 238-249.

BOCQUET, C., J. [H.] STOCK & G. KLEETON, 1963. Copépodes parasites d'invértebrés des côtes de la Manche. XI. Sur le développement de *Trochicola entericus* Dollfus, 1914, Copépode Cyclopoïde parasite de Trochidae. Archives de Zoologie Expérimentale et Générale, **102** (Notes et Revue) (2): 49-68.

BOERS, J. J. & J. C. H. CARTER, 1978. Instar development rates of *Diaptomus minutus* (Copepoda: Calanoida) in a small lake in Quebec. Canadian Journal of Zoology, **56**: 1710-1714.

—— & ——, 1978. The life history of *Cyclops scutifer* Sars (Copepoda: Cyclopoida) in a small lake off the Matamek River system, Quebec. Canadian Journal of Zoology, **56**: 2603-2607.

BONNET, D. & F. CARLOTTI, 2001. Development and egg production in *Centropages typicus* (Copepoda: Calanoida) fed different food types: a laboratory study. Marine Ecology Progress Series, **224**: 133-148.

BORUTZKY, E. V., 1925. Lichinki presnovodnykh Harpacticoida (Copepoda). [Die Larven der Süsswasser-Harpacticoida.] Trudy Kosinskoi Biologicheskoy Stantsii, **3**: 49-61. [In Russian, with German summary.]

——, 1926. Copepoda-Harpacticoida basseina r. Volgi. [Copepoda-Harpacticoida des Wolga Bassins.] Russiskii Gidrobiologicheskii Zhurnal, Saratov, **5**: 210-218. [In Russian with German summary.]

——, 1952. Freshwater Harpacticoida. Fauna SSSR, (Crustacea) **3/4**: 1-424.

BOULIGAND, Y., 1966. Recherches récentes sur les Copépodes associés aux Anthozoaires. Symposia of the Zoological Society of London, **16**: 267-306.

BOURGUET, J., 1986a. Contribution à l'étude de *Cletocamptus retrogressus* Schmankevitch, 1875 (Copepoda, Harpacticoida) I. Développement larvaire - stades copépodites. Crustaceana, **51**: 66-76.

——, 1986b. Contribution à l'étude de *Cletocamptus retrogressus* Schmankevitch, 1875 (Copepoda, Harpacticoida) II. Développement larvaire - stades naupliens. Crustaceana, **51**: 113-122.

BOWMAN, T. & L. KORNICKER, 1967. Two new crustaceans: the parasitic copepod *Sphaeronellopsis monothrix* (Choniostomatidae) and its myodocopid ostracod host *Parasterope pollex* (Cylindroleberidae) from the southern New England coast. Proceedings of the United States National Museum, **123**: 1-28 + 1 pl.

BOXASPEN, K. & T. NAESS, 2000. Development of eggs and the planktonic stages of salmon lice (*Lepeophtheirus salmonis*) at low temperatures. Bijdragen tot de Dierkunde, Amsterdam, **69**: 51-55.

BOXSHALL, G. A., 1974b. *Lepeophtheirus pectoralis* (O. F. Müller, 1776); a description, a review and some comparisons with the genus *Caligus* Müller, 1785. Journal of Natural History, London, **8**: 445-468.

——, 1979. The planktonic copepods of the northeastern Atlantic Ocean: Harpacticoida, Siphonostomatoida and Mormonilloida. Bulletin of the British Museum (Natural History), (Zoology) **35**: 201-264.

——, 1985. The comparative anatomy of two copepods, a predatory calanoid and a particle-feeding mormonilloid. Philosophical Transactions of the Royal Society of London, (B, Biological Sciences) **311**: 303-376.

BOXSHALL, G. A. & R. LINCOLN, 1983. Some new parasitic copepods (Siphonostomatoida: Nicothoidae) from deep-sea asellote isopods. Journal of Natural History, London, **17**: 891-900.

BOXSHALL, G. A., M. A. MONTU & A. SCHWARZBOLD, 1997. A new species of *Lernaea* L. (Copepoda: Cyclopoida) from Brazil, with notes on its ontogeny. Systematic Parasitology, **37**: 195-205.

BRADFORD, J. M., 1988. Review of the taxonomy of the Calanidae (Copepoda) and the limits to the genus *Calanus*. Hydrobiologia, **167/168**: 73-81.

BRADFORD, J., M. OHMAN & J. JILLETT, 1988. Larval morphology and development of *Neocalanus tonsus*, *Calanoides macrocarinatus* and *Calanus australis* (Copepoda; Calanoida) in the laboratory. New Zealand Journal of Marine and Freshwater Research, **22**: 301-320.

BRADY, G. S., 1883. Report on the Copepoda collected by H.M.S. Challenger during the years 1873-76. Report of the Scientific Results of the Exploring Voyage of H.M.S. Challenger during the years 1873-76, **8**: 1-142 + 55 pls.

——, 1894. On *Fucitrogus rhodymeniae*, a gall-producing copepod. Journal of the Royal Microscopical Society, **1894**: 168-170.

BRANCH, G., 1974. *Scutellidium patellarum* n. sp., a harpacticoid copepod associated with *Patella* spp. in South Africa, and a description of its larval development. Crustaceana, **26**: 179-200.

BRAND, G. W., 1985. Effect of crowding on larval viability in *Tisbe holothuriae* (Copepoda: Harpacticoida). Marine Biology, Berlin, **88**: 67-72.

BREITHOLTZ, M. & L. WOLLENBERGER, 2003. Effects of three PBDEs on development, reproduction and population growth rate of the harpacticoid copepod *Nitocra spinipes*. Aquatic Toxicology, **64**: 85-96.

BREITHOLTZ, M., L. WOLLENBERGER & L. DINAN, 2003. Effects of four synthetic musks on the life cycle of the harpacticoid copepod *Nitocra spinipes*. Aquatic Toxicology, **36**: 103-118.

BRESCIANI, J., 1961. Some features of the larval development of *Stenhelia* (*Delavalia*) *palustris* Brady, 1868 (Copepoda: Harpacticoida). Videnskabelige Meddelelser fra Dansk Naturhistorisk Forening, **123**: 237-247.

——, 1962. Parasitic copepods from the west coast of Sweden including some new or little known species. Videnskabelige Meddelelser fra Dansk Naturhistorisk Forening, **124**: 367-408.

BRESCIANI, J. & J. LÜTZEN, 1961a. The anatomy of a parasitic copepod, *Saccopsis steenstrupi* n. sp. Crustaceana, **3**: 9-24.

—— & ——, 1961b. *Gonophysema gullmarensis* (Copepoda parasitica). An anatomical and biological study of an endoparasite living in the ascidian *Ascidiella aspersa*. II. Biology and development. Cahiers de Biologie Marine, **2**: 347-371.

—— & ——, 1974. On the biology and development of *Aphanodomus* Wilson (Xenocoelomidae), a parasitic copepod of the polychaete *Thelepus cincinnatus*. Videnskabelige Meddelelser fra Dansk Naturhistorisk Forening, **137**: 25-63.

BRESCIANI, J. & K. W. OCKELMANN, 1966. *Axinophilus thyasirae* gen. et sp. nov., a parasitic copepod from the bivalves *Thyasira flexuosa* and *T. sarsi*. Ophelia, **3**: 179-182.

BRIAN, A., 1912. Copépodes parasites des poissons et des échinides provenant des campagnes scientifiques de S.A.S. le Prince Albert Ier de Monaco (1866-1910). Résultats des Campagnes Scientifiques du Prince Albert Ier de Monaco, **38**: 1-58.

——, 1919. Sviluppo larvale della *Psamathe longicauda* Ph. e della *Harpacticus uniremis* Kröyer (Copepodi Harpacticoidi) (descrizione della serie Copepodiforme). Atti di Società Italiano de Scienze Naturali, **58**: 29-58 + 2 pls.

——, 1921. I Copepodi Harpacticoidi del Golfo de Genova: 1-112 + 12 pls. (Genova).

——, 1922. L'*Alteutha depressa* Baird (copepodo Harpacticoide) e i suoi stadi larvali. Monitore Zoologico Italiano, **33**: 8-14.

BRIAN, A. & P. GRAY, 1928. Morphologie externe et interne d'un nouveau copépode parasite *Cardiodectes anchorellae,* n. sp. trouvé a Madras. Bollettino dei Musei e Laboratori di Zoologia e Anatomia Comparata della R. Università di Genova, **38**: 1-58.

BRIGGS, R., 1977. Larval stages of *Paranthessius anemoniae* Claus (Copepoda, Cyclopoida), an associate of the snakelocks anemone, *Anemonia sulcata* (Pennant). Crustaceana, **33**: 249-258.

BRODSKY, K., 1959. On phylogenetic relations of some *Calanus* (Copepoda) species of the northern and southern hemispheres. Zoologicheskii Zhurnal, **38**: 1537-1552.

——, 1972. Phylogeny of the family Calanidae (Copepoda) on the basis of a comparative morphological analysis of its characters. Geographical and seasonal variability of marine plankton. Studies of Marine Fauna, **12**: 1-128.

BRON, J. E., A. P. SHINN & C. SOMMERVILLE, 2000a. Moulting in the chalimus larva of the salmon louse *Lepeophtheirus salmonis* (Copepoda, Caligidae). Bijdragen tot de Dierkunde, Amsterdam, **69**: 31-38.

——, —— & ——, 2000b. Ultrastructure of the cuticle of the chalimus larva of the salmon louse *Lepeophtheirus salmonis* (Kroyer, 1837) (Copepoda: Caligidae). Bijdragen tot de Dierkunde, Amsterdam, **69**: 39-49.

BRON, J. E. & C. SOMMERVILLE, 1998. The functional and comparative morphology of the photoreceptors of the copepodid larva of the salmon louse *Lepeophtheirus salmonis* (Krøyer, 1837) (Crustacea: Copepoda, Caligidae). Zoologischer Anzeiger, **237**: 113-126.

BRON, J. E., C. SOMMERVILLE & G. H. RAE, 1993. Aspects of the behaviour of copepodid larvae of the salmon louse *Lepeophtheirus salmonis* (Krøyer, 1837). In: G. A. BOXSHALL & D. DEFAYE (eds.), Pathogens of wild and farmed fish: sea lice: 125-142. (Ellis Horwood, London).

BROWN, R. J., S. D. RUNDLE, T. H. HUTCHINSON, T. D. WILLIAMS & M. B. JONES, 2003. Small-scale detritus-invertebrate interactions: influence of detrital biofilm composition on development and reproduction in a meiofaunal copepod. Archiv für Hydrobiologie, **157**: 117-129.

——, ——, ——, —— & ——, 2005. A microplate freshwater copepod bioassay for evaluating acute and chronic effects of chemicals. Environmental Toxicology and Chemistry, **24**: 1528-1531.

BRUCET, S., D. BOIX, R. LÓPEZ-FLORES, A. BADOSA & X. D. QUINTANA, 2005. Ontogenetic changes of amino acid composition in planktonic crustacean species. Marine Biology, Berlin, **148**: 131-139.

BUCHHOLZ, R., 1869. Beiträge zur Kenntnis der innerhalb der Ascidien lebenden parasitischen Crustaceen des Mittelmeeres. Zeitschrift für Wissenschaftliche Zoologie, **19**: 99-155, pls. 5-11.

BURNS, C. W., 1979. Population dynamics and production of *Boeckella dilatata* (Copepoda: Calanoida) in Lake Hayes, New Zealand. Archiv für Hydrobiologie, (Supplement) **54**: 409-465.

——, 1981. Instar development rates and production of three generations of *Boeckella dilatata* (Copepoda: Calanoida) in a warm-monomictic lake. Verhandlungen der Internationale Vereinigung für Theoretische und Angewandte Limnologie, **21**: 1578-1583.

——, 1985. The effects of starvation on naupliar development and survivorship of three species of *Boeckella* (Copepoda: Calanoida). Ergebnisse der Limnologie, **21**: 297-309.

——, 1988. Starvation resistance among copepod nauplii and adults. Verhandlungen der Internationale Vereinigung für Theoretische und Angewandte Limnologie, **23**: 2087-2091.

——, 1992. Population dynamics of crustacean zooplankton in a mesotrophic lake, with emphasis on *Boeckella hamata* Brehm (Copepoda: Calanoida). Internationale Revue der Gesamten Hydrobiologie, **77**: 553-577.

BYRNES, E. F., 1921. The metamorphosis of *Cyclops americanus* Marsh and *Cyclops signatus* var. *tenuicornis*. Cold Spring Harbor Monograph IX. The Brooklyn Institute for Arts and Sciences: 3-53.

CABRAL, P., F. COSTE & A. RAIBAUT, 1984. Cycle évolutif de *Lernanthropus kroyeri* Van Beneden, 1851, Copépode branchial hématophage du loup *Dicentrachus labrax* (Linné, 1758) dans des populations naturelles et en élevage. Annales de Parasitologie Humaine et Comparée, **59**: 189-207.

CALBET, A. & M. ALCARAZ, 1997. Growth and survival rates of early developmental stages of *Acartia grani* (Copepoda: Calanoida) in relation to food concentration and fluctuations in food supply. Marine Ecology Progress Series, **147**: 181-186.

CALBET, A., I. TREPAT & L. ARIN, 2000. Naupliar growth versus egg production in the calanoid copepod *Centropages typicus*. Journal of Plankton Research, **22**: 1393-1402.

CALMAN, W., 1908. On a parasitic copepod from *Cephalodiscus*. Transactions of the South African Philosophical Society, **17**: 177-184 + 2 pls.

CAMPBELL, R. G., M. M. WAGNER, G. J. TEEGARDEN, C. A. BOUDREAU & E.G. DURBIN, 2001. Growth and development rates of the copepod *Calanus finmarchicus* reared in the laboratory. Marine Ecology Progress Series, **221**: 161-183.

CANU, E., 1890. Sur le développement des Copépodes. Comptes Rendus des Séances de la Société de Biologie, Paris, **111**: 1919-1920.

——, 1892. Les Copépodes du Boulonnais, morphologie, embryologie, taxonomie. Travaux du Laboratoire de Zoologie Maritime de Wimereux-Ambleteuse, **6**: 1-292 + 30 pls.

CAPACCIONI, R., D. GRAS & E. CARBONELL, 1993. A copepod of the genus *Rhodinicola* Levinsen, 1878 (Poecilostomatoida, Clausiidae) parasitic on *Clymenura clypeata* (Saint-Joseph, 1894) (Polychaeta, Maldanidae) from the Alfaques inlet (Ebro river delta, Spain, western Mediterranean). Crustaceana, **64**: 129-136.

CARAMUJO, M. J., E. VAN DER GRINTEN & W. ADMIRAAL, 2005. Trophic interactions between benthic copepods and algal assemblages: a laboratory study. Journal of the North American Benthological Society, **24**: 890-903.

CARLI, A. & C. L. BRUZZONE, 1974. Nota su *Kroyeria carchariaeglauci* (Copepoda: Eudactylinidae) Hesse, 1878. Parassita di *Prionace glauca* (L.) e sui primi stadi naupliari ottenuti dalla schiusa delle uova in laboratorio. Bollettino Pesca, Piscicoltura e Idrobiologia, (n. ser.) **28**: 101-109.

—— & ——, 1977. Développement larvaire du Copépode *Lepeophtheirus nordmanni* (Edwards) parasite de *Mola mola* (L.). Rapports et Procès-verbaux des Réunions de la Commission Internationale pour l'Exploration Scientifique de la Mer Méditerranée, **24**: 135-136.

CARLI, A. & A. FIORI, 1977. Sviluppo larvale del *Tigriopus fulvus* Fischer. In: Atti del IX Congresso della Società Italiana di Biologia marina, Lacco Ameno, Ischia: 181-190.

CARLOTTI, F., M. KRAUSE & G. RADACH, 1993. Growth and development of *Calanus finmarchicus* related to the influence of temperature: experimental results and conceptual model. Limnology and Oceanography, **38**: 1125-1134.

CARLOTTI, F. & P. NIVAL, 1992. Model of copepod growth and development: moulting and mortality in relation to physiological processes during an individual moult cycle. Marine Ecology Progress Series, **84**: 219-233.

CARLOTTI, F. & S. NIVAL, 1991. Individual variability of development in laboratory-reared *Temora stylifera* copepodites: consequences for the population dynamics and interpretation in the scope of growth and development rates. Journal of Plankton Research, **13**: 813-815.

—— & ——, 1992. Moulting and mortality rates of copepods related to age within stage: experimental results. Marine Ecology Progress Series, **84**: 235-243.

CARLOTTI, F. & A. SCIANDRA, 1989. Population dynamics model of *Euterpina acutifrons* (Copepoda: Harpacticoida) coupling individual growth and larval development. Marine Ecology Progress Series, **56**: 225-242.

CAROTENUTO, Y., 1999. Morphological analysis of larval stages of *Temora stylifera* (Copepoda, Calanoida) from the Mediterranean Sea. Journal of Plankton Research, **21**: 1613-1632.

CAROTENUTO, Y., A. IANORA, I. BUTTINO, G. ROMANO & A. MIRALTO, 2002. Is postembryonic development in the copepod *Temora stylifera* negatively affected by diatom diets? Journal of Experimental Marine Biology and Ecology, **276**: 49-66.

CARR, R. S., D. C. CHAPMAN, B. J. PRESLEY, J. M. BIEDENBACH, L. ROBERTSON, P. BOOTHE, R. KILADA, T. WADE & P. MONTAGNA, 1996. Sediment porewater toxicity assessment studies in the vicinity of offshore oil and gas production platforms in the Gulf of Mexico. Canadian Journal of Fisheries and Aquatic Sciences, **53**: 2618-2628.

CARTER, J. C. H., 1969. Life cycles of *Limnocalanus macrurus* and *Senecella calanoides*, and seasonal abundance and vertical distribution of various planktonic copepods, in Parry Sound, Georgian Bay. Journal of the Fisheries Research Board of Canada, **26**: 2543-2560.

——, 1974. Life cycles of three limnetic copepods in a beaver pond. Journal of the Fisheries Research Board of Canada, **31**: 421-434.

CARTER, M. E., 1944. Harpacticoid copepods of the region of Mountain Lake, Virginia (with a description of *Moraria virginiana*, n. sp.). Journal of the Elisha Mitchell Scientific Society, **60**: 158-166 + 3 pls.

CARTON, Y., 1968. Développement de *Cancerilla tubulata* Dalyell parasite de l'ophiure *Amphipholis squamata* Della Chiaje. Crustaceana, (Supplement) **1**: 11-28.

CASPERS, H., 1939. Über Vorkommen und Metamorphose von *Mytilicola intestinalis* Steuer (Copepoda paras.) in der südlichen Nordsee. Zoologischer Anzeiger, **126**: 161-171.

CASTEL, J., 1976. Développment larvaire et biologie de *Harpacticus littoralis* G. O. Sars, 1910 (Copépode, Harpacticoïde) dans les étangs saumâtres de la région d'Arcachon. Cahiers de Biologie Marine, **17**: 195-212.

CASTRO, L. R., P. A. BERNAL & V. A. TRONCOSO, 1993. Coastal intrusion of copepods: mechanisms and consequences on the population biology of *Rhincalanus nasutus*. Journal of Plankton Research, **15**: 501-515.

CAULLERY, M. & F. MESNIL, 1919. *Xenocoeloma brumpti* C. & M., Copépode parasite de *Polycirrus arenivorum* C. Bulletin Biologique de la France et de la Belgique, **53**: 161-233.

CHANDLER, G. T., T. L. CARY, A. C. BEJARANO, J. PENDER & J. L. FERRY, 2004. Population consequences of fipronil and degradates to copepods at field concentrations: an integration of life cycle testing with Leslie matrix population modeling. Environmental Science and Technology, **38**: 6407-6414.

CHANDLER, G. T., T. L. CARY, D. C. VOLZ, S. S. WALSE, J. L. FERRY & S. L. KLOSTERHAUS, 2004. Fipronil effects on estuarine copepod (*Amphiascus tenuiremis*) development, fertility, and reproduction: a rapid life-cycle assay in 96-well microplate format. Environmental Toxicology and Chemistry, **23**: 117-124.

CHANDRAN, A. & N. BALAKRISHNAN NAIR, 1980. Morphology of the copepodid stage of three marine lernaeopodids (Crustacea: Copepoda) from the south-west coast of India. Proceedings of the Indian Academy of Sciences, (Animal Sciences) **89**: 155-163.

CHANDRASEKHARA RAO, G., 1967. On the life-history of a new sand dwelling harpacticoid copepod. Crustaceana, **13**: 129-136.

CHANGEUX, J. P., 1957. Description de deux stades larvaires d'un Copépode Cancerillidae, parasite sur *Ophiopsila aranea* Forbes. Vie et Milieu, **8**: 297-311.

CHAPMAN, M. A. & C. W. BURNS, 1976. Sexual and other differences in copepodite stages of some New Zealand *Calamoecia* and *Boeckella* spp. (Copepoda, Calanoida). New Zealand Journal of Marine and Freshwater Research, **10**: 131-137.

CHAPPUIS, P.-A., 1916a. *Viguierella coeca* Maupas. Ein Beitrag zur Entwicklungsgeschichte der Crustaceen. Revue Suisse de Zoologie, **24**: 521-564.

——, 1916b. Die Metamorphose einiger Harpacticidengenera. Zoologischer Anzeiger, **48**: 20-31.

CHAPPUIS, P.-A. & F. KIEFER, 1952. Copépodes cavernicoles des Pyrénées. Notes Biospéologiques, **7**: 113-131.

CHATTON, E., 1909. Sur un nauplius double anadydne d'*Ophioseides joubini* Chatton. Comptes Rendus des Séances de la Société de Biologie, Paris, **67**: 482-484.

CHATTON, E. & E. BREMENT, 1915. Les oostégites, les ptérostégites et la cavité incubatrice des Ascidicolidae (Copépodes); développement, homologies, valeur phylogénétique et taxonomique. Bulletin de la Société Zoologique de France, **40**: 143-155.

CHIBA, T., 1956. Studies on the development and the systematics of Copepoda. Journal of the Shimonoseki College of Fisheries, **6**: 1-90.

CHISHOLM, L. A. & J. C. ROFF, 1990. Abundance, growth rates, and production of tropical neritic copepods off Kingston, Jamaica. Marine Biology, Berlin, **106**: 79-89.

CHRISTOU, E. D. & G. C. VERRIOPOULOS, 1993. Analysis of the biological cycle of *Acartia clausi* (Copepoda) in a meso-oligotrophic coastal area of the eastern Mediterranean Sea using time-series analysis. Marine Biology, Berlin, **115**: 643-651.

CHUA, T. E., 1975. The developmental stages of *Tisbe longisetosa* Gurney, 1927 (Copepoda, Harpacticoida). Crustaceana, **28**: 158-167.

CICCHINO, G. [N.], 1972. Desarollo postembrionario de *Notodiaptomus coniferoides* (Wright, 1927) y *Eucyclops neumanni* Pesta, 1927 (Crustacea, Copepoda). Physis, (B) **31**: 585-596.

——, 1974a. Desarrollo postembrionario de *Attheyella* (*Chappuisiella*) *huaronensis* (Delachaux, 1917) (Crustacea, Copepoda). Physis, (B) **33**: 11-18.

——, 1974b. La larva nauplius de Copepodos Ciclopoideos dulceacuicolas. Physis, (B) **33**: 19-44.

CIPÓLLI, M., 1975. Morfologia externa das fases de desenvolvimento de *Diaptomus corderoi* Wright, 1936 (Crustacea, Copepoda, Calanoida). Boletim de Zoologia e Biologia Marinha, (n. ser.) **30**: 567-612.

CLAUS, C., 1868a. Beobachtungen über *Lernaeocera*, *Peniculus* und *Lernaea*. Ein Beitrag zur Naturgeschichte der Lernaeen. Sitzungsberichte der Gesellschaft zur Beförderung der gesamten Naturwissenschaften zu Marburg, **2**: 1-32.

——, 1868b. Über die Metamorphose und systematischen Stellung der Lernaeen. Sitzungsberichte der Gesellschaft zur Beförderung der gesamten Naturwissenschaften zu Marburg, **1868**: 5-13.

——, 1886. Ueber *Lernaeascus nematoxys*, eine seither unbekannt gebliebene Lernaee. Anzeiger der Akademie der Wissenschaften in Wien, (Mathematisch-Naturwissenschaftliche Klasse) **23**: 231-233.

——, 1887a. On *Lernaeascus nematoxys*, a hitherto unknown Lernaean. [Translation of Claus, 1886]. Annals and Magazine of Natural History, **14**: 241-242.

——, 1887b. Über *Lernaeascus nematoxys* Cls. und die Familie der Philichthyden. Arbeiten aus dem Zoologischen Institut der Universität Wien und der Zoologischen Station in Triest, **7**: 1-35, pls. 1-4.

——, 1893a. Über die Entwicklung und das System der Pontelliden. Arbeiten aus dem Zoologischen Institut der Universität Wien und der Zoologischen Station in Triest, **10**: 233-282.

——, 1893b. Neue Beobachtung über die Organisation und Entwicklung von *Cyclops*. Ein Beitrag zur Systematik der Cyclopiden. Arbeiten aus dem Zoologischen Institut der Universität Wien und der Zoologischen Station in Triest, **10**: 283-356.

CLOGSTON, F. L., 1965. Postembryonic development of species of harpacticoid copepods from the Pacific coast of the United States and an application of developmental patterns to their systematics: 1-246. (Ph.D. Thesis, University of Washington, Seattle).

CODREANU, R. & V. MACK-FIRA, 1961. Sur un Copépode, *Sunaristes paguri* Hesse, 1867, et un Polychète, *Polydora ciliata* (Johnston, 1838), associés au Pagure, *Diogenes pugilator* (Roux), dans la mer Noire et la Méditerranée. La notion de cryptotropisme. Rapports et Procès-verbaux des Réunions de la Commission Internationale pour l'Exploration Scientifique de la Mer Méditerranée, **16**: 471-494.

COGNARD, C., 1973. Influence de la température sur les développements embryonnaire et post-embryonnaire du Copépode Diaptomide *Eudiaptomus gracilis* (Sars, 1863). Crustaceana, **24**: 323-331.

COKER, R., 1934. Contribution to the knowledge of North American freshwater harpacticoid Crustacea. Journal of the Elisha Mitchell Scientific Society, **50**: 75-141.

CONLEY, D. C. & M. A. CURTIS, 1993. Effects of temperature and photoperiod on the duration of hatching, swimming, and copepodid survival of the parasitic copepod *Salmincola edwardsii*. Canadian Journal of Zoology, **71**: 972-976.

CONNOLLY, C., 1929. A new copepod parasite *Choniosphaera cancrorum* gen. et sp. n. representing a new genus, and its larval development. Proceedings of the Zoological Society of London, **1929** (3): 415-427.

CONOVER, R. J., 1956. Biology of *Acartia clausi* and *A. tonsa*. Oceanography of Long Island Sound, 1952-1954, VI. Bulletin Bingham Oceanographic Collection, **15**: 156-233.

——, 1965. Notes on the molting cycle, development of sexual characters and sex ratio in *Calanus hyperboreus*. Crustaceana, **8**: 308-320.

——, 1988. Comparative life histories in the genera *Calanus* and *Neocalanus* in high latitudes of the northern hemisphere. Hydrobiologia, **167/168**: 127-142.

CONOVER, R. J. & M. HUNTLEY, 1991. Copepods in ice-covered seas. Distribution, adaptations to seasonally limited food, metabolism, growth patterns and life cycle strategies in Polar seas. Journal of Marine Systems, **2**: 1-42.

CONROY-DALTON, S., 2003. Systematics and phylogeny of the Ancorabolidae (Copepoda: Harpacticoida). IV. Redescription, ontogeny and position of *Echinopsyllus normani*. Cahiers de Biologie Marine, **44**: 153-169.

CONWAY, D. V. P., 2006. Identification of the copepodite developmental stages of twenty-six North Atlantic copepods. Occasional Publications, Marine Biological Association of the United Kingdom, **21**: 1-28.

COOLEY, J. M., 1973. The life history, population dynamics and production of *Leptodiaptomus minutus* Lillj. (Copepoda: Calanoida) in Bluff Lake, Nova Scotia. (Ph.D. Thesis, Dalhousie University, Halifax, Nova Scotia).

COONEY, J. D., D. L. BUNTING, II & C. W. GEHRS, 1984. Effects of acridine and temperature on the survival, development, growth, and reproduction of *Diaptomus clavipes* Schacht. Archives of Environmental Contamination and Toxicology, **13**: 665-674.

COONEY, J. D. & C. W. GEHRS, 1980. The relationship between egg size and naupliar size in the calanoid copepod *Diaptomus clavipes* Schacht. Limnology and Oceanography, **25**: 549-552.

—— & ——, 1984. Effects of temperature, feeding and acridine on development and mortality of eggs and nauplii of *Diaptomus clavipes* Schacht. Aquatic Toxicology, **5**: 197-209.

CORKETT, C. J., 1967. The copepodid stages of *Temora longicornis* (O. F. Müller, 1792) (Copepoda). Crustaceana, **12**: 261-273.

——, 1968. Observations sur les stades larvaires de *Pseudocalanus elongatus* Boeck et *Temora longicornis* Müller. Pelagos, **8**: 51-57.

——, 1981. The copepodid stages of the copepods *Acartia tonsa*, *A. clausii* and *Eurytemora herdmani* from the Annapolis River, Nova Scotia. Proceedings of the Nova Scotian Institute of Science, **31:** 173-179.

——, 1984. Observations on development in copepods. Crustaceana, (Supplement) **7**: 150-153.

CORKETT, C. J. & I. A. MCLAREN, 1970. Relationships between development rate of eggs and older stages of copepods. Journal of the Marine Biological Association of the United Kingdom, **50**: 161-168.

—— & ——, 1978. The biology of *Pseudocalanus*. Advances in Marine Biology, **15**: 1-231.

COSTANZO, G., 1968. Su *Lichomolgus canui* G.O. Sars (Copepoda, Cyclopoida) ed alcuni suoi copepoditi del Lago di Faro (Messina). Bollettino di Zoologia, **35**: 27-37.

COSTANZO, G. & N. CRESCENTI, 2001. Copepodid stages of laboratory-reared *Astericola asterinae* (Bocquet, 1952) (Copepoda, Poecilostomatoida, Lichomolgidae), a copepod associated with *Asterina gibbosa* (Pennant, 1777) of Lake Faro (Messina). Journal of Natural History, London, **35**: 1795-1805.

COSTANZO, G., N. CRESCENTI & N. CALAFIORE, 1996a. Copepodid stages of *Paranthessius anemoniae* Claus, 1889 (Copepoda, Poecilostomatoida, Sabelliphilidae), a copepod associated with *Aiptasia diaphana* (Rapp, 1829) of Lake Faro (Messina) reared in the laboratory. Crustaceana, **69**: 387-399.

——, —— & ——, 1996b. Copepodid stages of *Lichomolgus forficula* Thorell, 1859 (Copepoda, Poecilostomatoida, Lichomolgidae), a copepod associated with *Phallusia mammillata* (Cuvier, 1815). Journal of Natural History, London, **31**: 1019-1028.

——, —— & ——, 1996c. Copepodid development of *Pseudanthessius gracilis* Claus, 1889 (Copepoda, Poecilostomatoida, Pseudanthessiidae), a copepod associated with *Hydroides elegans* (Haswell, 1883) (Annelida, Polychaeta) of Lake Faro (Messina, Italy) as obtained in the laboratory. Crustaceana, **69**: 295-305.

COSTANZO, G., N. CRESCENTI & G. ZAGAMI, 2004. Postembryonic development of *Pseudocyclops umbraticus* Giesbrecht, 1893 (Copepoda, Calanoida) from coastal waters of Sicily. Crustaceana, **77**: 49-66.

COSTANZO, G., G. ZAGAMI, N. CRESCENTI & A. GRANATA, 2002. Postembryonic naupliar development of *Stephos longipes* (Giesbrecht, 1902) (Copepoda Calanoida) from the annual sea ice of Terra Nova Bay, Antarctica. Journal of Crustacean Biology, **22**: 855-860.

COULL, B. C. & B. W. DUDLEY, 1976. Delayed naupliar development of meiobenthic copepods. Biological Bulletin, Woods Hole, **150**: 38-46.

CRESSEY, R., 1967. Revision of the family Pandaridae (Copepoda: Caligoida). Proceedings of the United States National Musuem, **121**: 1-133.

——, 1968. Caligoid copepods parasitic on *Isurus oxyrinchus* with an example of habitat shift. Proceedings of the United States National Musuem, **125**: 1-26.

CRISAFI, P., 1960a. Note biologiche e stadi copepodiformi di *Centropages violaceus* Cls. (Copepoda, Calanoida). Atti della Società Peloritana di Scienze Fisiche, Matematiche e Naturali, **5**: 503-512.

——, 1960b. I Copepodi dello Stretto di Messina. Nota 1. Su alcuni stadi copepodiformi di *Monops regalis* Dana (Copepoda, Calanoida). Atti della Società Peloritana di Scienze Fisiche, Matematiche e Naturali, **6**: 279-288.

——, 1960c. I Copepodi dello Stretto di Messina. Nota II. *Labidocera brunescens* Czerniawsky e *Labidocera brunescens* var. *Dulzettoi* (Copepoda, Calanoida). Atti della Società Peloritana di Scienze Fisiche, Matematiche e Naturali, **6**: 289-292.

——, 1960d. I Copepodi dello Stretto di Messina. Nota III. Osservazioni su alcuni stadi copepodiformi di *Pontella mediterranea* Claus (Copepoda, Calanoida). Atti della Società Peloritana di Scienze Fisiche, Matematiche e Naturali, **6**: 293-299.

——, 1962. I Copepodi dello Stretto di Messina. Nota IV. *Pleuromamma abdominalis* (Lubbock) e *Pleuromamma gracilis* (Claus) (Copepoda, Calanoida). Morfologia, sviluppo e frequenza. Atti della Società Peloritana di Scienze Fisiche, Matematiche e Naturali, **8**: 475-486.

——, 1963a. I Copepodi dello Stretto di Messina. La famiglia Candaciidae nel quinquennio 1958-1962 (esame storico, morfometria, sviluppo, frequenza delle otto specie incontrate con notizie sul periodo riproduttivo). Atti della Società Peloritana di Scienze Fisiche, Matematiche e Naturali, **9**: 81-143.

——, 1963b. Les copépodes pélagiques du détroit de Messine. *Pleuromamma abdominalis* (Lubbock) et *Pleuromamma gracilis* (Claus) (Copepoda, Calanoida). Morphologie, développement et fréquence. Rapports et Procès-verbaux des Réunions de la Commission Internationale pour l'Exploration Scientifique de la Mer Méditerranée, **17**: 555-561.

——, 1965a. I Copepodi dello Stretto di Messina. Variabilità e sviluppo di *Euchaeta marina* (Prestandrea). Bollettino di Zoologia, **32**: 263-281.

——, 1965b. Les copépodes du détroit de Messine - Oeufs, stades naupliens et segmentation du corps du copépode pélagique *Pontella mediterranea* Claus. Rapports et Procès-verbaux des Réunions de la Commission Internationale pour l'Exploration Scientifique de la Mer Méditerranée, **18**: 412-416.

——, 1966. Caratteri morfologici, sviluppo e presenza nel Mediterreneo di Copepodi appartenenti al genere *Sapphirina* J.V. Thompson, 1829. Atti della Società Peloritana di Scienze Fisiche, Matematiche e Naturali, **12**: 651-663.

CRISAFI, P. & M. CRESCENTI, 1972. Comportamento, morfologia, sviluppo, misure, confronti e revisione di otto specie della famiglia Acartiidae, incontrate in acque marine inquinate sopra tutto da scarichi industriali. Bollettino Pesca, Piscicoltura e Idrobiologia, (n. ser.) **27**: 221-254.

CRUZ-PIZARRO, L., 1978. Comparative vertical zonation and diurnal migration among Crustacea and Rotifera in the small high mountain lake La Caldera (Granada, Spain). Verhandlungen der Internationale Vereinigung für Theoretische und Angewandte Limnologie, **20**: 1026-1032.

CURRIE, M. E., 1919. Exuviation and variation of plankton copepods with special reference to *Calanus finmarchicus*. Proceedings and Transactions of the Royal Society of Canada, (3) **12** (4): 207-233.

CZAIKA, S. C., 1982. Identification of nauplii N1-N6 and copepodids CI-CVI of the Great Lakes calanoid and cyclopoid copepods (Calanoida, Cyclopoida, Copepoda). Journal of Great Lakes Research, **8**: 439-469.

CZAIKA, S. C. & A. ROBERTSON, 1968. Identification of the copepodids of the Great Lakes species of *Diaptomus* (Calanoida, Copepoda). The University of Michigan, Great Lakes Research Division Contribution, **71**: 39-60.

DAHMS, H.-U., 1986. Zur Biologie von *Paramphiascella fulvofasciata* (Copepoda, Harpacticoida). Helgoländer Wissenschaftliche Meeresuntersuchungen, **40**: 267-277.

——, 1987b. Die Nauplius-Stadien von *Bryocamptus pygmaeus* (Sars, 1862) (Copepoda, Harpacticoida, Canthocamptidae). Drosera, **87**: 47-58.

——, 1988. Postembryonic development of Harpacticoida (Crustacea, Copepoda) and its significance for the reconstruction of phylogenetic relationships: 1-280. (Ph.D. Thesis, Universität Oldenburg, Oldenburg).

——, 1989b. First report of a lecithotrophic nauplius among Harpacticoida (Crustacea, Copepoda) from the eastern Weddell Sea. Polar Biology, **10**: 221-224.

——, 1991. Usefulness of postembryonic characters for phylogenetic reconstruction in Harpacticoida (Crustacea, Copepoda). Journal of Plankton Research, (Special Volume): 87-104.

——, 1992b. Die Naupliusentwicklung der Süsswassercopepoden. Mikrokosmos, **81**: 276-281.

——, 1994. Ontogenetic characters as a tool for the assessment of phylogenetic relationships among Crustacea. Verhandlungen der Deutschen Zoologischen Gesellschaft, **87**: 218.

DAHMS, H.-U. & G. DIECKMANN, 1987. *Drescheriella glacialis* gen. nov., sp. nov. (Copepoda, Harpacticoida) from Antarctic Sea ice. Polar Biology, **7**: 329-337.

DAHMS, H.-U. & P.-Y. QIAN, 2004. Life histories of the Harpaticoida (Copepoda, Crustacea): a comparison with meiofauna and macrofauna. Journal of Natural History, London, **38**: 1725-1734.

DAHMS, H.-U., N. V. SCHIZAS & T. C. SHIRLEY, 2005. Naupliar evolutionary novelties of *Stenhelia peniculata* (Copepoda, Harpacticoida) from Alaska affirming taxa belonging to different categorical rank. Invertebrate Zoology, **2**: 1-14.

DAMKAER, D. M., 1975. Calanoid copepods of the genera *Spinocalanus* and *Mimocalanus* from the central Arctic Ocean, with a review of the Spinocalanidae. National Oceanic and Atmospheric Administration Technical Reports, National Marine Fisheries Service Circular, **391**: 1-88.

DAVIS, C. C., 1943. The larval stages of calanoid copepod *Eurytemora hirundoide*s (Nordquist). Chesapeake Biological Laboratory, Publication, **58**: 1-52.

DEARBORN, J., F. D. FERRARI & K. EDWARDS, 1986. Can pelagic aggregations cause benthic satiation? Feeding biology of the Antarctic brittle star *Astrotoma agassizii* (Echinodermata: Ophiuroidea). Antarctic Research Series, **44**: 1-28.

DECHO, A. W. & J. W. FLEEGER, 1988. Ontogenetic feeding shifts in the meiobenthic harpacticoid copepod *Nitocra lacustris*. Marine Biology, Berlin, **97**: 191-197.

DEDIE, O., 1940. Etude de *Salminicola mattheyi* n. sp. Copépode parasite de l'ombre-chevalier (*Salmo salvelinus* L.). Revue Suisse de Zoologie, **47**: 1-63.

DELAMARE DEBOUTTEVILLE, C., 1958. Intérêt des transfuges en parasitologie. 83e Congrès des Sociétés Savantes: 563-570.

——, 1960. Biologie des eaux souterraines littorales et continentales. Vie et Milieu, (Supplément) **9**: 1-740.

DELAMARE DEBOUTTEVILLE, C. & L. P. NUNES-RUIVO, 1955. Remarques sur le développement de la femelle d'*Haemobaphoides ambiguus* (T. Scott), et analyse critique des genres *Haemobaphes* Steenstrup et Lütken, *Haemobaphoides* T. et A. Scott et *Collipravus* Wilson, (Crust. Copepoda). Bulletin de la Société Zoologique de France, **80**: 27-37.

DELGADO, L. E., R. JAÑA & V. H. MARÍN, 1998. Testing hypotheses on life-cycle models for Antarctic calanoid copepods, using qualitative, winter, zooplankton samples. Polar Biology, **20**: 74-76.

DEMOTT, W. R., 1986. The role of taste in food selection by freshwater zooplankton. Oecologia, **69**: 334-340.

DIAZ, W. & F. EVANS, 1983. The reproduction and development of *Microsetella norvegica* (Boeck) (Copepoda, Harpacticoida) in Northumberland coastal waters. Crustaceana, **45**: 113-130.

DIBBERN, S. & G. ARLT, 1989. Post-embryonic development of *Mesochra aestuarii* Gurney, 1921 (Copepoda, Harpacticoida). Crustaceana, **57**: 263-287.

DIEL, S., 1991. Zur Lebensgeschichte dominanter Copepodenarten (*Calanus finmarchicus*, *C. glacialis*, *C. hyperboreus*, *Metridia longa*) in der Framstrasse. [On the life history of dominant copepod species (*Calanus finmarchicus*, *C. glacialis*, *C. hyperboreus*, *Metridia longa*) in the Fram Strait.]. Berichte zur Polarforschung, **88**: 1-113. [In German, with English abstract.]

DIEL, S. & W. C. M. KLEIN BRETELER, 1986. Growth and development of *Calanus* spp. (Copepoda) during spring phytoplankton succession in the North Sea. Marine Biology, Berlin, **91**: 85-92.

DIGBY, P. S. B., 1950. The biology of the small planktonic copepods of Plymouth. Journal of the Marine Biological Association of the United Kingdom, **29**: 393-438.

DOUWE, C. VAN, 1899. Zur Morphologie des rudimentären Copepoden-Fusses. Zoologischer Anzeiger, **22**: 447-450.

DUKINA, V. V., 1956. Vidovye razlichiya lichinok tsiklopov. [Specific differences in cyclopoid larvae.]. Zoologicheskii Zhurnal, **35**: 680-690.

DURAN, M., 1974. Description des copépodites II à V de *Neocalanus gracilis* (Dana) (Copepoda Calanoida). Rapports et Procès-verbaux des Réunions de la Commission Internationale pour l'Exploration Scientifique de la Mer Méditerranée, **22**: 79-80.

DURBIN, E. G., P. R. GARRAHAN & M. C. CASAS, 2000b. Abundance and distribution of *Calanus finmarchicus* on the Georges Bank during 1995 and 1996. Journal of Marine Science, **57**: 1664-1685.

DURBIN, E. G., J. A. RUNGE, R. G. CAMPBELL, P. R. GARRAHAN, M. C. CASAS & S. PLOURDE, 1997. Late fall-early winter recruitment of *Calanus finmarchicus* on Georges Bank. Marine Ecology Progress Series, **151**: 103-114.

DUREN, L. A. VAN & J. J. VIDELER, 1995. Swimming behavior of developmental stages of the calanoid copepod *Temora longicornis* at different food concentrations. Marine Ecology Progress Series, **126**: 153-161.

DUSSART, B. [H.], 1969. Les Copépodes des eaux continentales d'Europe occidentale, **2**, Cyclopoïdes et biologie: 1-292. (N. Boubee et Cie, Paris).

——, 1982. Crustacés Copépodes des eaux intérieures. Faune de Madagascar, **58**: 1-146.

DUTZ, J., W. C. M. KLEIN BRETELER & G. KRAMER, 2005. Inhibition of copepod feeding by exudates and transparent exopolymer particles (TEP) derived from a *Phaeocystis globosa* dominated phytoplankton community. Harmful Algae, **4**: 929-940.

ECKSTEIN, H., 1964. Untersuchungen über die Einfluss des Rheinwassers auf die Limnologie des Schluchsees. IV. Untersuchungen über die Einbürgerung des *Eudiaptomus gracilis* Sars in den Schluchsee. Archiv für Hydrobiologie, **28**(5)(2-3): 119-182.

EDVARDSEN, A., D. SLAGSTAD, K. S. TANDE & P. JACCARD, 2003. Assessing zooplankton advection in the Barents Sea using underway measurements and modelling. Fisheries Oceanography, **12**: 61-74.

EINSLE, U. [K.], 1964a. Die Gattung *Cyclops* s. str. im Bodensee. Archiv für Hydrobiologie, **60**: 133-199.

——, 1964b. Larvalentwicklung von Cyclopiden und Photoperiodik. Naturwissenschaften, **51**: 345.

——, 1968. Die Gattung *Mesocyclops* im Bodensee. Archiv für Hydrobiologie, **60**: 131-169.

——, 1989. The identification of copepodid instars of some calanoid and cyclopoid copepods of Lake Konstanz, Federal Republic of Germany. Crustaceana, **57**: 79-87.

ELBOURN, C. A., 1966. The life cycle of *Cyclops strenuus* Fischer in a small pond. Journal of Animal Ecology, **35**: 333-347.

ELGMORK, K., 1955. A resting-stage without encystment in the annual cycle of the freshwater copepod *Cyclops strenuus strenuus*. Ecology, **36**: 739-743.

——, 1965. A triennal copepod (Crustacea) in the temperate zone. Nature, London, **205**: 413.

ELGMORK, K., J. P. NILSSEN, T. BROCH & R. OVREVIK, 1978. Life cycle strategies in neighbouring populations of the copepod *Cyclops scutifer* Sars. Verhandlungen der Internationale Vereinigung für Theoretische und Angewandte Limnologie, **20**: 2518-2523.

EL-MAGHRABY, A. M., 1964. The developmental stages and occurrence of the copepod *Euterpina acutifrons* Dana in the marine environment of Alexandria, United Arab Republic. Annals and Magazine of Natural History, **13**: 223-233.

ELMORE, J. L., 1982. The influence of food concentration and container volume on life history parameters of *Diaptomus dorsalis* Marsh from subtropical Florida. Hydrobiologia, **89**: 215-223.

ELOFSSON, R., 1966. The nauplius eye and frontal organs of the non-Malacostraca (Crustacea). Sarsia, **25**: 1-128.

ELSTER, H.-J., 1932a. Monographische Studien an *Hetercope weismanni* Imhof. I. Postembryonal-Entwicklung und Morphologie. Internationale Revue der Gesamten Hydrobiologie und Hydrographie, **27**: 1-101.

——, 1932b. Monographische Studien an *Hetercope weismanni* Imhof. I. Postembryonal-Entwicklung und Morphologie. 2 Teil. Internationale Revue der Gesamten Hydrobiologie und Hydrographie, **27**: 177-233.

——, 1954. Über die Populationsdynamik von *Eudiaptomus gracilis* Sars und *Heterocope borealis* Fischer im Bodensee-Obersee. Archiv für Hydrobiologie, (Suppl.) **20**: 546-614.

ELSTER, H.-J., R. SCHROEDER, M. HAWARY & I. SCHWOERBEL, 1960. Population dynamics of zooplankton in the Nozha-Hydrodrome near Alexandria. *Arctodiaptomus salinus* Daday and *Diaphanosoma excisum* Sars. Notes and Memoirs of the Hydrobiology Department, United Arab Republic, **50**: 1-27.

EPP, R. W. & W. M. LEWIS, JR., 1980. The nature and ecological significance of metabolic changes during the life history of copepods. Ecology, **61**: 259-264.

EPPACHER, T., 1968. Physiographie und Zooplankton des Gossenkollesees (2.413 m, Kuhtai, Österreich). Berichte des Naturwissenschaftlich-Medizinischer Vereins in Innsbruck, **56**: 31-123.

ESCRIBANO, R., 1998. Population dynamics of *Calanus chilensis* in the Chilean Eastern Boundary Humboldt Current. Fisheries Oceanography, **7**: 245-251.

ESCRIBANO, R., C. IRRIBARREN & L. RODRIGUEZ, 1997. Influence of food quantity and temperature on development and growth of the marine copepod *Calanus chilensis* from northern Chile. Marine Biology, Berlin, **128**: 281-288.

ESCRIBANO, R., V. H. MARIN & P. HIDALGO, 2001. The influence of coastal upwelling on the distribution of *Calanus chilensis* in the Mejillones Peninsula (northern Chile): implications for its population dynamics. Hydrobiologia, **453-454**: 143-151.

ESCRIBANO, R. & I. A. MCLAREN, 1999. Production of *Calanus chilensis* in the upwelling area of Antofagasta, northern Chile. Marine Ecology Progress Series, **177**: 147-156.

ESCRIBANO, R. & L. RODRIGUEZ, 1994. Life cycle of *Calanus chilensis* Brodsky in Bay of San Jorge, Antofagasta, Chile. Hydrobiologia, **292/293**: 289-294.

—— & ——, 1995. Seasonal size variation and growth of *Calanus chilensis* Brodsky in northern Chile. Revista Chilena de Historia Natural, **68**: 373-382.

ESCRIBANO, R., L. RODRIGUEZ & C. IRRIBARREN, 1998. Temperature-dependent development and growth of *Calanus chilensis* Brodsky from northern Chile. Journal of Experimental Marine Biology and Ecology, **229**: 19-34.

ESTERLY, C. O., 1905. The pelagic Copepoda of the San Diego Region. University of California Publications in Zoology, **2**: 113-233.

——, 1924. The free-swimming Copepoda of San Francisco Bay. University of California Publications in Zoology, **26**: 81-129.

EVANSON, M., E. A. BORNHOLD, R. H. GOLDBLATT, P. J. HARRISON & A. G. LEWIS, 2000. Temporal variation in body composition and lipid storage of the overwintering, subarctic copepod *Neocalanus plumchrus* in the Strait of Georgia, British Columbia (Canada). Marine Ecology Progress Series, **192**: 239-247.

EWERS, L. A., 1930. The larval development of freshwater Copepoda. Contributions of the Franz Theodore Stone Laboratory, Ohio State University, Columbus, **3**: 1-43.

FABER, D. J., 1966a. Free-swimming copepod nauplii of Narragansett Bay with a key to their identification. Journal of the Fisheries Research Board of Canada, **23**: 189-205.

FAIRBRIDGE, W., 1945a. West Australian fresh-water calanoids (Copepoda). I. Three new species of *Boeckella*, with an account of the developmental stages of *B. opaqua* n. sp. and a key to the genus. Journal of the Royal Society of Western Australia, **29**: 25-65.

FAIRBRIDGE, W., 1945b. West Australian fresh-water calanoids (Copepoda). II. Two new species of *Brunella*, with an account of the developmental stages of *B. subattenuata* n. sp. Journal of the Royal Society of Western Australia, **29**: 67-89.

FALKENHAUG, T., K. S. TANDE & T. SEMENOVA, 1997. Diel, seasonal and ontogenetic variations in the vertical distributions of four marine copepods. Marine Ecology Progress Series, **149**: 105-119.

FASTEN, N., 1913. The behaviour of a parasitic copepod, *Lernaeopoda Edwardsii* Olsson. Journal of Animal Behaviour, **3**: 36-60.

——, 1914. Fertilization in the parasitic copepod, *Lernaeopoda edwardsii* Olsson. Biological Bulletin, Woods Hole, **27**: 115-126.

——, 1919. Morphology and attached stages of first copepodid larva of *Salmincola edwardsii* (Olsson) Wilson. Publications of the Puget Sound Marine Biological Station, **2**: 153-181.

——, 1921. Studies on parasitic copepods of the genus *Salmincola*. American Naturalist, **55**: 449-456.

FELLER, R. J., 1977. Life history and production of meiobenthic harpacticoid copepods in Puget Sound: 1-249. (Ph.D. Thesis, University of Washington, Seattle).

——, 1980. Development of the sand-dwelling meiobenthic harpacticoid copepod *Huntemannia jadensis* Poppe in the laboratory. Journal of Experimental Marine Biology and Ecology, **46**: 1-15.

FERRARI, F. D., 1980. *Pseudochirella squalida* Grice & Hulsemann, 1967 (Copepoda: Calanoida) from continental slope waters off Delaware. Proceedings of the Biological Society of Washington, **93**: 536-550.

——, 1992. Development of copepod swimming legs with unequal numbers of ramal segments. Journal of Crustacean Biology, **12**: 392-396.

——, 1993. Exceptions to the rule of development that anterior is older among serially homologous segments of post-maxillipedal legs in copepods. Journal of Crustacean Biology, **13**: 763-768.

FERRARI, F. D. & J. DEARBORN, 1989. A second examination of predation on pelagic copepods by the brittle star *Astrotoma agassizii*. Journal of Plankton Research, **11**: 1315-1320.

FERRARI, F. D. & M. DOJIRI, 1987. The calanoid copepod *Euchaeta antarctica* from Southern Ocean Atlantic Sector midwater trawls, with observations on spermatophore dimorphism. Journal of Crustacean Biology, **7**: 458-480.

FERRARI, F. D. & L.-A. HAYEK, 1990. Monthly differences in distributions of sex and asymmetry in a looking-glass copepod, *Pleuromamma xiphias*, off Hawaii. Journal of Crustacean Biology, **10**: 114-127.

FIERS, F., 1990. *Abscondicola humesi* n. gen., n. sp. from the gill chambers of land crabs and the definition of the Cancrincolidae n. fam. (Copepoda, Harpacticoida). Bulletin Institut Royal des Sciences Naturelles de Belgique, (Biologie) **60**: 69-103.

——, 1991b. Allocation of *Laophonte trispinosa* Sewell to *Xanthilaophonte* gen. nov. and the description of *X. cancrincola* spec. nov. (Harpacticoida, Laophontidae). Zoologische Mededelingen, Leiden, **65**: 1-28.

——, 1991c. A revision of the genus *Laophontina* Norman & T. Scott (Copepoda, Harpacticoida). Bulletin Institut Royal des Sciences Naturelles de Belgique, (Biologie) **61**: 5-54.

——, 1997. The genera *Triathrix* Gee & Burgess and *Sphingothrix* gen. nov. (Copepoda, Cletodidae sensu Por) from the Bay of Campeche, Gulf of Mexico. Sarsia, **82**: 237-257.

——, 1998. Female leg 4 development in Laophontidae (Harpacticoida): a juvenile adaptation to precopulatory behaviour. Journal of Marine Systems, (Special Volume) **15**: 41-51.

FIERS, F. & M. DE TROCH., 2000. New Tetragonicipitidae (Copepoda, Harpacticoida) from the Indo-Pacific. Hydrobiologia, **434**: 97-144.

FIKSEN, O. & F. CARLOTTI, 1998. A model of optimal life history and diel vertical migration in *Calanus finmarchicus*. Sarsia, **83**: 129-147.

FILHOL, M. J., 1934. Embryologie et développement de *Lamproglena pulchella* Nordmann, description du mâle. Bulletin International de l'Académie Polonaise des Sciences et des Lettres, Classe des Sciences Mathématiques et Naturelles, (B, Sciences Naturelles) **2**: 225-232.

——, 1936. Nouvelles observations sur le développement et la biologie du *Lamproglena pulchella* Nordmann. Annales de Parasitologie Humaine et Comparée, **14**: 246-255.

FISCHER, W., 1956. Untersuchungen über einen für die Deutsche Bucht neuen parasitären Copepoden: *Lecithomyzon maenadis* Bloch & Gallien (Familie Choniostomatidae) an *Carcinus maenas* Pennant (Crustacea Decapoda). Helgoländer Wissenschaftliche Meeresuntersuchungen, **5**: 326-352.

FLAMARIQUE, I. N., H. I. BROWMAN, M. BELANGER & K. BOXASPEN, 2000. Ontogenetic changes in visual sensitivity of the parasitic salmon louse *Lepeophtheirus salmonis*. Journal of Experimental Biology, **203**: 1649-1657.

FLEEGER, J. W., 1979. Population dynamics of three estuarine meiobenthic harpacticoids (Copepoda) in South Carolina. Marine Biology, Berlin, **52**: 147-156.

FLEMINGER, A. & K. HULSEMANN, 1974. Systematics and distribution of the four sibling species comprising the genus *Pontellina* Dana (Copepoda, Calanoida). Fishery Bulletin, U.S., **72**: 63-120.

FONSECA, V. L. & M. S. DE ALMEIDA PRADO, 1979. The naupliar development of *Oithona oligohalina* Fonseca & Björnberg, 1976. Boletim do Instituto Oceanografico, São Paulo, **28**: 17-22.

FORGET, J., J. F. PAVILLON, M. R. MENASRIA & G. BOCQUENE, 1998. Mortality and LC50 values for several stages of the marine copepod *Tigriopus brevicornis* (Müller) exposed to the metals arsenic and cadmium and the pesticides atrazine, carbofuran, dichlorvos, and malathion. Ecotoxicology and Environmental Safety, **40**: 239-244.

FORGET-LERAY, J., I. LANDRIAU, C. MINIER & F. LEBOULENGER, 2004. Impact of endocrine toxicants on survival, development, and reproduction of the estuarine copepod *Eurytemora affinis* (Poppe). Ecotoxicology and Environmental Safety, **60**: 288-294.

FOSSHAGEN, A., 1970. *Thespesiopsyllus paradoxus* (Sars) (Copepoda, Cyclopoida) from western Norway. Sarsia, **42**: 33-40.

——, 1983. A new genus of calanoid copepod from the Norwegian Sea. Sarsia, **68**: 257-262.

FRANSZ, H. G. & S. R. GONZALEZ, 1995. The production of *Oithona similis* (Copepoda: Cyclopoida) in the Southern Ocean. Journal of Marine Science, **52**: 549-555.

FRASER, J. H., 1936. The occurrence, ecology and life history of *Tigriopus fulvus* (Fischer). Journal of the Marine Biological Association of the United Kingdom, **20**: 523-536.

FRATELLO, B., I. BENEDETTI, M. A. SABATINI & L. MOLA, 1985. Glycogen in a parasitic crustacean, *Lernaea cyprinacea* L., during development and in adult life. Atti della Accademia Nazionale dei Lincei, (Classe di Scienze Fisiche Matematiche e Naturali), Rendiconti, (8) **77**: 243-247.

FRIČ, A. J., 1882. Note préliminaire sur l'ontogénie de nos Copépodes d'eau douce. Zoologischer Anzeiger, **5**: 498-503.

FRIEND, G. F., 1941. The life-history and ecology of the salmon gill-maggot *Salmincola salmonea* (L.) (Copepoda Crustacea). Transactions of the Royal Society of Edinburgh, **60**: 503-541.

FRISCH, D. & B. SANTER, 2004. Temperature-induced responses of a permanent-pond and a temporary-pond cyclopoid copepod: a link to habitat predictability? Evolutionary Ecology Research, **6**: 541-553.

FROST, B. W., 1985. Food limitation of the planktonic marine copepods *Calanus pacificus* and *Pseudocalanus* sp. in a temperate fjord. Archiv für Hydrobiologie, (Beiheft, Ergebnisse der Limnologie) **21**: 1-13.

FRYD, M., O. H. HASLUND & O. WOHLGEMUTH, 1991. Development, growth and egg production of the two copepod species *Centropages hamatus* and *Centropages typicus* in the laboratory. Journal of Plankton Research, **13**: 683-690.

FRYER, G., 1956. A report on the parasitic Copepoda and Branchiura of the fishes of Lake Nyasa. Proceedings of the Zoological Society of London, **127**: 293-344.

——, 1978. Free-living stages of freshwater parasitic Copepoda. Das Zooplankton der Binnengewässer, (2, Binnengewässer) **26**: 344-367.

——, 1986. Structure, function and behaviour, and the elucidation of evolution in copepods and other crustaceans. Syllogeus, **58**: 150-157.

FULTON, J. D., 1973. Some aspects of the life history of *Calanus plumchrus* in the Strait of Georgia. Journal of the Fisheries Research Board of Canada, **30**: 811-815.

GAARD, E., 1996. Life cycle, abundance and transport of *Calanus finmarchicus* in Faroese waters. Ophelia, **44**: 59-70.

GAARD, E. & B. HANSEN, 2000. Variations in the advection of *Calanus finmarchicus* onto the Faroe Shelf. Journal of Marine Science, **57**: 1612-1618.

GADD, P., 1904. Parasit-Copepoder i Finland. Acta Societatis pro Fauna et Flora Fennica, **26**(8): 1-60.

GALLEGO, A., J. MARDALJEVIC, M. R. HEATH, D. HAINBUCHER & D. SLAGSTAD, 1999. A model of the spring migration into the North Sea by *Calanus finmarchicus* overwintering off the Scottish continental shelf. Fisheries Oceanography, **8** (Supplement 1): 107-125.

GAO, S. W. & Y. LIN, 1986. A study on the biology of copepodid from Jiao Zhou Bay. 1. The copepodid stages of *Paracalanus parvus* (Claus). Marine Science, **10**: 51-59. [In Chinese, with English summary.]

GARBER, B. I., 1941. Postembrional'noe razvitie *Epischura baikalensis* Sars. [Postembryonal development of *Epischura baikalensis*.]. Izvestiya Akademii Nauk SSSR, (Biologicheskikh Nauk) **1941**: 105-115. [In Russian, with English summary.]

GATTEN, R. R., E. D. S. CORNER, C. C. KILVINGTON & J. R. SARGENT, 1979. A seasonal survey of the lipids in *Calanus helgolandicus* Claus from the English Channel. In: E. NAYLOR & R. G. HARTNOLL (eds.), Cyclic phenomena in marine plants and animals: 275-284. (Proceedings XIII. European Marine Biology Symposium, Oxford).

GAUDY, R., 1961. Note sur les stades larvaires de *Temora stylifera* Dana. Recueil des Travaux de la Station Marine d'Endoume, **36**: 115-122.

GAUDY, R. & M. PAGANO, 1987. Biologie d'un Copépode des mares temporaires du littoral méditerranéen français: *Eurytemora velox*. III. Réproduction. Marine Biology, Berlin, **94**: 335-345.

GEE, J. M. & J. T. DAVEY, 1986. Stages in the life history of *Mytilicola intestinalis* Steuer, a copepod parasite of *Mytilus edulis* (L.), and the effect of temperature on their rates of development. Journal du Conseil International pour l'Exploration de la Mer, **42**: 254-264.

GEGENBAUR, C., 1870. Grundzüge der vergleichende Anatomie: i-xii, 1-892. (W. Engelmann, Leipzig).

GEHRS, C. W. & A. ROBERTSON, 1975. Use of life tables in analyzing the dynamics of copepod populations. Ecology, **56**: 665-672.

GEHRS, C. W., J. R. TRABALKA & E. A. BARDILL, 1975. Sensitivity of adult and embryonic calanoid copepods to acute ionizing radiation. Radiation Research, **63**: 382-385.

GEILING, W. T. & R. S. CAMPBELL, 1972. The effect of temperature on the development rate of the major life stages of *Diaptomus pallidus* Herrick. Limnology and Oceanography, **17**: 304-307.

GELMINI, G., 1928. Contributo alla conoscenza dello sviluppo larvale di *Cyclops leuckarti* Claus. Rivista di Scienze Naturali, Milano, **19**: 89-96.

GEORGE, C. L. & J. A. LINDLEY, 1997. Hatching nauplii of planktonic calanoid copepods from intertidal estuarine sediments. Journal of the Marine Biological Association of the United Kingdom, **77**: 899-902.

GEORGE, D. G., 1976. Life cycle and production of *Cyclops vicinus* in a shallow eutrophic reservoir. Oikos, **27**: 101-110.

——, 2000. The impact of regional-scale changes in the weather on the long-term dynamics of *Eudiaptomus* and *Daphnia* in Esthwaite Water, Cumbria. Freshwater Biology, **45**: 111-121.

GEORGE, V. S., 1985. Demographic evaluation of the influence of temperature and salinity on the copepod *Eurytemora herdmani*. Marine Ecology Progress Series, **21**: 145-152.

GHEERARDYN, H., F. FIERS, M. VINCX & M. DE TROCH, 2006. *Paralaophonte harpagone* sp. n. (Copepoda: Harpacticoida), a laophontid with an extremely specialised maxilliped. Organisms, Diversity and Evolution, (Electronic Supplement) **14**: 1-9.

GIARD, A., 1873. Recherches sur les Ascidies composées ou Synascidies. Archives de Zoologie Expérimentale et Générale, **1**: 501-687.

GIARD, A. & J. BONNIER, 1889. Note sur l'*Aspidoecia normani* et sur la famille des Choniostomatidae. Bulletin Scientifique de la France et la Belgique, **20**: 341-372, 2 pls.

GIBBONS, S. G., 1936. Early development stages of Copepoda. - 1. *Rhincalanus nasutus* and *Eucalanus elongatus*. Annals and Magazine of Natural History, (10) **18**: 384-392.

——, 1937. Variations in copepod development. Nature, London, **140**: 1064-1065.

——, 1938. Early developmental stages of Copepoda. - 2. *Metridia lucens* Boeck. Annals and Magazine of Natural History, (11) **2**: 493-497.

GIBBONS, S. G. & H. S. OGILVIE, 1933. The developmental stages of *Oithona helgolandica* and *Oithona spinirostris* with a note on the occurrence of body spines in cyclopoid nauplii. Journal of the Marine Biological Association of the United Kingdom, (new series) **18**: 529-550.

GIBSON, V. & G. GRICE, 1976. Developmental stages of *Pontella meadi* Wheeler (Copepoda: Calanoida). Journal of the Fisheries Research Board of Canada, **33**: 847-854.

—— & ——, 1977. The developmental stages of *Labidocera aestiva* Wheeler, 1900 (Copepoda, Calanoida). Crustaceana, **32**: 7-20.

—— & ——, 1978. The developmental stages of a species of *Corycaeus* (Copepoda: Cyclopoida) from Saanich Inlet, British Columbia. Canadian Journal of Zoology, **56**: 66-74.

GIESBRECHT, W., 1882. Beiträge zur Kenntnis einiger Notodelphyiden. Mittheilungen aus der Zoologischen Station zu Neapel, **3**: 293-372, pls. 22-24.

——, 1893b. Mittheilungen über Copepoden 1-6. 6. Zur Morphologie der Maxillipeden. Mittheilungen aus der Zoologischen Station zu Neapel, **11**: 56-106, pls. 5-7.

——, 1895. Mittheilungen über Copepoden 7-9. 7. Zur Morphologie des weiblichen Abdomens. Mittheilungen aus der Zoologischen Station zu Neapel, **11**: 631-694.

——, 1897. Zur Ontogenese der Monstrilliden. Zoologischer Anzeiger, **20** (526): 70-72.

——, 1899. Die Asterocheriden des Golfes von Neapel und der angrenzenden Meeres-Abschnitte. Fauna und Flora des Golfes von Neapel und der Angrenzenden Meeres-Abschnitte, Herausgegeben von der Zoologischen Station zu Neapel, **25**: 1-217, pls. 1-11.

GIRALDO, A., R. ESCRIBANO & V. MARIN, 2002. Spatial distribution of *Calanus chilensis* off Mejillones Peninsula (northern Chile): ecological consequences upon coastal upwelling. Marine Ecology Progress Series, **230**: 225-234.

GISLASON, A., 2003. Life-cycle strategies and seasonal migrations of oceanic copepods in the Irminger Sea. Hydrobiologia, **503**: 195-209.

GISLASON, A. & O. S. ASTTHORSSON, 1998. Variability in the population structure of *Calanus finmarchicus* in Icelandic waters in spring. Journal of Marine Science, **55**: 811-813.

—— & ——, 2000. Winter distribution, ontogenetic migration, and rates of egg production of *Calanus finmarchicus* southwest of Iceland. Journal of Marine Science, **57**: 1727-1739.

GLATZEL, T., 1991. Neue morphologische Aspekte und die Copepodid-Stadien von *Parastenocaris phyllura* Kiefer (Copepoda, Harpacticoida). Zoologica Scripta, **20**: 375-393.

GNADEBERG, W., 1949. Beiträge zur Biologie und Entwicklung des *Ergasilus sieboldii* v. Nordmann (Copepoda parasitica). Zeitschrift für Parasitenkunde, **14**: 103-180.

GNANAMUTHU, C. P., 1950. Sex differences in the chalimus and adult forms of *Caligus polycanthi* sp. nov. (Crustacea: Copepoda) parasitic on *Balistes maculatus* from Madras. Records of the Indian Museum, **47**: 159-170.

——, 1951. Notes on the life history of a parasitic copepod, *Lernaea chackoensis*. Parasitology, **41**: 148-155.

GOLEZ, M. S. N., T. TAKAHASHI, T. ISHIMARU & A. OHNO, 2004. Post-embryonic development and reproduction of *Pseudodiaptomus annandalei* (Copepoda: Calanoida). Plankton Biology and Ecology, **51**: 15-25.

GÓMEZ, S. & S. CONROY-DALTON, 2002. Description of *Ancorabolus hendrickxi* sp. nov. (Copepoda: Harpacticoida: Ancorabolidae) from the Neotropics and notes on caudal ramus development within oligoarthran harpacticoids. Cahiers de Biologie Marine, **43**: 111-129.

GONZALEZ, S. A., 1968. Desarollo larvario de *Diaptomus proximus* Kiefer (Copepoda, Calanoida). Hydrobiologia, **32**: 528-544.

GOODING, R. U., 1963. External morphology and classification of marine poecilostome copepods belonging to the families Clausidiidae, Clausiidae, Nereicolidae, Eunicicolidae, Synaptiphilidae, Catiniidae, Anomopsyllidae, and Echiurophilidae: 1-247, 26 pls. (Ph.D. Thesis, University of Washington, Seattle).

——, 1988. The *Saphirella* problem. Hydrobiologia, **167-168**: 363-366.

GOODSIR, H. D. S., 1842a. Note sur le développement des oeufs du *Caligus* et sur les métamorphoses que ce Crustacé éprouve. Annales des Sciences Naturelles, (Zoologie et Biologie Animale) **2** (18): 181-184.

——, 1842b. On a new genus, and on six new species of Crustacea, with observations on the development of the egg, and on the metamorphosis of *Caligus*, *Carcinus* and *Pagurus*. Edinburgh New Philosophical Journal, **33**: 174-192.

GOPHEN, M., 1976. Temperature effect on lifespan, metabolism and development time of *Mesocyclops leuckarti* (Claus). Oecologia, **25**: 271-277.

GOSWAMI, S. C., 1971. Metamorphosis of two species of the genus *Oithona* Baird (Copepoda). Indian Journal of Marine Science, **4**: 60-67.

——, 1972. Life history of *Cyclops buxtoni* Gurney (Copepoda). Indian Journal of Marine Science, **1**: 128-132.

——, 1977b. Development and predation in a calanoid copepod *Tortanus forcipatus* (Giesbrecht). Indian Journal of Marine Science, **6**: 154-159.

——, 1977c. Laboratory culture of a harpacticoid copepod *Laophonte setosa* (Boeck): 563-570. (National Institute of Oceanography, Goa, Special Publication).

——, 1978b. Developmental stages of calanoid copepods, *Labidocera pavo* Giesbrecht and *L. minuta* Giesbrecht. Indian Journal of Marine Science, **7**: 288-294.

GOTTO, R. V., 1957. The biology of a commensal copepod, *Ascidicola rosea* Thorell, in the ascidian *Corella parallelogramma* (Müller). Journal of the Marine Biological Association of the United Kingdom, **36**: 281-290.

——, 1979. The association of copepods with marine invertebrates. Advances in Marine Biology, **16**: 1-109.

GOUDEY-PERRIERE, F., 1975. Développement de *Amphiurophilus amphiurae* (Hérouard), Crustacé Copépode, parasite des bourses génitales de l'Ophiure *Amphipholis squamata* Della Chiaje, Echinoderme. Comptes Rendus Hebdomadaires des Séances de l'Académie des Sciences, (D) **280**: 857-860.

——, 1979. *Amphiurophilus amphiurae* (Hérouard), Crustacé Copépode parasite des bourses génitales de l'Ophiure *Amphipholis squamata* Della Chiaje, Echinoderme: morphologie des adultes et étude des stades juveniles. Cahiers de Biologie Marine, **20**: 201-230.

GOUVÊA, E. P. DE, 1978. Estágios de desenvolvimento pós embrionário de *Thermocyclops minutus* (Lowndes) e de *Notodiaptomus conifera* (Sars) (Crustacea, Copepoda) da Lagoa do Abaeté (Salvador - Ba.): 1-131. (Ph.D. Thesis, Universidade de São Paulo).

——, 1980. Estágios do desenvolvimento naupliar de *Notodiaptomus conifer* (Sars, 1901) (Copepoda, Calanoida). Ciência e Cultura, **32**: 1047-1059.

GRABDA, J., 1956. Badania nad rozwojem *Lernaea esocina* (Burm., 1835) i *Lernaea cyprinacea* L., 1758. [Research on the development of *Lernaea esocina* (Burmeister, 1835) and *Lernaea cyprinacea* L., 1758.]. Wiadomosci Parazytologiczne, (Supplement) **5** (2): 257-258. [In Polish, with English summary.]

——, 1958. Cykl rozwojowy *Lernaea cyprinacea* L. [Developmental cycle of *Lernaea cyprinacea* L.]. Wiadomosci Parazytologiczne, **4**: 633-636. [In Polish, with English summary.]

——, 1971. Pasozyty kraglovstych i ryb. Parasiti cyclostomatorum et piscium. Katalog Fauny Pasozytniczej Polski, **2**: 1-304.

——, 1973. Contribution to knowledge of biology of *Cecrops latreillii* Leach, 1816 (Caligoida: Cercopidae) the parasite of the ocean sunfish *Mola mola* (L.). Acta Ichthyologica et Piscatoria, **3**: 61-74.

GRAINGER, J. N. R., 1951. Notes on the biology of the copepod *Mytilicola intestinalis* Steuer. Parasitology, **41**: 135-142.

GRANDORI, R., 1912. Studi sullo sviluppo larvale dei copepodi pelagici. Redia, **8**: 360-437.

——, 1925. Sullo sviluppo larvale di *Centropages typicus* Krøyer. Rivista di Biologia, Milano, **7**: 137-145.

GRAS, R., 1971. Poids individuel, durée de développement et production des différents stades de *Tropodiaptomus incognitus* (Crustacés Copépodes). Cahiers de l'Office de la Recherche Scientifique et Technique Outre-Mer, (Hydrobiologie) **4**: 63-70.

GRAS, R. & L. SAINT-JEAN, 1981. Durée du développement juvénile de quelques Copépodes planctoniques du lac Tchad. Revue d'Hydrobiologie Tropicale, **14**: 39-52.

GRAVIER, C. J., 1912. Sur quelques Crustacés parasites annélidicoles provenant de la seconde Expédition Antarctique Française. Note (1). Présentée par M. E. L. Bouvier. Comptes Rendus de l'Académie des Sciences, **154**: 830-832.

GRAY, P., 1933a. The nauplii of *Notodelphis agilis* Thorell and *Doropygus porcicauda* Brady. Journal of the Marine Biological Association of the United Kingdom, **18**: 519-522.

——, 1933b. *Mycophilus rosocula* n. sp., a notodelphoid copepod parasitic within *B.* (*Botrylloides*) *leachii* Sav., with a description of the nauplius and notes on the habits. Journal of the Marine Biological Association of the United Kingdom, **18**: 523-527.

GREEN, E. P., R. P. HARRIS & A. DUNCAN, 1992. The production and ingestion of faecal pellets by nauplii of marine calanoid copepods. Journal of Plankton Research, **14**: 1631-1643.

——, —— & ——, 1993. The seasonal abundance of the copepodite stages of *Calanus helgolandicus* and *Pseudocalanus elongatus* off Plymouth. Journal of the Marine Biological Association of the United Kingdom, **73**: 109-122.

GREEN, J., 1961. A new species of *Sabellacheres* (Crustacea: Copepoda) parasitic on the fan worm *Eudistylia polymorpha* (Johnson). Proceedings of the Zoological Society of London, **136**: 623-628.

GREENE, C. H. & M. R. LANDRY, 1985. Patterns of prey selection in the cruising calanoid predator *Euchaeta elongata*. Ecology, **66**: 1408-1416.

GRESTY, K. A., G. A. BOXSHALL & K. NAGASAWA, 1993. Antennulary sensors of the infective copepodid larva of the salmon louse, *Lepeophtheirus salmonis* (Copepoda: Caligidae). In: G. A. BOXSHALL & D. DEFAYE (eds.), Pathogens of wild and farmed fish. Sea lice: 83-98 (Ellis Horwood, London).

GRICE, G. D., 1971. The developmental stages of *Eurytemora americana* Williams, 1906, and *Eurytemora herdmani* Thompson & Scott, 1897 (Copepoda, Calanoida). Crustaceana, **20**: 145-158.

GRICE, G. D. & V. R. GIBSON, 1982. The developmental stages of the calanoid copepod *Labidocera wollastoni* (Lubbock) with observations on its eggs. Cahiers de Biologie Marine, **23**: 215-225.

GRIGA, R. E., 1960. Razvitie nekotorykh Harpacticoida Chernogo morya. [Development of some Black Sea harpacticoids.]. Trudy Sevastopol Biologicheskoy Stantsii, **13**: 68-77. [In Russian, with English summary.]

GRIGG, H. & S. J. BARDWELL, 1982. Seasonal observations on moulting and maturation in stage V copepodites of *Calanus finmarchicus* from the Firth of Clyde. Journal of the Marine Biological Association of the United Kingdom, **62**: 315-327.

GRIGG, H., S. J. BARDWELL & L. J. HOLMES, 1987. Comparative observations on the biometry and development of *Calanus finmarchicus* and *C. helgolandicus* in copepodite stage V, with comments on other Calanidae. Marine Biology, Berlin, **96**: 253-262.

GRIGG, H., L. J. HOLMES & S. J. BARDWELL, 1988. Patterns of variation in the dry body weight of *Calanus finmarchicus* in copepodite stage V during autumn and winter in the Firth of Clyde. Journal of the Marine Biological Association of the United Kingdom, **69**: 101-122.

GROBBEN, K., 1892. Zur Kenntnis des Stammbaumes und des Systems der Crustaceen. Sitzungsberichte der Kaiserlichen Akademie der Wissenschaften in Wien, (Mathematik-naturwissenschaft.) (1) **101**: 237-274.

GRYGIER, M. J., 1994. Identity of *Thaumatoessa* (= *Thaumaleus*) *typica* Krøyer, the first described monstrilloid copepod. Sarsia, **78**: 235-242.

GUIGLIA, D., 1926. Sullo sviluppo larvale de *Tigriopus fulvus* (Fischer). Bollettino dei Musei di Zoologia e Anatomia Comparata della Roy. Università di Genova, (2) **6**(3): 11-16.

GUISANDE, C. & R. HARRIS, 1995. Effect of total organic content of eggs on hatching success and naupliar survival in the copepod *Calanus helgolandicus*. Limnology and Oceanography, **40**: 476-482.

GUNNERØD, T. B., 1971. The life history and population dynamics of *Cyclops scutifer* Sars in Pedro Bay, Iliamna Lake, Alaska: 1-136. (Ph.D. Thesis, University of Washington, Seattle).

GURNEY, R., 1913. Some notes on the parasitic copepod *Thersitina gasterostei* Pagenstecher. Annals and Magazine of Natural History, **12**: 415-424, 4 pls.

——, 1923. Notes on some Copepoda from Plymouth. Journal of the Marine Biological Association of the United Kingdom, **19**: 299-304.

——, 1930b. The larval stages of the copepod *Longipedia*. Journal of the Marine Biological Association of the United Kingdom, **16**: 461-474.

——, 1933b. British freshwater Copepoda, **3**. Cyclopoida: 1-384. (The Ray Society, London).

——, 1934b. The development of *Rhincalanus*. Discovery Reports, **9**: 207-214.

——, 1945. Some notes on the development and classification of parasitic Copepoda. Annals and Magazine of Natural History, (11) **12**: 121-127.

——, 1947. Some notes on parasitic Copepoda. Journal of the Marine Biological Association of the United Kingdom, **27**: 133-137.

HADA, A., 1991. Effect of cannibalism on the laboratory cultured population of the brackish-water copepod *Sinocalanus tenellus*. Bulletin of the Plankton Society of Japan, **38**: 43-52.

HADA, A., S. I. UYE & T. ONBÉ, 1986. The seasonal life cycle of *Sinocalanus tenellus* (Copepoda: Calanoida) in a brackish-water pond. Bulletin of the Plankton Society of Japan, **33**: 29-41.

HADDON, K., 1912. *Herpyllobius arcticus*. Quarterly Journal of Microscopical Science, (New Series) **58**: 385-410.

HALISCH, W., 1939. Anatomie und Biologie von *Ergasilus minor* Halisch. Zeitschrift für Parasitenkunde, **11**: 284-330.

HALL, L. W., JR., M. C. ZIEGENFUSS, R. D. ANDERSON & D. P. TIERNEY, 1995. The influence of salinity on the chronic toxicity of atrazine to an estuarine copepod: implications for development of an estuarine chronic criterion. Archives of Environmental Contamination and Toxicology, **28**: 344-348.

HALLBERG, E. & H.-J. HIRCHE, 1980. Differentiation of mid-gut in adults and over-wintering copepodids of *Calanus finmarchicus* (Gunnerus) and *C. helgolandicus*. Journal of Experimental Marine Biology and Ecology, **48**: 283-295.

HALSBAND-LENK, C., H. J. HIRCHE & C. F. JUERGEN, 2002. Temperature impact on reproduction and development of congener copepod populations. Journal of Experimental Marine Biology and Ecology, **271**: 121-153.

HALVORSEN, E. & K. S. TANDE, 1999. Physical and biological factors influencing the seasonal variation in distribution of zooplankton across the shelf at Nordvestbanken, northern Norway, 1994. Sarsia, **84**: 279-292.

HALVORSEN, E., K. S. TANDE, A. EDVARDSEN, D. SLAGSTAD & O. P. PEDERSEN, 2003. Habitat selection of overwintering *Calanus finmarchicus* in the NE Norwegian Sea and shelf waters off northern Norway in 2000-02. Fisheries Oceanography, **12**: 339-351.

HAMON, M., 1952. Un nouveau Lernaeidae, parasite larvaire du poisson *Serranus cabrilla* L. Annales des Sciences Naturelles, (Zoologie et Biologie Animale) (11) **14**: 207-218.

HANAOKA, T., 1944. Nauplius of the fresh-water genus, *Cyclops*. Suisan Gakkai Ho, **9**: 109-137.

——, 1952a. Study of free-living Copepoda-Nauplius. Bulletin of the Naikai Regional Fishery Research Laboratory, **1**: 1-36. [In Japanese, with English summary.]

——, 1952b. On nauplius of *Oncaea mediterranea* and *Corycaeus* sp. Bulletin of the Naikai Regional Fishery Research Laboratory, **1**: 37-41. [In Japanese, with English summary.]

HANSEN, A. M., 1996. Variable life history of a cyclopoid copepod: the role of food availability. Hydrobiologia, **320**: 223-227.

HANSEN, A. M. & N. G. HAIRSTON, JR., 1998. Food limitation in a wild cyclopoid copepod population: direct and indirect life history responses. Oecologia, **115**: 320-330.

HANSEN, A. M. & E. JEPPESEN, 1992. Life cycle of *Cyclops vicinus* in relation to food availability, predation, diapause and temperature. Journal of Plankton Research, **14**: 591-605.

HANSEN, A. M. & B. SANTER, 1995. The influence of food resources on the development, survival and reproduction of the two cyclopoid copepods: *Cyclops vicinus* and *Mesocyclops leuckarti*. Journal of Plankton Research, **17**: 631-646.

—— & ——, 2003. The life cycle of *Cyclops vicinus* in Lake Sobygård: new aspects derived from sediment analyses. Hydrobiologia, **510**: 17-21.

HANSEN, B. W., T. MARKER, P. ANDREASSEN, E. ARASHKEWICH, F. CARLOTTI, P. LINDEQUE, K. S. TANDE & M. WAGNER, 2003. Differences in life-cycle traits of *Calanus finmarchicus* originating from 60°N and 69°N, when reared in mesocosms at 69°N. Marine Biology, Berlin, **142**: 877-893.

HANSEN, H.-J., 1897. The Choniostomatidae, a family of Copepoda, parasites of Crustacea Malacostraca: 1-205. (Andr. Fred. Host & Son, Copenhagen).

HAQ, S. M., 1965a. The larval development of *Oithonina nana*. Journal of Zoology, London, **146**: 555-566.

——, 1965b. Development of the copepod *Euterpina acutifrons* with special reference to dimorphism in the male. Proceedings of the Zoological Society of London, **144**: 175-201.

HARRIS, J. R. W., 1983. The development and growth of *Calanus* copepodites. Limnology and Oceanography, **28**: 142-147.

HARRIS, R. P., X. IRIGOIEN, R. N. HEAD, C. REY, B. H. HYGUM, B. W. HANSEN, B. NIEHOFF, H. B. MEYER & F. CARLOTTI, 2000. Feeding, growth, and reproduction in the genus *Calanus*. Journal of Marine Science, **57**: 1708-1726.

HARRIS, R. P. & G. A. PAFFENHÖFER, 1976. Feeding growth and reproduction of the marine planktonic copepod *Temora longicornis*. Journal of the Marine Biological Association of the United Kingdom, **56**: 675-690.

HART, R. C., 1991. Food and suspended sediment influences on the naupliar and copepodid durations of freshwater copepods: comparative studies on *Tropodiaptomus* and *Metadiaptomus*. Journal of Plankton Research, **13**: 645-660.

——, 1996. Naupliar and copepodite growth and survival of two freshwater calanoids at various food levels: demographic contrasts, similarities, and food needs. Limnology and Oceanography, **41**: 648-658.

——, 1998a. Copepod equiproportional development: experimental confirmation of its independence of food supply level, and a conceptual model accounting for apparent exceptions. Hydrobiologia, **380**: 77-85.

——, 1998b. Comparative functional feeding responses of two seasonally alternating African calanoids: the importance of juvenile instars in inter-specific differentiation. Verhandlungen der Internationale Vereinigung für Theoretische und Angewandte Limnologie, **26**: 1945-1951.

HART, R. C. & B. R. ALLANSON, 1976. The distribution and diel vertical migration of *Pseudodiaptomus hessei* (Mrázek) (Calanoida: Copepoda) in a subtropical lake in southern Africa. Freshwater Biology, **6**: 183-198.

HART, R. C., K. IRVINE & R. WAYA, 1995. Experimental studies on food dependency of development times and reproductive effort (fecundity and egg size) of *Tropodiaptomus cunningtoni* in relation to its natural distribution in Lake Malawi. Archiv für Hydrobiologie, **133**: 23-47.

HARTMANN, R., 1870. Beiträge zur anatomischen Kenntnis der Schmarotzerkrebse. 2. *Lernaeocera barnimii* mihi. Archiv für Anatomie, Physiologie, und Wissenschaftliche Medizin, **1870**: 726-757, pls. 17-18.

HARTOG, M., 1888. The morphology of *Cyclops* and the relations of the copepods. Transactions of the Linnean Society of London, **5**: 1-46.

HEATH, M. R., 1999. The ascent migration of *Calanus finmarchicus* from overwintering depths in the Faroe-Shetland Channel. Fisheries Oceanography, **8** (Supplement 1): 84-99.

HEATH, M. R., O. S. ASTTHORSSON, J. DUNN, B. ELLERTSEN, E. GAARD, A. GISLASON, W. S. C. GURNEY, A. T. HIND, X. IRIGOIEN, W. MELLE, B. NIEHOFF, K. OLSEN, S. SKRESLET & K. S. TANDE, 2000. Comparative analysis of *Calanus finmarchicus* demography at locations around the northeast Atlantic. Journal of Marine Science, **57**: 1562-1580.

HEATH, M. R., J. O. BACKHAUS, K. RICHARDSON, E. MCKENZIE, D. SLAGSTAD, D. BEARE, J. DUNN, J. G. FRASER, A. GALLEGO, D. HAINBUCHER, S. HAY, S. JONASDOTTIR, H. MADDEN, J. MARDALJEVIC & A. SCHACHT, 1999. Climate fluctuations and the spring invasion of the North Sea by *Calanus finmarchicus*. Fisheries Oceanography, **8** (Supplement 1): 163-176.

HEATH, M. R., P. R. BOYLE, A. GISLASON, W. S. C. GURNEY, S. J. HAY, E. J. H. HEAD, S. HOLMES, A. INGVARSDÓTTIR, S. H. JÓNASDÓTTIR, P. LINDEQUE, R. T. POLLARD, J. RASMUSSEN, K. RICHARDS, K. RICHARDSON, G. SMERDON & D. SPEIRS, 2004. Comparative ecology of overwintering *Calanus finmarchicus* in the northern North Atlantic, and implications for life cycle patterns. Journal of Marine Science, **61**: 698-708.

HEATH, M. R., J. G. FRASER, A. GISLASON, S. J. HAY, S. H. JÓNASDÓTTIR & K. RICHARDSON, 2000. Winter distribution of *Calanus finmarchicus* in the Northeast Atlantic. Journal of Marine Science, **57**: 1628-1635.

HEATH, M. R. & S. H. JÓNASDÓTTIR, 1999. Distribution and abundance of overwintering *Calanus finmarchicus* in the Faroe-Shetland Channel. Fisheries Oceanography, **8** (Supplement 1): 40-60.

HECKMAN, C. W., 1986. The anadromous migration of a calanoid copepod, *Eurytemora affinis* (Poppe, 1880) in the Elbe estuary. Crustaceana, **50**: 176-181.

HEEGAARD, P. E., 1947. Contribution to the phylogeny of the Arthropods. Copepoda. Spolia Zoologica Musei Hauniensis, **8**: 1-236.

HELLE, K., 2000. Distribution of the copepodite stages of *Calanus finmarchicus* from Lofoten to the Barents Sea in July 1989. Journal of Marine Science, **57**: 1636-1644.

HENDERSON, J. T., 1926. Description of a copepod gill parasite of pike perches in lakes of northern Quebec, including an account of the free swimming male and some developmental stages. Contributions to Canadian Biology and Fisheries, (New Series) **3**: 237-245.

HERRICK, C. L., 1883. Types of animal life, selected for laboratory use in inland districts. Part I. Arthropoda: 1-33, 7 pls. (L. Kimball & Co., Minneapolis).

——, 1884. Final report on the Crustacea of Minnesota, included in the orders Cladocera and Copepoda, together with a synopsis of the described species in North America, and keys to the known species of the more important genera. Reports of the Geological and Natural History Survey of Minnesota, **12**(5): 1-192, pls. 1-29.

HERRING, P. J., 1988. Copepod luminescence. Hydrobiologia, **167-168**: 183-195.

HERRINGTON, W. C., H. M. BEARSE & F. E. FIRTH, 1939. Observations on the life history, occurrence, and distribution of the redfish parasite *Sphyrion lumpi*. Special Scientific Report of the United States Bureau of Fisheries, **5**: 1-18.

HESSE, E., 1858. Mémoire sur les moyens à l'aide desquels certains Crustacés parasites assurent la conservation de leurs espèces. Annales des Sciences Naturelles, (Zoologie et Biologie Animale) (4) **9**: 120-125.

——, 1877. Description des mâles, non encore connus, des *Lernanthropes* de Gisler et de Krøyer, ainsi que de la femelle d'une espèce nouvelle, dessinés et peints d'après des individus vivants. Revue des Sciences Naturelles, Montpellier, **6**: 252-260.

HEUCH, P. A., 1995. Experimental evidence for aggregation of salmon louse copepodids (*Lepeophtheirus salmonis*) in steep salinity gradients. Journal of the Marine Biological Association of the United Kingdom, **75**: 927-939.

HEUCH, P. A. & H. E. KARLSEN, 1997. Detection of infrasonic water oscillations by copepodids of *Lepeophtheirus salmonis* (Copepoda: Caligida). Journal of Plankton Research, **19**: 735-747.

HEUCH, P. A., A. PARSONS & K. BOXASPEN, 1995. Diel vertical migration: a possible host-finding mechanism in salmon louse (*Lepeophtheirus salmonis*) copepodids? Canadian Journal of Fisheries and Aquatic Sciences, **52**: 681-689.

HEWITT, G. C., 1964. The postchalimus development of *Lepeophtheirus polyprioni* Hewitt, 1963 (Copepoda: Caligidae). Transactions of the Royal Society of New Zealand, (Zoology) **4**: 157-159.

HIDALGO, P., R. ESCRIBANO & C. E. MORALES, 2005. Ontogenetic vertical distribution and diel migration of the copepod *Eucalanus inermis* in the oxygen minimum zone off northern Chile (20-21°S). Journal of Plankton Research, **27**: 519-529.

HIPEAU-JACQUOTTE, R., 1978. Relation entre âge de l'hôte et type de développement chez un Copépode ascidicole Notodelphyidae. Comptes Rendus de l'Académie des Sciences, Paris, (D) **287**: 1207-1210.

——, 1980. Le développement atypique du Copépode ascidicole Notodelphyidae *Pachypygus gibber* (Thorell, 1859). Archives de Zoologie Expérimentale et Générale, **121**: 29-47.

HIRAKAWA, K., 1974. Biology of a pelagic harpacticoid copepod, *Microsetella norvegica* Boeck in Oshoro Bay, Hokkaido I. Life history. Bulletin of the Plankton Society of Japan, **21**: 41-54.

HIRAKAWA, K. & A. IMAMURA, 1993. Seasonal abundance and life history of *Metridia pacifica* (Copepoda: Calanoida) in Toyama Bay, southern Japan Sea. Bulletin of the Plankton Society of Japan, **40**: 41-54.

HIRCHE, H. J., 1980. The cultivation of *Calanoides carinatus* Kroyer (Copepoda: Calanoida) under different temperature and food conditions - with a description of eggs and nauplii. Journal of the Marine Biological Association of the United Kingdom, **60**: 115-125.

——, 1996. The reproductive biology of the marine copepod, *Calanus finmarchicus*: a review. Ophelia, **44**: 111-128.

——, 1997. Life cycle of the copepod *Calanus hyperboreus* in the Greenland Sea. Marine Biology, Berlin, **128**: 607-618.

HIRCHE, H. J. & K. KOSOBOKOVA, 2003. Early reproduction and development of dominant calanoid copepods in the sea ice zone of the Barents Sea-need for a change of paradigms? Marine Biology, Berlin, **143**: 769-781.

HIRCHE, H. J. & S. KWASNIEWSKI, 1997. Distribution, reproduction and development of *Calanus* species in the Northeast Water in relation to environmental conditions. Journal of Marine Systems, **10** (Special Volume): 299-317.

HIRST, A. G. & M. SHEADER, 1997. Are in situ weight-specific growth rates body-size independent in marine planktonic copepods? A re-analysis of the global syntheses and a new empirical model. Marine Ecology Progress Series, **154**: 155-165.

HIRST, A. G. & R. S. LAMPITT, 1998. Towards a global model of in situ weight-specific growth in marine planktonic copepods. Marine Biology, Berlin, **132**: 247-257.

HJORT, J., 1895. Zur Anatomie und Entwicklungsgeschichte einer im Fleisch von Fischen schmarotzenden Crustacee (*Sarcotaces arcticum*, Collett). Skrifter udgivne av Videnskabsselskabet i Christiania, (Math.-Nat. Klasse) **1895** (2): 1-14.

HO, J.-S., 1966. Larval stages of *Cardiodectes* sp. (Caligoida: Lernaeoceriformes) a copepod parasitic on fishes. Bulletin of Marine Science, **16**: 159-199.

——, 1984. New family of poecilostomatoid copepods (Spiophanicolidae) parasitic on polychaetes from southern California, with a phylogenetic analysis of nereicoliform families. Journal of Crustacean Biology, **4**: 134-146.

——, 1987. Larval stages of *Ismaila occulta* Ho, 1981 and the affinity of Splanchnotrophidae (Copepoda, Poecilostomatoida). Researches on Crustacea, **16**: 67-83.

HO, J.-S. & W. WARDLE, 1992. *Pholadicola intestinalis*, new genus and species of clausidiid copepod parasitic in a deep-burrowing clam from Texas. Bulletin of Marine Science, **51**: 37-44.

HOCKLEY, A. R., 1951. On the biology of *Mytilicola intestinalis* (Steuer). Journal of the Marine Biological Association of the United Kingdom, **30**: 223-232.

HOEK, P. C., 1877. Zur Entwicklungsgeschichte der Entomostraken. II. Zur Embryologie der freilebenden Copepoden. Niederländisches Archiv für Zoologie, **4**: 55-74.

HOGANS, W. E. & D. J. TRUDEAU, 1989. Preliminary studies on the biology of sea lice, *Caligus elongatus*, *Caligus curtus* and *Lepeophtheirus salmonis* (Copepoda: Caligoida) parasitic on cage-cultured salmonids in the lower Bay of Fundy. Canadian Technical Report of Fisheries and Aquatic Sciences, **1715**: 1-14.

HOPCROFT, R. R., C. CLARKE, R. J. NELSON & K. A. RASKOFF, 2005. Zooplankton communities of the Arctic's Canada Basin: the contribution by smaller taxa. Polar Biology, **28**: 198-206.

HOPCROFT, R. R. & J. C. ROFF, 1998. Zooplankton growth rates: the influence of female size and resources on egg production of tropical marine copepods. Marine Biology, Berlin, **132**: 79-86.

—— & ——, 1998. Zooplankton growth rates: the influence of size in nauplii of tropical marine copepods. Marine Biology, Berlin, **132**: 87-96.

HOPP, U. & G. MAIER, 2005. Survival and development of five species of cyclopoid copepods in relation to food supply: experiments with algal food in a flow-through system. Freshwater Biology, **50**: 1454-1463.

HOPPENHEIT, M., 1976. Zur Dynamik exploitierter Populationen von *Tisbe holothuriae* (Copepoda, Harpacticoida). III. Reproduktion, Geschlechtsverhältnis, Entwicklungsdauer und Überlebenszeit. Helgoländer Wissenschaftliche Meeresuntersuchungen, **28**: 109-137.

HUANG, C., S.-I. UYE & T. ONBÉ, 1992. Ontogenetic diel vertical migration of the planktonic copepod *Calanus sinicus* in the Inland Sea of Japan. II. Late fall and early spring. Marine Biology, Berlin, **113**: 391-400.

——, —— & ——, 1993a. Geographic distribution, seasonal life cycle, biomass and production of a planktonic copepod *Calanus sinicus* in the Inland Sea of Japan and its neighboring Pacific Ocean. Journal of Plankton Research, **15**: 1229-1246.

——, —— & ——, 1993b. Ontogenetic diel vertical migration of the planktonic copepod *Calanus sinicus* in the inland Sea of Japan. 3. Early summer and overall seasonal pattern. Marine Biology, Berlin, **117**: 289-299.

HUIZINGA, H. W., 1971. Cultivation, life history and salinity tolerance of the tidepool copepod *Tigriopus californicus* Baker, 1912, in artificial sea water. Transactions of the Illinois State Academy of Science, **64**: 230-236.

HULSEMANN, K., 1965. A new genus and species of siphonostome cyclopoid copepod from deep North Atlantic waters. Crustaceana, **9**: 45-50.

HUMES, A. G., 1955. The postembryonic developmental stages of a fresh-water calanoid copepod, *Epischura massachusettsensis* Pearse. Journal of Morphology, **96**: 441-472.

——, 1960. The harpacticoid copepod *Sacodiscus* (= *Unicalteutha*) *ovalis* (C. B. Wilson, 1944) and its copepodid stages. Crustaceana, **1**: 279-294.

——, 1969. *Aspidomolgus stoichactinus* n. gen., n. sp. (Copepoda, Cyclopoida) associated with an actiniarian in the West Indies. Crustaceana, **16**: 225-242.

HUMES, A. G. & M. S. WILSON, 1951. The last copepodid instar of *Diaptomus sanguineus* Forbes (Copepoda). Journal of the Washington Academy of Sciences, **41**: 395-399.

HUNTLEY, M. E. & E. R. BROOKS, 1982. Effects of age and food availability on diel vertical migration of *Calanus pacificus*. Marine Biology, Berlin, **71**: 23-31.

HUNTLEY, M. E. & F. ESCRITOR, 1991. Dynamics of *Calanoides acutus* (Copepoda: Calanoida) in Antarctic coastal waters. Deep-Sea Research, (A) **38**: 1145-1167.

—— & ——, 1992. Ecology of *Metridia gerlachei* Giesbrecht in the western Bransfield Strait, Antarctica. Deep-Sea Research, (A) **39**: 1027-1055.

HUNTLEY, M. E., M. ZHOU & M. D. G. LOPEZ, 1994. *Calanoides acutus* in Gerlache Strait, Antarctica, 2. Solving an inverse problem in population dynamics. Deep-Sea Research, (II) **41**: 209-227.

HUTCHINSON, T. H., N. A. POUNDS, M. HAMPEL & T. D. WILLIAMS, 1999. Life-cycle studies with marine copepods (*Tisbe battagliai*) exposed to 20-hydroxyecdysone and diethylstilbestrol. Environmental Toxicology and Chemistry, **18**: 2914-2920.

HUYS, R., P.-J. LOPEZ-GONZALEZ, E. ROLDAN & A.-A. LUQUE, 2002. Brooding in cocculiniform limpets (Gastropoda) and familial distinctiveness of the Nucellicolidae (Copepoda): misconceptions reviewed from a chitonophilid perspective. Biological Journal of the Linnean Society, London, **75**: 187-217.

HWA, T. K., 1965. Studies on the life-history of a fish-louse (*Caligus orientalis* Gussev). Acta Zoologica Sinica, **17**: 48-63.

HYGUM, B. H., C. REY, B. W. HANSEN & F. CARLOTTI, 2000. Rearing cohorts of *Calanus finmarchicus* (Gunnerus) in mesocosms. Journal of Marine Science, **57**: 1740-1751.

IANORA, A., A. MIRALTO, S. A. POULET, Y. CAROTENUTO, I. BUTTINO, G. ROMANO, R. CASOTTI, G. POHNERT, T. WICHARD, L. COLUCCI-D'AMATO, G. TERRAZZANO & V. SMETACEK, 2004. Aldehyde suppression of copepod recruitment in blooms of a ubiquitous planktonic diatom. Nature, London, **429**: 403-407.

IANORA, A., B. SCOTTO DI CARLO, M. G. MAZZOCCHI & P. MASCELLARO, 1990. Histomorphological changes in the reproductive condition of parasitized marine planktonic copepods. Journal of Plankton Research, **12**: 249-258.

IGARASHI, S., 1963. Developmental cycle of *Tigriopus japonicus* Mori. Science Reports of the Tohoku University, (4, Biology) **29**: 59-72.

IKEDA, T. & K. HIRAKAWA, 1996. Early development and estimated life cycle of the mesopelagic copepod *Pareuchaeta elongata* in the southern Japan Sea. Marine Biology, Berlin, **126**: 261-270.

ILLG, P. & P. DUDLEY, 1980. The family Ascidicolidae and its subfamilies (Copepoda, Cyclopoida), with descriptions of new species. Mémoires du Muséum National d'Histoire Naturelle, (n. ser. A, Zoologie) **117**: 1-192.

INCZE, L. S. & T. AINAIRE, 1994. Distribution and abundance of copepod nauplii and other small (40-300 μm) zooplankton during spring in Shelikof Strait, Alaska. Fishery Bulletin, United States National Marine Fisheries Service, **92**: 67-78.

INFANTE, A. DE, 1981. Natural food of copepod larvae from Lake Valencia, Venezuela. Verhandlungen der Internationale Vereinigung für Theoretische und Angewandte Limnologie, **21**: 709-714.

INGVARSDÓTTIR, A., D. F. HOULIHAN, M. R. HEATH & S. J. HAY, 1999. Seasonal changes in respiration rates of copepodite stage V *Calanus finmarchicus* (Gunnerus). Fisheries Oceanography, **8** (Supplement 1): 73-83.

IRIGOIEN, X., 2000. Vertical distribution and population structure of *Calanus finmarchicus* at station India (59°N, 19°W) during the passage of the great salinity anomaly, 1971-1975. Deep-Sea Research, (I, Oceanographic Research Papers) **47**: 1-26.

IRIGOIEN, X. & R. P. HARRIS, 2003. Interannual variability of *Calanus helgolandicus* in the English Channel. Fisheries Oceanography, **12**: 317-326.

IRVINE, K., D. SNOOK & B. MOSS, 1995. Life histories of *Neomysis integer*, and its copepod prey, *Eurytemora affinis*, in a eutrophic and brackish shallow lake. Hydrobiologia, **304**: 59-76.

ISHIKAWA, A., S. BAN & N. SHIGA, 1999. Effects of salinity on survival, and embryonic and postembryonic development of *Eurytemora affinis* from a freshwater lake. Plankton Biology and Ecology, **46**: 113-119.

ITÔ, T., 1972. On the fifth copepodid stage of a harpacticoid copepod of the genus *Cletopsyllus* from Sagami Bay. Journal of the Faculty of Science, Hokkaido University, (6, Zoology) **18**: 406-410.

——, 1974. Descriptions and records of marine harpacticoid copepods from Hokkaido V. Journal of the Faculty of Science Hokkaido University, (6, Zoology) **19**: 546-640.

——, 1977. New species of marine harpacticoid copepods of the genera *Harpacticella* and *Tigriopus* from the Bonin Islands, with reference to the morphology of copepodid stages. Journal of the Faculty of Science Hokkaido University, (6, Zoology) **21**: 61-91.

——, 1984. A phylogenetic study of the family Harpacticidae (Harpacticoida): some problems in character differentiation processes through the copepodid stages. Crustaceana, (Supplement) **7**: 267-278.

ITÔ, T. & M. FUKUCHI, 1978. *Harpacticus furcatus* Lang from the Antarctic Peninsula with reference to the copepodid stages (Copepoda: Harpacticoida). Antarctic Research, **61**: 40-64.

ITOH, H., 2006. Parasitic and commensal copepods occurring as planktonic organisms with special reference to *Saphirella*-like copepods. Bulletin of the Plankton Society of Japan, **53**: 53-63. [In Japanese, with title and abstract in English.]

IVANENKO, V. N., 1998. Deep-sea hydrothermal vent Copepoda (Siphonostomatoida: Dirivultidae) in plankton over the Mid-Atlantic Ridge (29°N); morphology of their first copepodid stage. Zoologicheskii Zhurnal, **77**: 1249-1256. [In Russian with English abstract.]

IVANENKO, V. N., D. DEFAYE & R. HUYS, 2005. A new species of *Ambilimbus* nom. nov., a replacement name for *Amphicrossus* Huys, 1991 (Copepoda, Cyclopoida, Erebonasteridae), from the Mid-Atlantic Ridge at 36 degrees N (hydrothermal vent site Rainbow). Marine Biology Research, **1**: 233-241.

IZAWA, K., 1975b. A new and a known chondracanthid copepods parasitic on fishes from Tanabe Bay. Annotationes Zoologicae Japonenses, **48**: 108-118.

——, 1976a. Two new parasitic copepods (Cyclopoida: Myicolidae) from Japanese gastropod molluscs. Publications of the Seto Marine Biological Laboratory, **23**: 213-227.

——, 1976b. A new parasitic copepod, *Philoblenna arabici* gen. et sp. nov., from a Japanese gastropod, with proposal of a new family Philoblennidae (Cyclopoida: Poecilostoma). Publications of the Seto Marine Biological Laboratory, **23**: 229-235.

——, 2004. The copepodid and two chalimus stages of *Caligus latigenitalis* Shiino, 1954 (Copepoda, Siphonostomatoida, Caligidae), parasitic on Japanese black sea bream, *Acanthopagrus schlegeli*. Contributions to the Biological Laboratory, Kyoto University, **29**: 329-341.

JACOBS, M., 1925. Entwicklungsphysiologische Untersuchungen am Copepodenei (*Cyclops viridis* Jurine). Zeitschrift für Wissenschaftliche Zoologie, **124**: 487-541.

JACOBS, R. P. W. M. & A. M. J. BOUWHUIS, 1979. The year cycle of *Eudiaptomus vulgaris* (Schmeil, 1896) (Copepoda, Calanoida) in a small, acid water body during 1973, development in the natural habitat and relationships between temperature and duration of developmental stages. Hydrobiologia, **64**: 17-36.

JAMIESON, C. D., 1980. The predatory feeding of copepodid stages III to adult *Mesocyclops leuckarti* (Claus). In: W. C. KERFOOT (ed.), Evolution and ecology of zooplankton communities: 518-537. (American Society of Limnology and Oceanography, Hanover, New Hampshire).

——, 1980. Observations on the effect of diet and temperature on rate of development of *Mesocyclops leuckarti* (Claus) (Copepoda, Cyclopoida). Crustaceana, **38**: 145-154.

——, 1986a. A method for distinguishing the females and late copepodites of three species of *Boeckella*. New Zealand Journal of Marine and Freshwater Research, **20**: 153-156.

——, 1986b. The effects of temperature and food on naupliar development, growth and metabolism in three species of *Boeckella* (Copepoda: Calanoida). Hydrobiologia, **139**: 277-286.

JAMIESON, C. D. & C. W. BURNS, 1988. The effect of temperature and food on copepodite development, growth and reproduction in three species of *Boeckella* (Copepoda: Calanoida). Hydrobiologia, **164**: 235-257.

JAMIESON, C. D. & B. SANTER, 2003. Maternal aging in the univoltine freshwater copepod *Cyclops kolensis*: variation in egg sizes, egg development times, and naupliar development times. Hydrobiologia, **510**: 75-81.

JERLING, H. L. & T. H. WOOLDRIDGE, 1989. The developmental stages of *Pseudodiaptomus hessei* (Copepoda: Calanoida). South African Journal of Science, **24**: 139-145.

—— & ——, 1991. Population dynamics and estimates of production for the calanoid copepod *Pseudodiaptomus hessei* in a warm temperate estuary. Estuarine, Coastal and Shelf Science, **33**: 121-135.

JERSABEK, C. D., A. GOLDSCHMIDT & R. SCHABETSBERGER, 1996. Naupliar development of *Acanthodiaptomus denticornis* (Wierzejski, 1887) and *Arctodiaptomus alpinus* (Imhof,

1885) (Copepoda: Calanoida) and a comparison with other Diaptomidae. Journal of Plankton Research, **18**: 2027-2061.

JIANG, H. & G. A. PAFFENHÖFER, 2004. Relation of behavior of copepod juveniles to potential predation by omnivorous copepods: an empirical-modeling study. Marine Ecology Progress Series, **278**: 225-239.

JIMÉNEZ-MELERO, R., B. SANTER & F. GUERRERO, 2005. Embryonic and naupliar development of *Eudiaptomus gracilis* and *Eudiaptomus graciloides* at different temperatures: comments on individual variability. Journal of Plankton Research, **27**: 1175-1187.

JOHANNESSEN, A., 1978. Early stages of *Lepeophtheirus salmonis* (Copepoda, Caligidae). Sarsia, **63**: 169-176.

JOHNSON, C. L., 2003. Ecdysteroids in the oceanic copepod *Calanus pacificus*: variation during molt cycle and change associated with diapause. Marine Ecology Progress Series, **257**: 159-165.

JOHNSON, G. F., 1969. A contribution to the biology of *Corycaeus anglicus*, a marine cyclopoid copepod. (M.Sc. Thesis, University of California, San Diego).

JOHNSON, M. W., 1934a. The life history of the copepod *Tortanus discaudatus* (Thompson and Scott). Biological Bulletin, Woods Hole, **67**: 182-200.

——, 1934b. The developmental stages of the copepod *Epilabidocera amphitrites* McMurrich. Biological Bulletin, Woods Hole, **67**: 466-483.

——, 1935. The developmental stages of *Labidocera*. Biological Bulletin, Woods Hole, **68**: 397-421.

——, 1953. The copepod *Cyclops dimorphus* Kiefer from the Salton Sea. American Midland Naturalist, **49**: 188-192.

——, 1957. The copepod *Choniosphaera cancrorum* parasitizing a new host, the green crab *Carcinides maenas*. Journal of Parasitology, **43**: 470-473.

——, 1965. The nauplius larvae of *Pontellopsis occidentalis* Esterly (Copepoda, Calanoida). Transactions of the American Microscopical Society, **84**: 43-48.

——, 1966. The nauplius larvae of *Eurytemora herdmani* Thompson and Scott, 1897 (Copepoda, Calanoida). Crustaceana, **11**: 307-313.

JOHNSON, M. W. & J. OLSON, 1948. The life history and biology of a marine harpacticoid copepod, *Tisbe furcata* (Baird). Biological Bulletin, Woods Hole, **95**: 320-332.

JOHNSON, S. C., 1993. A comparison of development and growth rates of *Lepeophtheirus salmonis* (Copepoda: Caligidae) on native Atlantic (*Salmo salar*) and chinook (*Oncorhynchus tshawytscha*) salmon. In: G. A. BOXSHALL & D. DEFAYE (eds.), Pathogens of wild and farmed fish: sea lice: 68-80. (Ellis Horwood, London).

JOHNSON, S. C. & L. J. ALBRIGHT, 1991. The development stages of *Lepeophtheirus salmonis* (Krøyer, 1837) (Copepoda: Caligidae). Canadian Journal of Zoology, **69**: 929-950.

JONES, D. H. & J. B. L. MATTHEWS, 1968. On the development of *Sphyrion lumpi* (Krøyer). Crustaceana, (Supplement) **1**: 177-185.

JUDAY, C., 1904. The diurnal movement of plankton Crustacea. Transactions of the Wisconsin Academy of Sciences, Arts, and Letters, **14**: 534-568.

JUNGERSEN, H., 1911. On a new gymnoblastic hydroid (*Ichthyocodium sarcotretis*) epizoic on a new parasitic copepod (*Sarcotretes scopeli*) infesting *Scopelus glacialis* Rhdt. Videnskabelige Meddelelser fra Dansk Naturhistorisk Forening, **64**: 1-33, 2 pls.

KABATA, Z., 1958. *Lernaeocera obtusa* n. sp., its biology and its effects on the haddock. Marine Research, **1958**(3): 1-26.

——, 1959. Ecology of the genus *Acanthochondria* Oakley (Copepoda Parasitica). Journal of the Marine Biological Association of the United Kingdom, **38**: 249-261.

——, 1963. The free-swimming stage of *Lernaeenicus* (Copepoda, Parasitica). Crustaceana, **5**: 181-187.

——, 1964. On the adult and juvenile stages of *Vanbenedenia chimaerae* (Heegaard, 1962) (Copepoda: Lernaeopodiae) from Australian waters. Proceedings of the Linnean Society of New South Wales, **89**: 254-267.

——, 1966a. Copepoda parasitic on Australian fishes. V. Genus *Dissonus* (Diosonidae). Annals and Magazine of Natural History, (13) **9**: 211-226.

——, 1966b. The genus *Haemobaphes* (Copepoda: Lernaeoceridae) in the waters of British Columbia. Canadian Journal of Zoology, **45**: 853-875.

——, 1976. Early stages of some copepods (Crustacea) parasitic on marine fishes of British Columbia. Journal of the Fisheries Research Board of Canada, **33**: 2507-2525.

——, 1981. Copepoda (Crustacea) parasitic on fishes: problems and perspectives. Advances in Parasitology, **19**: 1-71.

KABATA, Z. P. & O. A. KHODOREVSKY, 1977. Kopepoditnaya stadiya *Dichelesthium oblongum* (Abildgaard, 1794) paraziticheskoi kopepody osetrovykh. [The copepodid stage of *Dichelesthium oblongum* (Abildgaard, 1794), a copepod parasite of acipenserid fishes.]. Parazitologiya, **11**: 236-240. [In Russian, with English summary.]

KAMAL, A. & K. ARMITAGE, 1967. External morphology of adult and copepodid stages of *Diaptomus clavipes* Schacht, 1897. The University of Kansas Science Bulletin, **47**: 559-573.

KANE, J., 2003. Spatial and temporal abundance patterns for the late stage copepodites of *Metridia lucens* (Copepoda: Calanoida) in the US northeast continental shelf ecosystem. Journal of Plankton Research, **25**: 151-167.

KARANAS, J. J., R. C. WORREST & N. VAN DIJKE, 1981. Impact of UV-B radiation on the fecundity of the copepod *Acartia clausii*. Marine Biology, Berlin, **65**: 125-134.

KASAHARA, S., 1962. Studies on the biology of the parasitic copepod *Lernaea cyprinacea* Linnaeus and the method for controling this parasite in fish-culture ponds. Contributions to the Fisheries Laboratory, Faculty of Agriculture, University of Tokyo, **3**: 103-196. [In Japanese, with English summary.]

KATONA, S. K., 1970. Growth characteristics of the copepods *Eurytemora affinis* and *E. herdmani* in laboratory cultures. Helgoländer Wissenschaftliche Meeresuntersuchungen, **20**: 373-384.

——, 1971. The developmental stages of *Eurytemora affinis* (Poppe, 1880) (Copepoda, Calanoida) raised in laboratory cultures, including a comparison with the larvae of *Eurytemora americana* Williams, 1906, and *Eurytemora herdmani* Thompson & Scott, 1897. Crustaceana, **21**: 5-20.

KATONA, S. K. & C. F. MOODIE, 1969. Breeding of *Pseudocalanus elongatus* in the laboratory. Journal of the Marine Biological Association of the United Kingdom, **49**: 743-747.

KATTNER, G., H. J. HIRCHE & M. KRAUSE, 1989. Spatial variability in lipid composition of calanoid copepods from Fram Strait, the Arctic. Marine Biology, Berlin, **102**: 473-480.

KAWABATA, K., 1987. Ecology of large phytoplankters in Lake Biwa: population dynamics and food relations with zooplankters. Bulletin of the Plankton Society of Japan, **34**: 165-172.

——, 1987. Abundance and distribution of *Eodiaptomus japonicus* (Copepoda: Calanoida) in Lake Biwa. Bulletin of the Plankton Society of Japan, **34**: 173-183.

——, 1989. Natural development time of *Eodiaptomus japonicus* (Copepoda: Calanoida) in Lake Biwa. Journal of Plankton Research, **11**: 1261-1272.

——, 1991. Ontogenetic changes in copepod behaviour: an ambush cyclopoid predator and a calanoid prey. Journal of Plankton Research, **13**: 27-34.

——, 1995. Ontogenetic niches of a planktonic copepod in Lake Biwa studied on a fine temporal scale. Ecological Research, **10**: 207-215.

——, 2006. Clearance rate of the cyclopoid copepod *Mesocyclops dissimilis* on the calanoid copepod *Eodiaptomus japonicus*. Plankton & Benthos Research, **1**: 68-70.

KAWABATA, K. & T. NARITA, 2003. Feeding rate of naupliar *Eodiaptomus japonicus*. Plankton Biology and Ecology, **50**: 27-29.

KAWATOW, K., K. MUROGA, K. IZAWA & S. KASAHARA, 1980. Life cycle of *Alella macrotrachelus* (Copepoda) parasitic on cultured black sea-bream. Journal of Faculty of Applied Science, Hiroshima University, **19**: 199-214. [In Japanese.]

KENDALL, A. W., JR., L. S. INCZE, P. B. ORTNER, S. R. CUMMINGS & P. K. BROWN, 1994. The vertical distribution of eggs and larvae of walleye pollock, *Theragra chalcogramma*, in Shelikof Strait, Gulf of Alaska. Fishery Bulletin, U.S. National Marine Fisheries Service, **92**: 540-554.

KESTEVEN, H. L., 1913. A new endoparasitic copepod: morphology and development. Proceedings of the Linnean Society of New South Wales, **37**: 673-688.

KHAN, M. F., 1965. The effect of constant and varying temperatures on the development of *Acanthocyclops viridis* (Jurine). Proceedings of the Royal Irish Academy. (B) **64**: 117-130.

KIEFER, F., 1973. Vergleichende Studien an Nauplien verschiedener Cyclopiden. Memorie dell'Istituto Italiano di Idrobiologia, **30**: 45-60.

KIM, I.-H., 1993. Developmental stages of *Caligus punctatus* Shiino, 1955 (Copepoda: Caligidae). In: G. A. BOXSHALL & D. DEFAYE (eds.), Pathogens of wild and farmed fish: sea lice: 16-29. (Ellis Horwood, London).

KIMOTO, K., 1988. Segregation of vertical distribution of calanoid copepod *Acartia omorii* depending on the developmental stages in Shijiki Bay, western Kyushu, Japan. Bulletin of the Seikai Regional Fisheries Research Laboratory, **66**: 35-39.

KIMOTO, K., S. I. UYE & T. ONBÉ, 1986. Growth characteristics of a brackish-water calanoid copepod *Sinocalanus tenellus* in relation to temperature and salinity. Bulletin of the Plankton Society of Japan, **33**: 43-57.

KIØRBOE, T. & M. SABATINI, 1995. Scaling of fecundity, growth and development in marine planktonic copepods. Marine Ecology Progress Series, **120**: 285-298.

KIØRBOE, T., E. SAIZ & A. W. VISSER, 1999. Hydrodynamic signal perception in the copepod *Acartia tonsa*. Marine Ecology Progress Series, **179**: 97-111.

KIRTISINGHE, P., 1950. Parasitic copepods of fish from Ceylon. III. Parasitology, **40**: 77-86.

KITAJIMA, C., 1973. Experimental trials on mass culture of copepods. Bulletin of the Plankton Society of Japan, **20**: 54-60. [In Japanese, with English summary.]

KLEIN BRETELER, W. C. M., 1982. The life stages of four pelagic copepods (Copepoda, Calanoida), illustrated by a series of photographs. Netherlands Institute of Sea Research, Publication Series, **6**: 1-32.

KLEIN BRETELER, W. C. M., H. G. FRANSZ & S. R. GONZALEZ, 1982. Growth and development of four calanoid copepod species under experimental and natural conditions. Netherlands Journal of Sea Research, **16**: 195-207.

KLEIN BRETELER, W. C. M. & S. R. GONZALEZ, 1986. Culture and development of *Temora longicornis* (Copepoda, Calanoida) at different conditions of temperature and food. Syllogeus, **58**: 71-84.

—— & ——, 1988. Influence of temperature and food concentration on body size, weight and lipid content of two calanoid copepod species. Hydrobiologia, **167/168**: 201-210.

KLEIN BRETELER, W. C. M., R. GONZALEZ & N. SCHOGT, 1995. Development of *Pseudocalanus elongatus* (Copepoda, Calanoida) cultured at different temperature and food conditions. Marine Ecology Progress Series, **119**: 99-110.

KLEIN BRETELER, W. C. M. & M. KOSKI, 2003. Development and grazing of *Temora longicornis* (Copepoda, Calanoida) nauplii during nutrient limited *Phaeocystis globosa* blooms in mesocosms. Hydrobiologia, **491**: 185-192.

KLEIN BRETELER, W. C. M. & N. SCHOGT, 1994. Development of *Acartia clausi* (Copepoda, Calanoida) cultured at different conditions of temperature and food. Hydrobiologia, **292/293**: 469-479.

KLEIN BRETELER, W. C. M., N. SCHOGT & S. R. GONZALEZ, 1990. On the role of food quality in grazing and development of life stages, and genetic change of body size during cultivation of pelagic copepods. Journal of Experimental Marine Biology and Ecology, **135**: 177-189.

KLEIN BRETELER, W. C. M., N. SCHOGT & J. VAN DER MEER, 1994. The duration of copepod life stages estimated from stage-frequency data. Journal of Plankton Research, **16**: 1039-1057.

KLEIN BRETELER, W. C. M., N. SCHOGT & S. RAMPEN, 2005. Effect of diatom nutrient limitation on copepod development: the role of essential lipids. Marine Ecology Progress Series, **291**: 125-133.

KLINTZ, J. H., 1907. Versuche über das geringe Regenerationsvermögen der Cyclopiden. Archiv für Entwicklungsmechanik der Organismen, **25**: 125-134.

KNUCKEY, R. M., G. L. SEMMENS, R. J. MAYER & M. A. RIMMER, 2005. Development of an optimal microalgal diet for the culture of the calanoid copepod *Acartia sinjiensis*: effect of algal species and feed concentration on copepod development. Aquaculture, **249**: 339-351.

KÔ, Y., 1961. The biology of the commensal copepods in the Japanese clam, *Tapes japonica* Deshayes. Records of Oceanographic Works in Japan, (Special Number) **5**: 129-141.

———, 1969a. On the reproduction and metamorphosis of a commensal copepod, *Ostrincola koe*, in the Japanese clam, *Tapes japonica* (preliminary note). Bulletin of the Faculty of Fisheries, Nagasaki University, **27**: 1-7. [In Japanese, with English summary.]

———, 1969b. The life-history of a commensal copepod, *Ostrincola koe*, in Japanese marine bivalves. Bulletin of the Faculty of Fisheries, Nagasaki University, **28**: 73-82. [In Japanese, with English summary.]

———, 1969c. On the occurrence of four commensal copepods in the Japanese clam, *Tapes japonica*. Bulletin of the Faculty of Fisheries, Nagasaki University, **28**: 83-91. [In Japanese, with English summary.]

KÔ, Y., Y. MURAKAMI & K. DIAKU, 1962. The biology of the commensal copepods in Japanese marine bivalves. Records of Oceanographic Works in Japan, (Special Number) **6**: 113-119.

KÔ, Y. & K. YOSHIKOSHI, 1974a. Laboratory rearing of *Ostrincola koe* (Copepoda, Cyclopoida), an associate of marine pelecypods, using the gills of a host clam as food. Bulletin of the Faculty of Fisheries, Nagasaki University, **38**: 95-100. [In Japanese, with English summary.]

——— & ———, 1974b. Consecutive breeding *Ostrincola koe* (Copepoda, Cyclopoida), an associate of marine pelecypods, through its over-wintering period using the gills of a host clam as food. Bulletin of the Faculty of Fisheries, Nagasaki University, **38**: 101-107. [In Japanese, with English summary.]

KÔ, Y., K. YOSHIKOSHI & N. ITÔ, 1974. External anatomy of a commensal copepod, *Ostrincola koe* - II. Nauplius stage. Bulletin of the Faculty of Fisheries, Nagasaki University, **38**: 87-93.

KOBARI, T. & S. BAN, 1998. Life cycles of two limnetic cyclopoid copepods, *Cyclops vicinus* and *Thermocyclops crassus*, in two different habitats. Journal of Plankton Research, **20**: 1073-1086.

KOBARI, T. & T. IKEDA, 1999. Vertical distribution, population structure and life cycle of *Neocalanus cristatus* (Crustacea: Copepoda) in the Oyashio region, with notes on its regional variations. Marine Biology, Berlin, **134**: 683-696.

—— & ——, 2001. Life cycle of *Neocalanus flemingeri* (Crustacea: Copepoda) in the Oyashio region, western Subarctic Pacific, with notes on its regional variations. Marine Ecology Progress Series, **209**: 243-255.

KOBARI, T., T. NAGAKI & K. TAKAHASHI, 2004. Seasonal changes in abundance and development of *Calanus pacificus* (Crustacea: Copepoda) in the Oyashio-Kuroshio Mixed Region. Marine Biology, Berlin, **144**: 713-722.

KOGA, F., 1960a. The developmental stages of nauplius larvae of *Pareuchaeta russelli* (Farran). Bulletin of the Japanese Society of Scientific Fisheries, **26**: 792-796.

——, 1960b. The nauplius larvae of *Centropages abdominalis* Sato. Bulletin of the Japanese Society of Scientific Fisheries, **26**: 877-881.

——, 1968. On the nauplius of *Undinula vulgaris* (Dana) (Copepoda, Calanoida). Journal of the Oceanographical Society of Japan, **24**: 173-177.

——, 1970a. On the life history of *Tigriopus japonicus* Mori (Copepoda). Journal of the Oceanographical Society of Japan, **26**: 11-21.

——, 1970b. On the nauplius of *Centropages yamadai* Mori, Copepoda. Journal of the Oceanographical Society of Japan, **26**: 195-202.

——, 1973. Life history of copepods especially of nauplius larvae ascertained mainly with cultivation of animals. Bulletin of the Plankton Society of Japan, **20**: 30-40.

——, 1984a. The developmental stages of *Temora turbinata* (Copepoda: Calanoida). Bulletin of the Plankton Society of Japan, **31**: 43-52.

——, 1984b. Morphology, ecology, classification and specialization of copepods nauplius. Bulletin of the Nansei Regional Fisheries Research Laboratory, **16**: 95-229.

KOLLAR, V., 1835. Beiträge zur Kenntniss der Lernaenartigen Crustaceen. Annalen des Wiener Museums, **1**: 81-90, 2 pls.

KOMAI, T., 1924. Notes on *Sarcotaces pacificus* n. sp. with remarks on its systematic position. Memoirs of the College of Science, Kyoto Imperial University, (B) **1**: 265-271, 1 pl.

KOSKI, M., W. [C. M.] KLEIN BRETELER & N. SCHOGT, 1998. Effect of food quality on rate of growth and development of the pelagic copepod *Pseudocalanus elongatus* (Copepoda, Calanoida). Marine Ecology Progress Series, **170**: 169-187.

KOTANI, Y., 2006. Lipid content and composition of dominant copepods in the Oyashio waters analyzed by the thin layer chromatography flame ionization detection method. Plankton and Benthos Research, **1**: 85-90.

KRAUSE, E. P. & A. G. LEWIS, 1979. Ontogenetic migration and the distribution of *Eucalanus bungii* (Copepoda; Calanoida) in British Columbia inlets. Canadian Journal of Zoology, **57**: 2211-2222.

KRAUSE, M. & G. RADACH, 1980. On the succession of developmental stages of herbivorous zooplankton in the northern North Sea during FLEX '76. 1. First statement about the main groups of the zooplankton community. Meteor Forschungsergebnisse, (D, Biologie) **22**: 133-149.

KRAUSE, M. & J. TRAHMS, 1982. Vertical distribution of copepods (all developmental stages) and other zooplankton during spring bloom in the Fladen Ground area of the North Sea. Netherlands Journal of Sea Research, **16**: 217-230.

KRISHNASWAMY, S., 1950. Larval stages of some copepods in the Madras plankton and their seasonal fluctuation. Journal of the Madras University, (B) **19**: 33-58.

——, 1951. Development of a harpacticoid copepod *Macrosetella gracilis* (Dana). Journal of the Madras University, (B) **21**: 256-271.

——, 1955. Life history of the psammophilous copepod *Leptastacus euryhalinus*. Journal of the Madras University, (B) **25**: 353-359.

KRØYER, H., 1837. Om Snyltekrebsene, især med Hensyn til den danske Fauna. [Concerning parasitic Crustacea with special reference to the Danish fauna.]. Naturhistorisk Tidsskrift, **1** (2): 172-208, 2 pls.; **1** (3): 252-304, pl. 3; **1** (5): 476-506, pl. 5; **1** (6): 605-628, pl. 6.

——, 1838. Om Snyltekrebsene, især med Hensyn til den danske Fauna. [Concerning parasitic Crustacea with special reference to the Danish fauna.]. Naturhistoriske Tidsskrift, **2** (1): 8-52, pl. 1; **2** (1): 131-57, pl. 3.

KUEI, H. Y. N. & T. K. S. BJÖRNBERG, 2002. Developmental stages of *Oncaea curta* Sars, 1916 (Copepoda, Poecilostomatoida). Nauplius, **10**: 1-14.

KUITUNEN-EKBAUM, E., 1949. The occurrence of *Sarcotaces* in Canada. Journal of the Fisheries Research Board of Canada, **7**: 505-512.

KUPERMAN, B. I. & R. E. SHUL'MAN, 1977. O vliyanii nekotorykh abioticheskikh faktorov na razvitie *Ergasilus sieboldi* (Crustacea, Copepoda). [On the influence of some abiotic factors on the development of *Ergasilus sieboldi* (Crustacea, Copepoda).]. Parazitologiya, **11**: 117-121.

KURIHARA, H., S. SHIMODE & Y. SHIRAYAMA, 2004. Effects of raised CO_2 concentration on the egg production rate and early development of two marine copepods (*Acartia steueri* and *Acartia erythraea*). Marine Pollution Bulletin, **49**: 721-727.

KURZ, W., 1877. *Eunicicola clausii*, ein neuer Anneliden parasit. Sitzungsberichten Akademie der Wissenschaften in Wien, (Mathematisch-Naturwissenschaftliche Klasse) (1) **75**: 21-28.

KUWAHARA, A. & S. SUZUKI, 1984. Diurnal changes in vertical distributions of anchovy eggs and larvae in the western Wakasa Bay. Bulletin of the Japanese Society of Scientific Fisheries, **50**: 1285-1292.

LAHAV, M., S. SARIG & M. SHILO, 1964. The eradication of *Lernaea* in storage ponds of carps through destruction of the copepodidal stage by Dipterex. Bamidgeh, Bulletin for Fish Culture in Israel, **16**: 87-94.

LAI, H. C. & J. C. H. CARTER, 1970. Life cycle of *Diaptomus oregonensis* Lilljeborg in Sunfish Lake, Ontario. Canadian Journal of Zoology, **48**: 1299-1302.

LANDRY, M. R., 1975. The relationship between temperature and the development of life stages of the marine copepod *Acartia clausi* Giesbr. Limnology and Oceanography, **20**: 854-857.

——, 1978b. Predatory feeding behavior of a marine copepod, *Labidocera trispinosa*. Limnology and Oceanography, **23**: 1103-1113.

——, 1983. The development of marine calanoid copepods with comment on the isochronal rule. Limnology and Oceanography, **28**: 614-624.

LANDRY, M. R., V. L. FAGERNESS & W. K. PETERSON, 1991. Ontogenetic patterns in the distribution of *Pseudocalanus* spp. during upwelling off the coast of Washington, USA. Bulletin of the Plankton Society of Japan, (Special Volume): 451-466.

LANG, K., 1948a. Monographie der Harpacticiden, **1-2**: 1-1682. (Hakan Ohlsson's Boktryckeri, Lund).

——, 1948b. Copepoda 'Notodelphyoida' from the Swedish west-coast with an outline on the systematics of the Copepoda. Arkiv för Zoologi, (A) **40** (14): 1-36.

——, 1949. On some Swedish marine semi-parasitic and parasitic copepods. Arkiv för Zoologi, (A) **42** (22): 1-10.

LAUBIER, L., 1961. *Phyllodicola petiti* (Delamare et Laubier, 1960) et la famille des Phyllodicolidae, Copépodes parasites d'Annélides polychètes en Méditerranée occidentale. Crustaceana, **2**: 228-242.

LAVERACK, M. S. & M. Q. HULL, 1993. Sensory innervation of the antenna of the pre-adult male *Caligus elongatus*. In: G. A. BOXSHALL & D. DEFAYE (eds.), Pathogens of wild and farmed fish: sea lice: 114-122. (Ellis Horwood, London).

LAWSON, T., 1973. The developmental stages of *Paracalanus crassirostris* Dahl, 1894 (Copepoda, Calanoida). Crustaceana, **24**: 43-56.

LAWSON, T. & G. D. GRICE, 1970. The developmental stages of *Centropages typicus* Kröyer (Copepoda, Calanoida). Crustaceana, **18**: 187-208.

LEBOUR, M. V., 1916. Stages in the life history of *Calanus finmarchicus* (Gunnerus), experimentally reared by Mr. L. R. Crawshay in the Plymouth Laboratory. Journal of the Marine Biological Association of the United Kingdom, **11**: 1-17.

LEE, C. E. & C. H. PETERSEN, 2002. Genotype-by-environment interaction for salinity tolerance in the freshwater-invading copepod *Eurytemora affinis*. Physiological and Biochemical Zoology, **75**: 335-344.

LEE, H. W., S. BAN, T. IKEDA & T. MATSUISHI, 2003. Effect of temperature on development, growth and reproduction in the marine copepod *Pseudocalanus newmani* at satiating food condition. Journal of Plankton Research, **25**: 261-271.

LEE, S. S., 1986. Morphological studies of developmental copepodid stages of calanoids in the southern coastal waters of Korea. Bulletin of Fisheries Research and Development Agency, **37**: 45-150.

LEIDY, J., 1889. A parasitic copepod. Proceedings of the Academy of Natural Sciences of Philadelphia, **1889**: 95.

LEIGH-SHARPE, W. H., 1918. *Clavella sciatherica,* n. sp., a parasitic copepod of *Gadus morrhua*. Parasitology, **11**: 118-126.

——, 1920. *Clavella iadda* n. sp. a parasitic copepod of *Gadus morrhua*, including some further remarks upon *C. sciatherica*, with a detailed account of the renal excretory system. Journal of the Marine Biological Association of the United Kingdom, (New Series) **12**: 332-338.

——, 1925. A revision of the British species of *Clavella* (Crustacea, Copepoda) with a diagnosis of new species: *C. devastatrix* and *C. invicta*. Parasitology, **17**: 194-200.

——, 1926a. *Nicothoe astaci* (Copepoda), with a revision of its appendages. Parasitology, **18**: 148-153.

——, 1926b. A list of parasitic Copepoda found at Plymouth, with a note on the bulla of *Clavella devastatrix*. Parasitology, **18**: 384-386.

——, 1933. A second list of parasitic Copepoda of Plymouth with a description of three new species. Parasitology, **25**: 113-118.

——, 1934. The Copepoda of the Siboga Expedition. Part II. Commensal and parasitic Copepoda. Siboga Expeditie, Monograph, **29b**: 1-43, figs. 1-39.

LEPEZ, I. M., 1974. Descripción de algunos estadios de nauplios y copepoditos de *Porcellidium rubrum* Pallares, 1966 (Copepoda, Harpacticoida). Boletin de la Sociedad de Biologia de Concepción, **48**: 265-272.

LESCHER-MOUTOUÉ, F., 1973. Sur la biologie et l'écologie des Copépodes Cyclopides hypogés (Crustacés). Annales de Spéléologie, **28**: 429-502.

——, 1974. Sur le nombre de stades naupliens chez *Eucyclops serrulatus*. Annales de Limnologie, **10**: 263-274.

LEVINSEN, G. M. R., 1878. Om nogle parasitiske Krebstyr, der snylte hos Annelider. Videnskabelige Meddelelser fra Dansk Naturhistoriske Forening, **1877**: 351-380, fig. 1, pl. 6.

LEWIS, A. G., L. CHATTERS & M. RAUDSEP, 1998. Feeding structures and their functions in adult and preadult *Tigriopus californicus* (Copepoda: Harpacticoida). Journal of the Marine Biological Association of the United Kingdom, **78**: 451-466.

LEWIS, A. G. & A. RAMNARINE, 1969. Some chemical factors affecting the early developmental stages of *Euchaeta japonica* (Crustacea: Copepoda: Calanoida) in the laboratory. Journal of the Fisheries Research Board of Canada, **26**: 1347-1362.

LEWIS, B. G., S. LUFF & J. W. WHITEHOUSE, 1971. Laboratory culture of *Cyclops abyssorum* Sars, 1863 (Copepoda, Cyclopoida). Crustaceana, **21**: 176-182.

LI, S. J. & J. C. FANG, 1983a. The developmental stages of *Temora turbinata* (Dana). Journal of the Xiamen University of Natural Sciences, **22**: 96-101. [In Chinese, with English summary.]

——, 1983b. The developmental stages of *Calanus sinicus* Brodsky in the Taiwan Strait. Journal of the Xiamen University of Natural Sciences, **2**: 110-118. [In Chinese, with English summary.]

——, 1983c. The developmental stages of *Labidocera euchaeta* Giesbrecht. Journal of the Xiamen University of Natural Sciences, **22**: 375-381. [In Chinese, with English summary.]

——, 1984. The developmental stages of *Calanopia thompsoni* A. Scott. Journal of Xiamen University of Natural Sciences, **23**: 391-397. [In Chinese, with English summary.]

——, 1990. Larval stages of marine planktonic copepods of China: 1-141. (China Ocean Press, Beijing). [In Chinese.]

LIANG, D. & S.-I. UYE, 1996a. Population dynamics and populations of the planktonic copepods in a eutrophic inlet of the Inland Sea of Japan. II. *Acartia omorii*. Marine Biology, Berlin, **125**: 109-117.

——, 1996b. Population dynamics and production of the planktonic copepods in a eutrophic inlet of the Inland Sea of Japan. III. *Paracalanus* sp. Marine Biology, Berlin, **127**: 219-227.

—— & ——, 1997. Population dynamics and production of the planktonic copepods in a eutrophic inlet of the Inland Sea of Japan. IV. *Pseudodiaptomus marinus*, the egg-carrying calanoid. Marine Biology, Berlin, **128**: 415-421.

LIANG, D., S.-I. UYE & T. ONBÉ, 1996. Population dynamics and production of the planktonic copepods in a eutrophic inlet of the Inland Sea of Japan. I. *Centropages abdominalis*. Marine Biology, Berlin, **124**: 527-536.

LILLJEBORG, W., 1902. Tres species novae generis Canthicampti, e Novaja Semlja et Sibiria Boreali, sive Trenne nya Arter of Slägtet *Canthocamptus* från Novaja Semlja och Norra Sibirien. Bihang till Kongliga Svenska Vetenskaps-Akademiens Handlingar, (4) **28** (9): 1-20, pls. 1-3.

LIN, C. L. & J. S. HO, 1993. Life history of *Caligus epidemicus* Hewitt, parasitic on the tilapia (*Oreochromis mossambica*) cultured in brackish water. In: G. A. BOXSHALL & D. DEFAYE (eds.), Pathogens of wild and farmed fish: sea lice: 5-15. (Ellis Horwood, London).

LIN, Y. S. & S. J. LI, 1984. A preliminary study of the life cycle of *Calanus sinicus* (Brodsky) in Xiamen harbour. Journal of the Xiamen University of Natural Sciences, **23**: 111-117. [In Chinese, with English summary.]

LINCOLN, R. & G. [A.] BOXSHALL, 1983. Deep-sea asellote isopods of the northeast Atlantic: the family Dendrotionidae and some new ectoparasitic copepods. Zoological Journal of the Linnean Society, London, **79**: 297-318.

LINDQUIST, A., 1959. Studien über das Zooplankton der Bottensee. I. Nauplien und Copepoditen von *Limnocalanus grimaldii* (de Guerne) (Copepoda, Calanoida). Reports of the Fisheries Research Board of Sweden, Institute of Marine Research Lysekil, (Biologie) **10**: 1-19.

LINE, R. Y., 1980. Some observations on the development cycle of *Temora longicornis* and *Centropages hamatus* in the Baltic Sea. Rybokhozyaistvennye Issledovaniya v Basseine Baltiiskogo Morya, **15**: 71-75.

LIST, J. H., 1890. Das Genus *Gastrodelphys*. Zeitschrift für Wissenschaftliche Zoologie, **49**: 71-146.

LONGHURST, A. R., 1986. Instar increments in copepod growth. Canadian Journal of Fisheries and Aquatic Sciences, **43**: 1671-1674.

LOPEZ, G., 1976. Redescription and ontogeny of *Lepeophtheirus kareii* Yamaguti, 1936 (Copepoda, Caligoida). Crustaceana, **31**: 203-207.

LOPEZ, G. W., 1980. Description of the larval stages of *Tisbe cucumariae* (Copepoda: Harpacticoida) and comparative development within the genus *Tisbe*. Marine Biology, Berlin, **57**: 61-71.

— —, 1982. Short-term population dynamics of *Tisbe cucumariae* (Copepoda: Harpacticoida). Marine Biology, Berlin, **68**: 333-341.

LOPEZ, M. D. G., 1991. Molting and mortality depend on age and stage in naupliar *Calanus pacificus*: implication for development time of field cohorts. Marine Ecology Progress Series, **75**: 79-89.

— —, 1996. Effect of starvation on development and survivorship of naupliar *Calanus pacificus* (Brodsky). Journal of Experimental Marine Biology and Ecology, **203**: 133-146.

LOPEZ, M. D. G. & M. E. HUNTLEY, 1995. Feeding and diel vertical migration cycles of *Metridia gerlachei* (Giesbrecht) in coastal waters of the Antarctic Peninsula. Polar Biology, **15**: 21-30.

LÓPEZ-GONZÁLEZ, P. J., J. BRESCIANI & M. CONRADI, 1998. *Tarificola bulbosus*, new genus, new species, a highly transformed parasitic copepod, with information on its parasitism and larval development. Journal of Crustacean Biology, **18**: 581-589.

LORENZEN, S., 1969. Harpacticoiden aus dem lenitischen Watt und den Salzwiesen der Nordseeküste. Kieler Meeresforschungen, **25**: 215-223.

LOTUFO, G. R. & J. W. FLEEGER, 1997. Effects of sediment-associated phenanthrene on survival, development and reproduction of two species of meiobenthic copepods. Marine Ecology Progress Series, **151**: 91-102.

LOVEGROVE, T., 1956. Copepod nauplii, II. Fiches d'Identification du Zooplancton, **63**: 1-4. (Conseil Permanent International pour l'Exploration de la Mer, Charlottenlund Slot).

LOWNDES, A. G., 1928. The results of breeding experiments and other observations on *Cyclops vernalis* Fischer and *Cyclops robustus* G.O. Sars. Internationale Revue der Gesamten Hydrobiologie und Hydrographie, **21**: 171-188.

— —, 1932a. The results of breeding experiments on the genus *Leptocyclops* G.O. Sars, with some general notes on the results of culture experiments. Annals and Magazine of Natural History, **10**: 45-80.

——, 1932b. The results of further breeding experiments on four species of *Cyclops*. Annals and Magazine of Natural History, **10**: 265-297.

LUČIĆ, D., J. NJIRE, M. MOROVIČ, R. PRECALI, D. FUKS & J. BOLOTIN, 2003. Microzooplankton in the open waters of the northern Adriatic Sea from 1990 to 1993: the importance of copepod nauplii densities. Helgoland Marine Research, **57**: 73-81.

LUCKS, R., 1929. *Cyclops phaleratus* Koch. Ein Beitrag zu seiner Entwicklungsgeschichte. Berichte des Westpreussischen Botanisch-Zoologischen Vereins, Danzig, **51**: 9-33.

LUGER, M. S., R. SCHABETSBURGER, C. D. JERSABEK & A. GOLDSCHMIDT, 2000. Life cycles, size and reproduction of the two coexisting calanoid copepods *Arctodiaptomus alpinus* (Imhof, 1885) and *Mixodiaptomus laciniatus* (Lilljeborg, 1889) in a small high-altitude lake. Archiv für Hydrobiologie, **148**: 161-185.

LÜTZEN, J., 1964. A revision of the family Herpyllobiidae (parasitic copepods) with notes on hosts and distribution. Ophelia, **1**: 241-274.

——, 1968b. *Heliograbalus phascolia,* sp.n., an ectoparasitic copepod of *Phascolion strombi* (Montagu) of the coast of New England. Videnskabelige Meddelelser fra Dansk Naturhistorisk Forening, **131**: 209-216.

MACKINNON, B. M., 1993. Response of the copepodid larvae of *Caligus elongatus* to light, and the ultrastructure of the eyes. Canadian Journal of Fisheries and Aquatic Sciences, **50**: 793-799.

MACLELLAN, D. & C. SHIH, 1974. Description of copepodite stages of *Chiridius gracilis* Farran, 1908 (Crustacea, Copepoda). Journal of the Fisheries Research Board of Canada, **31**: 1337-1349.

MACKAS, D. L., R. GOLDBLATT & A. G. LEWIS, 1998. Interdecadal variation in developmental timing of *Neocalanus plumchrus* populations at Ocean Station P in the Subarctic North Pacific. Canadian Journal of Fisheries and Aquatic Sciences, **55**: 1878-1893.

MACKAS, D. L., H. SEFTON, C. B. MILLER & A. RAICH, 1993. Vertical habitat partitioning by large calanoid copepods in the oceanic Subarctic Pacific during spring. Progress in Oceanography, **32**: 259-294.

MACKAS, D. L. & A. TSUDA, 1999. Mesozooplankton in the eastern and western Subarctic Pacific: community structure, seasonal life histories, and interannual variablility. Progress in Oceanography, **43**: 335-363.

MAGADZA, C. H. D. & P. Z. MUKWENA, 1979. Determination of the post-embryonic development period in *Thermocyclops neglectus* (Sars) using cohort analysis in batch cultures. Transactions of the Rhodesia Scientific Association, **59**: 41-45.

MCCLENDON, J. F., 1906. On the development of the parasitic copepods. Part 1. Biological Bulletin, Woods Hole, **12**: 37-52.

——, 1907. On the development of parasitic copepods. Part II. Biological Bulletin, Woods Hole, **12**: 57-88.

MCLAREN, I. A., 1965. Some relationships between temperature and egg size, body size, development rate and fecundity, of the copepod *Pseudocalanus*. Limnology and Oceanography, **10**: 528-538.

MCLAREN, I. A. & C. J. CORKETT, 1978. Unusual genetic variation of body size, development times, oil storage, and survivalship in the marine copepod *Pseudocalanus*. Biological Bulletin, Woods Hole, **155**: 347-359.

—— & ——, 1981. Temperature-dependent growth and production by a marine copepod. Canadian Journal of Fisheries and Aquatic Sciences, **38**: 77-83.

MCLAREN, I. A., E. LABERGE, C. J. CORKETT & J. M. SÉVIGNY, 1989. Life cycles of four species of *Pseudocalanus* in Nova Scotia. Canadian Journal of Zoology, **67**: 552-558.

MADSEN, S. D., T. G. NIELSEN & B. W. HANSEN, 2001. Annual population development and production by *Calanus finmarchicus*, *C. glacialis* and *C. hyperboreus* in Disko Bay, western Greenland. Marine Biology, Berlin, **139**: 75-93.

MAIER, G., 1989a. The effect of temperature on the development times of eggs, naupliar and copepodite stages of five species of cyclopoid copepods. Hydrobiologia, **184**: 79-88.

——, 1989b. The seasonal cycle of *Metacyclops gracilis* (Lilljeborg) in a shallow pond. Archiv für Hydrobiologie, **115**: 97-110.

——, 1989c. The seasonal cycle of *Thermocyclops crassus* (Fischer, 1853) (Copepoda: Cyclopoida) in a shallow, euthrophic lake. Hydrobiologia, **178**: 43-58.

——, 1990a. Reproduction and life history of a Central European planktonic *Diacyclops bicuspidatus* population. Archiv für Hydrobiologie, **117**: 485-495.

——, 1990b. The life history of two copepods with special reference to *Eudiaptomus vulgaris* Schmeil, 1898. Crustaceana, **59**: 204-212.

——, 1992. *Metacyclops minutus* (Claus, 1863), population dynamics and life history characteristics of a rapidly developing copepod. Internationale Revue der Gesamten Hydrobiologie, **77**: 455-466.

MAKINO, W. & S. BAN, 2000. Response of life history traits to food conditions in a cyclopoid copepod from an oligotrophic environment. Limnology and Oceanography, **45**: 396-407.

MAKINO, W. & S. BAN, 2003. Fecal pellet production between molts in a cyclopoid copepod: patterns, individual variability and implications for growth and development. Hydrobiologia, **501**: 101-107.

MALAQUIN, A., 1896. Parasitisme et évolution de deux Monstrillides (*Thaumaleus filogranarum* n. sp., *Haemocera* n.g., *Danae* Clapd.) a l'intérieur du système vasculaire des Filogranes et des Salmacynes. Comptes Rendus Hebdomadaires des Séances de l'Académie des Sciences, **123**: 1316-1318.

——, 1897. Evolution des Monstrillides (*Haemocera* n.g., *danae* Clpd. et *Haemocera filogranarum* n. sp.). Comptes Rendus Hebdomadaires des Séances de l'Académie des Sciences, **124**: 99-102.

——, 1900. Nouvelles recherches sur l'évolution des Monstrillides. Comptes Rendus Hebdomadaires des Séances de l'Académie des Sciences, **130**: 427-430.

——, 1901. Le parasitisme évolutif des Monstrillides (Crustacés Copépodes). Archives de Zoologie Expérimentale et Générale, (3) **9**: 81-232, figs. 1-8, pls. 2-8.

MALONEY, M. T. & W. L. TRESSLER, 1942. The diurnal migration of certain species of zooplankton in Caroga Lake, New York. Transactions of the American Microscopical Society, **61**: 40-52.

MANFREDI, P., 1923. Etude sur le développement larvaire de quelques espèces du genre *Cyclops*. Annales de Biologie Lacustre, Bruxelles, **12**: 272-303.

——, 1925. Etude sur le développement larvaire de quelques espèces du genre *Cyclops* (2me note). Annales de Biologie Lacustre, Bruxelles, **14**: 111-129.

——, 1930. Contributa alla conoscenza dello sviluppo larvale di *Canthocamptus trispinosus* (Copepodi Harpacticidi). Bollettino di Pesca, Piscicoltura e Idrobiologia, **6**: 232-235.

MARCIAL, H. S., A. HAGIWARA & T. W. SNELL, 2003. Estrogenic compounds affect development of harpacticoid copepod *Tigriopus japonicus*. Environmental Toxicology and Chemistry, **22**: 3025-3030.

MARGULIES, D., J. B. WEXLER, K. T. BENTLER, J. M. SUTER, S. MASUMA, N. TEZUKA, K. TERUYA, M. OKA, M. KANEMATSU & H. NIKAIDO, 2001. Food selection of yellowfin tuna, *Thunnus albacares*, larvae reared in the laboratory. Inter American Tropical Tuna Commission Bulletin, **22**: 9-51.

MARIN, V., 1987. The oceanographic structure of the eastern Scotia Sea – IV. Distribution of copepod species in relation to hydrography in 1981. Deep-Sea Research, **34**: 105-121.

——, 1988. Independent life cycles: an alternative to the asynchronism hypothesis for Antarctic calanoid copepods. Hydrobiologia, **167/168**: 161-168.

——, 1988. Qualitative models of the life cycles of *Calanoides acutus*, *Calanus propinquus*, and *Rhincalanus gigas*. Polar Biology, **8**: 439-446.

——, 1997. A simple-biology, stage-structured population model of the spring dynamics of *Calanus chilensis* at Mejillones del Sur Bay, Chile. Ecological Modelling, **105**: 65-82.

MARKEVICH, A. P., 1936. *Cardiodectes hardenbergi*, ein neuer parasitischer Copepode aus der Java See. Treubia, Bogor, **15**: 407-411.

——, 1939. Evolyutsiya kinshchivok v ontogenezi Lernaeopodidae. [Evolution of the appendages in the ontogeny of Lernaeopodidae.]. Trudy Institutu Zoolohiyi ta Biologiyi, Kyyiv [Kiev], (Biologicheskoi) **5**: 147-161.

MARKHASEVA, E., 1980. Calanoida of the genus *Jaschnovia*, nom. n. (*Derjuginia* Jaschnov, nom. praeocc.) (Calanoida, Aetideidae). Explorations of the Fauna of the Seas, **25**: 63-76. [In Russian, with English summary.]

MARSH, C. D., 1900. On some points in the structure of the larva of *Epischura lacustris* Forbes. Transactions of the Wisconsin Academy of Sciences, Arts, and Letters, **12**: 544-549.

——, 1933. Synopsis of the calanoid crustaceans, exclusive of the Diaptomidae, found in fresh and brackish waters, chiefly of North America. Proceedings of the United States National Museum, **82** (2959): 1-58, pls. 1-24.

MARSHALL, S. M. & A. P. ORR, 1956. On the biology of *Calanus finmarchicus*. IX. Feeding and digestion in the young stages. Journal of the Marine Biological Association of the United Kingdom, **35**: 587-603.

—— & ——, 1972. The biology of a marine copepod: i-viii, 1-195. (The 1955 ed. reprinted by Springer-Verlag, New York).

MARTINS, M. L. & F. L. SOUZA, JR., 1996. Infestação experimental em girinos de *Rana catesbeiana* Shaw por copepoditos de *Lernaea cyprinacea* Linnaeus (Copepoda, Lernaeidae). [Experimental infestation of *Rana catesbeiana* Shaw tadpoles by copepodids of *Lernaea cyprinacea* Linnaeus (Copepoda, Lernaeidae.).] Revista Brasileira de Zoologia, **12**: 619-625. [In Portuguese, with English abstract.]

MATIAS-PERALTA, H., F. M. YUSOFF, M. SHARIFF & A. ARSHAD, 2005. Effects of some environmental parameters on the reproduction and development of a tropical marine harpacticoid copepod *Nitocra affinis* f. *californica* Lang. Marine Pollution Bulletin, **51**: 722-728.

MATSUDAIRA, C., 1957. Culturing of a Copepoda, *Sinocalanus tenellus*. Information Bulletin on Planktology in Japan, **5**: 1-6. [In Japanese.]

MAUCHLINE, J. & T. NEMOTO, 1977. The occurrence of integumental organs in copepodid stages of calanoid copepods. Bulletin of the Plankton Society of Japan, **24**: 108-114.

MAZEPOVA, G. F., 1960. Morfologiya stadii metamorfoza u *Cyclops kolensis* iz ozera Baikal. Izvestiya Sibirskogo Otdeleniya Akademii Nauk SSSR, (Biologicheskikh Nauk) **6**: 103-115.

MAZZA, J., 1964. Le développement de quelques Copépodes en Méditerranée. I. Les stades jeunes d'*Euchaeta acuta* Giesbrecht et d'*E. spinosa* Giesbrecht. Revue et Travaux de l'Institut des Pêches Maritimes, **28**: 271-292.

——, 1965. Le développement de quelques Copépodes en Méditerranée. II. Les stades jeunes de *Gaetanus kruppi* Giesb., *Euchirella messinensis* Cl., *Chiridius poppei* Gies.,

Pseudaetideus armatus (Boeck) et *Heterorhabdus spinifrons* Cl. Revue des Travaux de l'Institut des Pêches Maritimes, **29**: 285-320.

McKinnon, A. D. & G. Arnott, 1985. The developmental stages of *Gladioferens pectinatus* (Brady, 1899) (Copepoda: Calanoida). New Zealand Journal of Marine and Freshwater Research, **19**. 21-42.

McLaren, I. A., 1964. Zooplankton of Lake Hazen, Ellesmere Island, and a nearby pond, with special reference to the copepod *Cyclops scutifer* Sars. Canadian Journal of Zoology, **42**: 613-629.

McLaren, I. A., E. Head & D. D. Sameoto, 2001. Life cycles and seasonal distributions of *Calanus finmarchicus* on the central Scotian Shelf. Canadian Journal of Fisheries and Aquatic Sciences, **58**: 659-670.

McLaren, I. A., E. Laberge, C. J. Corkett & J. M. Sevigny, 1989. Life cycles of four species of *Pseudocalanus* in Nova Scotia. Canadian Journal of Zoology, **67**: 552-558.

Meeüs, T. De, A. Raibaut & F. Renaud, 1993. Comparative life history of two species of sea lice. In: G. A. Boxshall & D. Defaye (eds.), Pathogens of wild and farmed fish: sea lice: 61-67. (Ellis Horwood, London).

Meise, C. J. & J. E. O'Reilly, 1996. Spatial and seasonal patterns in abundance and age-composition of *Calanus finmarchicus* in the Gulf of Maine and on Georges Bank: 1977-1987. Deep-Sea Research, (II) **43**: 1473-1502.

Melle, W. & H. R. Skjøldal, 1998. Reproduction and development of *Calanus finmarchicus*, *C. glacialis* and *C. hyperboreus* in the Barents Sea. Marine Ecology Progress Series, **169**: 211-228.

Meskova, T. M., 1952. Zooplankton ozera Sevan. Trudy Sevanskoj Gidrobiologicheskoi Stantsii, **13**: 5-170.

Metzger, A., 1868. Über das Männchen und Weibchen der Gattung *Lernaea* vor dem Eintritt der sogen. Rückschreitenden Metamorphose. Nachrichten von der Königlichen Gesellschaft der Wissenschaft, Göttingen, **2**: 31-36.

Michaud, M., F. Thomas, S. Becheikh, A. Raibaut, J. A. Shykoff & F. Renaud, 1999. Stage-dependent decisions in a parasitic copepod practising environmental sex determination. Marine Ecology Progress Series, **185**: 189-193.

Mielke, W., 1976. Ökologie der Copepoda eines Sandstrandes der Nordseeinsel Sylt. Mikrofauna des Meeresbodens, **59**: 1-86.

Miller, C. B., 1993. Development of large copepods during spring in the Gulf of Alaska. Progress in Oceanography, **32**: 295-317.

Miller, C. B. & M. J. Clemons, 1988. Revised life history analysis for large grazing copepods in the Subarctic Pacific Ocean. Progress in Oceanography, **20**: 293-313.

Miller, C. B., B. W. Frost, H. P. Batchelder, M. J. Clemons & R. E. Conway, 1984. Life histories of large, grazing copepods in a Subarctic ocean gyre: *Neocalanus plumchrus*, *Neocalanus cristatus* and *Eucalanus bungii* in the northeast Pacific. Progress in Oceanography, **13**: 201-243.

Miller, C. B., J. K. Johnson & D. R. Heinle, 1977. Growth rules in the marine copepod genus *Acartia*. Limnology and Oceanography, **22**: 326-335.

Miller, C. B. & R. D. Nielsen, 1988. Development and growth of large, calanid copepods in the Subarctic Pacific Ocean, May 1984. Progress in Oceanography, **20**: 275-292.

Miller, C. B. & K. S. Tande, 1993. Stage duration estimation for *Calanus* populations, a modelling study. Marine Ecology Progress Series, **102**: 15-34.

Miller, C. B. & M. Terazaki, 1989. The life histories of *Neocalanus flemingeri* and *Neocalanus plumchrus* in the Sea of Japan. Bulletin of the Plankton Society of Japan, **36**: 27-41.

MILSTEIN, A., 1979. Vertical distribution of *Paracalanus crassirostris* (Copepoda, Calanoida): analysis by the general linear model. Boletim do Instituto Oceanográfico, **28**: 65-78.

MIRZO'EVA, L. M., 1969. Nauplial'ni stadii *Sinergasilus lieni* Yin, 1949 (Crustacea, Copepoda parasitica). [Nauplius stages of *Sinergasilus lieni* Yin, 1949 (Crustacea: Copepoda parasitica).]. Dopovidi Akademiyi Nauk Ukrayins'koiy RSR, Kiev, (B) **1969**: 753-755. [In Ukrainian, with English summary.]

——, 1972. Zhiznennyi tsikl i biologiya *Sinergasilus lieni* Yin, 1949 (Crustacea, Copepoda parasitica). [The life cycle and biology of *Sinergasilus lieni* Yin, 1949 (Crustacea, Copepoda parasitica).]. Parazitologiya, **6**: 252-258. [In Russian, with English summary.]

——, 1973. Zhinennyi tsikl i morfologiya *Sinergasilus lieni*, Yin, 1949 (Copepoda, Parasitica). [Life cycle and morphology of *Sinergasilus lieni* Yin, 1949 (Copepoda parasitica).]. Trudy Vsesoyuznogo Nauchno-issledovatel'skogo Instituta Prudovogo Rybnogo Khozyaistva, **22**: 143-158. [In Russian.]

MITTELHOLZER, E., 1970. Populationsdynamik und Produktion des Zooplanktons im Greifensee und im Vierwaldstättersee. Schweizerische Zeitschrift für Hydrologie, **32**: 90-149.

MONOD, T., 1923. Notes carcinologiques (parasites et commensaux). Bulletin de l'Institut Océanographique, Monaco, **427**: 1-23.

MONOD, T. & R. P. DOLLFUS, 1934. Des Copépodes parasites de mollusques (deuxième supplément). Annales de Parasitologie Humaine et Comparée, **12**: 309-321.

—— & ——, 1938. Pandarinés peu connus (genres *Phyllothyreus* Norman, 1903 et *Gangliopus* Gerstaecker, 1854). Annales de Parasitologie Humaine et Comparée, **16**: 196-209.

MORRIS, J. T. & B. C. COULL, 1992. Population dynamics, numerical production, and potential predation impact on a meiobenthic copepod. Canadian Journal of Fisheries and Aquatic Sciences, **49**: 609-616.

MÜLLER, F., 1852. Eine Beobachtung uber die Beziehung der Gattungen *Caligus* und *Chalimus*. Archiv für Naturgeschichte, **18**: 91-92.

——, 1864. Für Darwin: 1-91. (Wilhelm Engelmann Verlag, Leipzig).

MULLIN, M. M., 1997. The demography of *Calanus pacificus* during winter-spring Californian El Niño conditions, 1992-1993: implications for anchovy? Fisheries Oceanography, **6**: 10-18.

MULLIN, M. M. & E. R. BROOKS, 1967. Laboratory culture, growth rate, and feeding behaviour of a planctonic marine copepod. Limnology and Oceanography, **12**: 657-666.

—— & ——, 1970. Production of the planktonic copepod, *Calanus helgolandicus*. Bulletin of Scripps Institution of Oceanography, (Technical Series) **17**: 89-103.

—— & ——, 1972. The vertical distribution of juvenile *Calanus* (Copepoda) and phytoplankton within the upper 50 m of water off La Jolla, California. In: A. Y. TAKENOUTI (ed.), Biological oceanography of the northern north Pacific Ocean: 347-354. (Idemitsu Shoten, Tokyo).

MURDOCH, R. C., 1990. Diet of hoki larvae *Macruronus novaezelandiae* off Westland New Zealand. New Zealand Journal of Marine and Freshwater Research, **24**: 519-528.

MURPHY, H. E., 1923. The life cycle of *Oithona nana*, reared experimentally. University of California Publications in Zoology, **22**: 449-454.

MURNANE, J., 1969. Postembryonic development and systematics of siphonostome copepods (Cyclopoida) associated with sponges from the northeastern coast of the United States: 1-260. (Ph.D. Thesis, Boston University, Boston).

MUSSELIUS, V. A., 1967. K biologii i spetsifichnosti *Sinergasilus lieni* Yin, 1949 (Crustacea, Copepoda). [On the biology and specificity of *Sinergasilus lieni* Yin.]. Parazitologiya, **1**: 158-160. [In Russian, with English summary.]

NÆSS, T. & J. P. NILSSEN, 1991. Life cycle dynamics of a *Cyclops strenuus* (Crustacea, Copepoda) population with unusual diapause and reproductive characteristics. Archiv für Hydrobiologie, **122**: 323-334.

NAGARAJ, M., 1988. Combined effects of temperature and salinity in the complete development of *Eurytemora velox* (Crustacea: Calanoidea). Marine Biology, Berlin, **99**: 353-358.

NAGASAWA, T., 1999. Early life history of yoroi-mebaru, *Sebastes hubbsi*, in the Sado Strait, Sea of Japan. Bulletin of the Japan Sea National Fisheries Research Institute, **49**: 97-108.

NAKAI, N., 1927. On the development of a parasitic copepod, *Lernaea elegans* Leigh-Sharpe, infesting on *Cyprinis carpio* L. Journal of the Imperial Fisheries Institute, **29**: 39-58, 3 pls.

NAKAI, N. & E. KOKAI, 1931. On the biological study of a parasitic copepod, *Lernaea elegans* Leigh-Sharpe, infesting on Japanese fresh-water fishes. Journal of the Imperial Fisheries Experimental Station, **2**: 93-128.

NAKATANI, T., 1991. Copepod nauplii as food of walleye pollock larvae (Pisces: Gadidae) in Funka Bay, Japan. Bulletin of the Plankton Society of Japan, (Special Volume): 515-525.

——, 1995. Monthly changes in density and size structure of copepod nauplii as the primary food for fish larvae in winter in Funka Bay and surrounding vicinity in Hokkaido. Fisheries Science, **61**: 382-386.

NAKATANI, T. & T. MAEDA, 1993. Early life history of walleye pollock. Scientific Reports of Hokkaido Fisheries Experimental Station, **42**: 15-22.

NASSOGNE, A., 1970. Influence of food organisms on the development and culture of pelagic copepods. Helgoländer Wissenschaftliche Meeresuntersuchungen, **20**: 333-345.

NICHOLLS, A. G., 1934. The developmental stages of *Euchaeta norvegica* Boeck. Proceedings of the Royal Society of Edinburgh, **54**: 31-50.

——, 1935. The larval stages of *Longipedia coronata* Claus, *L. scotti* G. O. Sars and *L. minor* T. & A. Scott, with a description of the male of *L. scotti*. Journal of the Marine Biological Association of the United Kingdom, (n. ser.) **20**: 29-45.

——, 1941. The developmental stages of *Metis jousseaumei* (Richard) (Copepoda, Harpacticoida). Annals and Magazine of Natural History, (11) **7**: 317-328.

NILSSEN, J. P. & K. ELGMORK, 1977. *Cyclops abyssorum* - life cycle dynamics and habitat selection. Memorie dell'Istituto Italiano di Idrobiologia, **34**: 197-237.

NILSSEN, J. P. & S. B. WÆRVÅGEN, 2000. Superficial ecosystem similarities vs. autecological stripping: the 'twin species' *Mesocyclops leuckarti* (Claus) and *Thermocyclops oithonoides* (Sars) - seasonal habitat utilisation and life history traits. Journal of Limnology, **59**: 79-102.

NOGUTI-KUEI, H. Y., 1987. Estágios de desenvolvimento de *Oncaea curta* Sars, 1916 (Crustacea, Copepoda, Poecilostomatoida): 1-64. (M.Sc. Thesis, Institute of Biology, Universidade de São Paulo, São Paulo).

NORRBIN, M. F., 1991. Gonad maturation as an indication of seasonal cycles for several species of small copepods in the Barents Sea. Polar Research, **10**: 421-432.

NORRBIN, M. F., R. E. OLSEN & K. S. TANDE, 1990. Seasonal variation in lipid class and fatty acid composition of two small copepods in Balsfjorden, northern Norway. Marine Biology, Berlin, **105**: 205-211.

O'DOHERTY, E. C., 1985. Stream-dwelling copepods: their life history and ecological significance. Limnology and Oceanography, **30**: 554-564.

O'NEIL, J. M. & M. R. ROMAN, 1994. Ingestion of the cyanobacterium *Trichodesmium* spp. by pelagic harpacticoid copepods *Macrosetella*, *Miracia*, and *Oculosetella*. Hydrobiologia, **292/293**: 235-240.

OGAWA, K., K. MATSUZAKI & H. MISAKI, 1997. A new species of *Balaenophilus* (Copepoda: Harpacticoida), an ectoparasite of a sea turtle in Japan. Zoological Science, **14**: 691-700.

OGILVIE, H. S., 1953. Copepod nauplii, I. Fiches d'Identification du Zooplancton, **50**: 1-4. (Conseil Permanent International pour l'Exploration de la Mer, Charlottenlund Slot).

OHMAN, M. D. & A. W. TOWNSEND, 1998. Egg strings in *Euchirella pseudopulchra* (Aetideidae) and comments on constraints on egg brooding in planktonic marine copepods. Journal of Marine Systems, (Special Volume) **15**: 61-69.

OHNO, A. & Y. OKAMURA, 1988. Propagation of the calanoid copepod, *Acartia tsuensis*, in outdoor tanks. Aquaculture, **70**: 39-51.

OHTSUKA, S., G. A. BOXSHALL & S. HARADA, 2005. A new genus and species of nicothoid copepod (Crustacea: Copepoda: Siphonostomatoida) parasitic on the mysid *Siriella okadai* Ii from off Japan. Systematic Parasitology, **62**: 65-81.

OHTSUKA, S. & R. HUYS, 2001. Sexual dimorphism in calanoid copepods: morphology and function. Hydrobiologia, **453/454**: 441-466.

OHTSUKA, S., S. OHAYE, A. TANIMURA, M. FUKUCHI, H. HATTORI, H. SASAKI & O. MATSUDA, 1993. Feeding ecology of copepodid stages of *Eucalanus bungii* in the Chukchi and northern Bering Seas in October 1988. Proceedings of the National Institute of Polar Research, **6**: 27-37.

OKAMOTO, K., 1984. Size, selective feeding of *Daphnia longispina hyalina* and *Eodiaptomus japonicus* on a natural phytoplankton assemblage with the fractionizing method. Memoirs of the Faculty of Science, Kyoto University, (Biology) **9**: 23-40.

OLSSON, P., 1869. Prodromus Faunae Copepodorum Parasitantium Scandinaviae. Acta Universitatis Lundensis, [1868] **5**(3) (8): 1-49.

ONBÉ, T., T. HOTTA & S. OHTSUKA, 1988. The developmental stages of the marine calanoid copepod *Labidocera rotunda* Mori. Journal of the Faculty of Applied Biological Science, Hiroshima University, **27**: 79-91.

OOISHI, S., 1980. The larval development of some copepods of the family Ascidicolidae, subfamily Haplostominae, symbionts of compound ascidians. Publications of the Seto Marine Biological Laboratory, **25**: 253-292.

——, 1999. Female and male *Botryllophilus ruber* (Copepoda: Cyclopoida) associated with the compound ascidian *Botryllus schlosseri*. Journal of Crustacean Biology, **19**: 556-577.

ØSTERGAARD, P., 1998. Anatomy and development of an endoparasitic copepod *Parachordeumium amphiurae* (Hérouard) (Cyclopidae, Chordeumiidae) living in the brittlestar *Amphipholis squamata* (Delle Chiaje). Zoologischer Anzeiger, **236**: 189-202.

OZAKI, K. & T. IKEDA, 1999. Vertical distribution, population structure and life cycle of the mesopelagic copepod *Paraeuchaeta elongata* off Cape Esan, southwestern Hokkaido, Japan. Plankton Biology and Ecology, **46**: 48-53.

—— & ——, 1998. Development time and generation length of the mesopelagic copepod *Paraeuchaeta elongata* reared in the laboratory. Plankton Biology and Ecology, **45**: 75-78.

PADMAVATI, G. & T. IKEDA, 2002. Development of *Metridia pacifica* (Crustacea: Copepoda) reared at different temperatures in the laboratory. Plankton Biology and Ecology, **49**: 93-96.

PADMAVATI, G., T. IKEDA & A. YAMAGUCHI, 2004. Life cycle, population structure and vertical distribution of *Metridia* spp. (Copepoda: Calanoida) in the Oyashio region (NW Pacific Ocean). Marine Ecology Progress Series, **270**: 181-198.

PAFFENHÖFER, G. A., 1971. Grazing and ingestion rates of nauplii, copepodids and adults of the marine planktonic copepod *Calanus helgolandicus*. Marine Biology, Berlin, **11**: 286-298.

——, 1984. Food ingestion by the marine planktonic copepod *Paracalanus* in relation to abundance and size distribution of food. Marine Biology, Berlin, **80**: 323-333.

PAFFENHÖFER, G. A. & P. A. LOYD, 2000. Ultrastructure of cephalic appendage setae of marine planktonic copepods. Marine Ecology Progress Series, **203**: 171-180.

PAIVA CARVALHO, J. DE, 1957. Sobre estadios larvales de *Lernaeenicus longiventris* Wilson (Crustacea-Copepoda). Boletim do Instituto Oceanografico, São Paulo, **8**: 241-254.

PALLARES, R. E., 1980. Observaciones sobre el ciclo biologico y comportamiento 'in vitro' de un copepodo harpacticoide litoral *Tigriopus californicus* (Backer). Contribuciones Cientificas del Centro de Investigaciones de Biología Marina (CIBIMA), Buenos Aires, **169**: 5-38.

PARAMESWARAN PILLAI, P., 1972. On the postnaupliar development of the calanoid copepod *Labidocera pectinata* Thompson and Scott (1903). Journal of the Marine Biological Association of India, **13**: 66-77.

——, 1975. Post-naupliar development of the calanoid copepod *Temora turbinata* (Dana), with remarks on the distribution of the species of the genus *Temora* in the Indian Ocean. Journal of the Marine Biological Association of India, **17**: 87-95.

PARK, C. W., 1976. The larval development of *Tisbe holothuriae* Humes (Copepoda, Harpacticoida) under the laboratory conditions. Publication of the Institute of Marine Sciences, National Fisheries University of Busan, **9**: 33-48.

PASTERNAK, A., E. ARASHKEVICH, K. TANDE & T. FALKENHAUG, 2001. Seasonal changes in feeding, gonad development and lipid stores in *Calanus finmarchicus* and *C. hyperboreus* from Malangen, northern Norway. Marine Biology, Berlin, **138**: 1141-1152.

PASTERNAK, A. F., K. PULKKINEN, V. N. MIKHEEV, T. HASU & E. T. VALTONEN, 1999. Factors affecting abundance of *Triaenophorus* infection in *Cyclops strenuus*, and parasite-induced changes in host fitness. International Journal for Parasitology, **29**: 1793-1801.

PATERSON, N. F., 1958. External features and life cycle of *Cucumaricola notabilis* nov. gen. et sp., a copepod parasite of the holothurian, *Cucumaria*. Parasitology, **48**: 269-290.

PEDASHENKO, D. D., 1897. Razvitie nervnoi sistemy, polovykh' kletok' i spinnoi organ' *Lernaea branchialis*. [Ueber die Entwicklung des Nervensystems und die Genitalzellen und die Dorsalorgane von *Lernaea branchialis* L.]. Raboty Proizvedemiya v' Laboratorii Zootomicheskogo Kabineta, **27**: 187-194.

——, 1898. Istoriya embrional'nago razvitiya i metamorfos' *Lernaea branchialis*. [Die Embryonal-Entwicklung und Metamorphose von *Lernaea branchialis*.]. Raboty Proizvedemiya v' Laboratorii Zootomicheskogo Kabineta, **26**: 247-307.

PEDERSEN, G., K. S. TANDE & G. O. OTTESEN, 1995. Why does a component of *Calanus finmarchicus* stay in the surface waters during the overwintering period in high latitudes? Journal of Marine Science, **52**: 523-531.

PEDERSEN, O. P., K. S. TANDE, A. TIMONIN & T. SEMENOVA, 2000. A possible connection between hydrography and the distribution of *Calanus finmarchicus* on the Norwegian midshelf in 1997. Journal of Marine Science, **57**: 1645-1655.

PENNAK, R. W., 1944. Diurnal movements of zooplankton organisms in some Colorado mountain lakes. Ecology, **25**: 387-403.

PERRYMAN, J. C., 1961. The functional morphology of the skeleto-musculature system of the larval and adult stages of the copepod *Calanus*, together with an account of the changes undergone by this system during larval development: 1-97. (Ph.D. Thesis, University of London, London).

PERTSOVA, N. M., 1974. Zhiznennyi tsikl i ekologiya teplovodnoi kopepody *Centropages hamatus* v Belom More. [Life cycle and ecology of a thermophilous copepod *Centropages hamatus* in the White Sea.]. Zoologicheskii Zhurnal, **53**: 1013-1022. [In Russian, with English summary.]

——, 1974. Raspredelenie izhiznennyi tsikl *Metridia longa* Lubbock v Belom More. [Distribution and life history of *Metridia longa* Lubbock in the White Sea.]. Trudy Belomorskoi Biologicheskoi Stantsii Moskovskogo Gosudarstvennogo Universiteta, **4**: 14-31.

——, 1981. Kolichestvo generatsii i ikh prodolzhitel'nost' u *Pseudocalanus elongatus* (Copepoda, Calanoida) v Belom More. [Number of generations and their span in *Pseudocalanus elongatus* (Copepoda, Calanoida) in the White Sea.]. Zoologicheskii Zhurnal, **60**: 673-684.

PERUEVA, E. G., 1976. O pitanii iv kopepoditnoi stadii *Calanus glacialis* Yashnov iz Belogo Morya kolonial'noi vodorosl'yu *Chaetoceros crinitus* Schütt. [On feeding of IV copepodid stage of *Calanus glacialis* Jaschnov from the White Sea on colonial alga *Chaetoceros crinitus* Schütt.]. Okeanologiya, **16**: 1087-1091. [In Russian, with English Summary.]

PESTA, O., 1907. Die Metamorphose von *Mytilicola intestinalis* Steuer. Zeitschrift für Wissenschaftliche Zoologie, **88**: 78-98, 1 pl.

——, 1909. Beiträge zur Kenntnis parasitischer Copepoden. Denkschriften der Akademie der Wissenschaften zu Wien, (Mathematisch-Naturwissenschaftliche Klasse) **84**: 257-268.

PETERSON, W. T., 1986. Development, growth, and survivorship of the copepod *Calanus marshallae* in the laboratory. Marine Ecology Progress Series, **29**: 61-72.

PETERSON, W. T. & C. B. MILLER, 1978. Seasonal cycle of zooplankton abundance and species composition along the central Oregon coast. Fishery Bulletin, United States Fish and Wildlife Service, **75**: 717-724.

PETERSON, W. T., C. B. MILLER & A. HUTCHINSON, 1979. Zonation and maintenance of copepod populations in the Oregon upwelling zone. Deep Sea Research, (A, Oceanographic Research Papers) **26**: 467-494.

PETERSON, W. T., S. J. PAINTING & L. HUTCHINGS, 1990. Diel variations in gut pigment content, diel vertical migration and estimates of grazing impact for copepods in the Southern Benguela upwelling region in October 1987. Journal of Plankton Research, **12**: 259-281.

PIASECKI, W., 1985. Badania wstepne nad cyklem Rozwojowym *Tracheliastes maculatus* Kollar, 1835 (Copepoda: Lernaeopodidae). [Introductory examinations of *Tracheliastes maculatus* Kollar, 1855 (Copepoda, Lernaeopodidae) life cycle.]. Wiadomosci Parazytologiczne, **13**: 587-596.

——, 1989. Life cycle of *Tracheliastes maculatus* Kollar, 1835 (Copepoda, Siphonostomatoida, Lernaeopodidae). Polskie Towarzystwo Parazytologiczne, **35**: 187-245.

——, 1993. Interantennary organ of *Achtheres percarum* Von Nordmann, 1832 (Copepoda, Siphonostomatoida, Lernaeopodidae) and some other new details of its external morphology. Acta Parasitologica, **36**: 36-40.

PIASECKI, W. & B. M. MACKINNON, 1993. Changes in structure of the frontal filament in sequential developmental stages of *Caligus elongatus* Von Nordmann, 1832 (Crustacea: Copepoda: Siphonostomatoida). Canadian Journal of Zoology, **71**: 889-895.

—— & ——, 1995. Life cycle of a sea louse, *Caligus elongatus* Von Nordmann, 1832 (Copepoda, Siphonostomatoida, Caligidae). Canadian Journal of Zoology, **73**: 74-82.

PIKE, A. W., K. MACKENZIE & A. ROWAND, 1993. Ultrastructure of the frontal filament in chalimus larvae of *Caligus elongatus* and *Lepeophtheirus salmonis* from Atlantic salmon, *Salmo salar*. In: G. A. BOXSHALL & D. DEFAYE (eds.), Pathogens of wild and farmed fish: sea lice: 99-113. (Ellis Horwood, London).

PIKE, A. W., A. J. MORDUE & G. RITCHIE, 1993. The development of *Caligus elongatus* Nordmann from hatching to copepodid in relation to temperature. In: G. A. BOXSHALL & D. DEFAYE (eds.), Pathogens of wild and farmed fish: sea lice: 51-60. (Ellis Horwood, London).

PINCHUK, A. I., 1997. Naupliar development of *Metridia pacifica* (Copepoda: Calanoida) reared in the laboratory. Journal of Crustacean Biology, **17**: 343-351.

——, 1997. Early life history of *Metridia pacifica* Brodsky, (Copepoda: Calanoida) from the southeastern Bering Sea and Gulf of Alaska: 1-65 (M.Sc. Thesis, University of Alaska, Fairbanks, Alaska).

PINERO DI VERDINELLI, M., 1981. Sistematica e ecologia das especies de *Metridia* (Copepoda, Calanoida) do Mar de Weddell, com considerações sobre a hidrografia e o fitoplancton: 1-196. (Ph.D. Thesis, Instituto Oceanografico da Universidade de São Paulo, São Paulo).

PINTO, C. S. C., L. P. SOUZA-SANTOS & P. J. P. SANTOS, 2001. Development and population dynamics of *Tisbe biminiensis* (Copepoda: Harpacticoida) reared on different diets. Aquaculture, **198**: 253-267.

PLEW, W. F. & R. W. PENNAK, 1949. A seasonal investigation of the vertical movement of zooplankton in an Indiana lake. Ecology, **30**: 93-100.

POLI, J. M. & J. CASTEL, 1984. Cycle biologique en laboratoire d'un Copépode planctonique de l'estuaire de la Gironde: *Eurytemora hirundoides* (Nordquist, 1888). Vie et Milieu, **33**: 79-86. [In French, with English summary.]

PÓNYI, J. E., I. H. PETER & N. P. ZÁNKAI, 1982. Daily changes in population structure and production of *Eudiaptomus gracilis* (G.O. Sars) (Copepoda, Calanoida) during summer in a shallow lake (Balaton, Hungary). Journal of Plankton Research, **4**: 913-926.

POULET, S. A., M. LAABIR, A. IANORA & A. MIRALTO, 1995. Reproductive response of *Calanus helgolandicus*. I. Abnormal embryonic and naupliar development. Marine Ecology Progress Series, **129**: 85-95.

POWLIK, J. J., A. G. LEWIS & M. SPAETH, 1997. Development, body length, and feeding of *Tigriopus californicus* (Copepoda, Harpacticoida) in laboratory and field populations. Crustaceana, **70**: 324-343.

PU, X. M., S. SUN, B. YANG, G. T. ZHANG & F. ZHANG, 2004. Life history strategies of *Calanus sinicus* in the southern Yellow Sea in summer. Journal of Plankton Research, **26**: 1059-1068.

PUGLIESI, E., 1914. Sullo sviluppo larvale di *Harpacticus gracilis* Cls. (serie copepodiforme). Atti dell'Accademia Gioenia di Scienze Naturali, **7**: 81-97.

PULKKINEN, K., A. F. PASTERNAK, T. HASU & E. T. VALTONEN, 2000. Effect of *Triaenophorus crassus* (Cestoda) infection on behavior and susceptibility to predation of the first intermediate host *Cyclops strenuus* (Copepoda). Journal of Parasitology, **86**: 664-670.

PURASJOKI, K. J., 1945. Quantitative Untersuchungen über die Mikrofauna des Meeresbodens in der Umgebung der Zoologischen Station Tvärminne an der Südküste Finnlands. Commentationes Biologicae, Societas Scientiarum Fennica, Helsinki, **9** (14): 1-24.

QUIDOR, A., 1922. Sur "*Mesoglicola delagei*" et son hôte. Annales des Sciences Naturelles, (Zoologie et Biologie Animale) (10) **5**: 77-81.

RAIBAUT, A., 1963. Le développement larvaire de *Laophonte commensalis* Raibaut (Copepoda, Harpacticoida). Crustaceana, **5**: 112-118.

——, 1966. Actions de la température et de la chlorinité sur le développement larvaire de deux Copépodes Harpacticoïdes. Comptes Rendus des Séances de l'Académie des Sciences, Paris, (D) **263**: 416-418.

——, 1985. Les cycles évolutifs des Copépodes parasites et les modalités de l'infestation. Année Biologique, (4) **24** (3): 233-274. [With English summary.]

RAMA DEVI, C., 1966. Seasonal occurrence and abundance of free-swimming copepod nauplii in Narragansett Bay. Journal of the Fisheries Research Board of Canada, **23**: 415-422.

RAMA DEVI, C. & Y. RANGA REDDY, 1989a. The complete postembryonic development of *Paradiaptomus greeni* (Gurney, 1906) (Copepoda, Calanoida) reared in the laboratory. Crustaceana, **56**: 141-161.

—— & ——, 1989b. The complete postembryonic development of *Allodiaptomus raoi* Kiefer, 1936 (Copepoda, Calanoida) reared in the laboratory. Crustaceana, **56**: 246-266.

—— & ——, 1990a. The complete postembryonic development of *Tropodiaptomus informis* Kiefer, 1936 (Copepoda: Calanoida) reared in the laboratory. Journal of Plankton Research, **12**: 55-75.

—— & ——, 1990b. Larval stages of *Neodiaptomus lindbergi* Brehm, 1953 (Copepoda: Calanoida). Hydrobiologia, **206**: 99-114.

RAMAMOHAMA RAO, V., 1958a. Development of a cyclopoid copepod *Oithona rigida* (Giesbrecht). Andhra University Memoirs in Oceanography, **2**: 128-131.

——, 1958b. The development of a marine copepod, *Euterpina acutifrons* (Dana). Andhra University Memoirs in Oceanography, **2**: 132-136.

RANGA REDDY, Y. & C. RAMA DEVI, 1989. The complete postembryonic development of *Heliodiaptomus contortus* (Gurney, 1907) (Copepoda, Calanoida) reared in the laboratory. Crustaceana, **57**: 113-133.

—— & ——, 1990a. Postembryonic development of *Phyllodiaptomus blanci* (Guerne & Richard, 1896) (Copepoda, Calanoida). Hydrobiologia, **190**: 155-170.

—— & ——, 1990b. The complete postembryonic development of *Heliodiaptomus cinctus* (Gurney, 1907) (Copepoda, Calanoida), reared in the laboratory. Crustaceana, **58**: 45-66.

RATHKE, H., 1843. Beiträge zur Fauna Norwegens. Nova Acta Academiae Caesareae Leopoldino-Carolinae Naturae Curiosorum, **20**: 1-264.

RAVERA, O., 1953. Gli stadi di sviluppo dei copepodi pelagici del Lago Maggiore. Memorie dell'Istituto Italiano di Idrobiologia, **7**: 129-151.

REDDIAH, K., 1965. The copepod fauna of Assam (India). 4. The last copepodid instar in three diaptomid genera. Crustaceana, **8**: 174-180.

REED, E. B. & J. G. ARONSON, 1989. Seasonal variation in length of copepodids and adults of *Diacyclops thomasi* (Forbes) in two Colorado montane reservoirs (Copepoda). Journal of Crustacean Biology, **9**: 67-76.

REY, C., R. HARRIS, X. IRIGOIEN, R. HEAD & F. CARLOTTI, 2001. Influence of algal diet on growth and ingestion of *Calanus helgolandicus* nauplii. Marine Ecology Progress Series, **216**: 151-165.

REY-RASSAT, C., X. IRIGOIEN, R. HARRIS, R. HEAD & F. CARLOTTI, 2002. Growth and development of *Calanus helgolandicus* reared in the laboratory. Marine Ecology Progress Series, **238**: 125-138.

RICHIARDI, S., 1876. Intorno al *Peroderma cylindricum* dell' Heller e sopra due specie nuove del genere *Philichthys*. Atti della Società Toscana di Scienze Naturali, Processi Verbali e Memorie, (B) **2** (2): 1-14.

RICHMOND, C., N. H. MARCUS, C. SEDLACEK, G. A. MILLER & C. OPPERT, 2006. Hypoxia and seasonal temperature: Short-term effects and long-term implications for *Acartia tonsa* Dana. Journal of Experimental Marine Biology and Ecology, **328**: 177-196.

RINGUETTE, M., L. FORTIER, M. FORTIER, J. A. RUNGE, P. LAROUCHE, J. M. WESLAWSKI & S. KWASNIEWSKI, 2002. Advanced recruitment and accelerated population development in Arctic calanoid copepods of the North Water. Deep Sea Research, (II, Topical Studies in Oceanography) **49**: 5081-5099.

RITCHIE, L., 1975. A new genus and two new species of Choniostomatidae (Copepoda) parasitic on two deep sea isopods. Zoological Journal of the Linnean Society, London, **57**: 155-178.

ROFF, J. C., J. T. TURNER, M. K. WEBBER & R. R. HOPCROFT, 1995. Bacterivory by tropical copepod nauplii: extent and possible significance. Aquatic Microbial Ecology, **9**: 165-175.

ROKNEDDINE, A., 2004. Influence de la salinité et de la température sur la reproduction d'*Arctodiaptomus salinus* (Daday, 1885) (Copepoda, Calanoida), du marais temporaire salé, 'La Sebkha Zima' (Maroc). [The influence of salinity and temperature on the reproduction of *Arctodiaptomus salinus* (Daday, 1885) (Copepoda, Calanoida) in the temporary salt marsh, 'La Sebkha Zima' (Morocco).]. Crustaceana, **77**: 923-940.

ROSE, M. & M. HAMON, 1952. Sur un Copépode parasite de certains Mollusques Ptéropodes Thécosomes. Annales des Sciences Naturelles, (Zoologie et Biologie Animale) (11) **14**: 219-230.

—— & ——, 1953. A propos de *Pennella varians* Steenstrup et Lütken, 1861, parasite des branchies de Céphalopodes. Bulletin de la Société d'Histoire Naturelle de l'Afrique du Nord, **44**: 172-183.

ROSENFIELD, D., 1967. The external morphology of the developmental stages of some diosaccid harpacticoid copepods (Crustacea) from Massachusetts Bay: 1-299. (Ph.D. Thesis, Boston University, Boston).

ROUCH, R., 1967. Nombre et taille des oeufs chez quelques Harpacticides épigés et hypogés (Copépodes). Spelunca Memoires, (4) **5**: 300-304.

——, 1968. Contribution à la connaissance des Harpacticides hypogés (Crustacés - Copépodes). Annales de Spéléologie, **23**: 5-167.

RUNGE, J. A., I. A. MCLAREN, C. J. CORKETT, R. N. BOHRER & J. A. KOSLOW, 1985. Molting rates and cohort development of *Calanus finmarchicus* and *C. glacialis* in the sea off southwest Nova Scotia. Marine Biology, Berlin, **86**: 241-246.

RUSSELL, F. S., 1925. A new species of *Caligus* from Egypt, *Caligus pageti* sp. n. Annals and Magazine of Natural History, (9) **15**: 611-618.

——, 1933. On the occurrence of young stages of Caligidae on pelagic young fish in the Plymouth area. Journal of the Marine Biological Association of the United Kingdom, (n. ser.) **18**: 551-553.

RYLOV, V. M., 1928. Ueber eine neue *Epischura*-Art aus dem fernen Osten (*Epischura chankensis* sp. nov.; Copepoda, Calanoida). Zoologischer Anzeiger, **77**: 125-137.

——, 1948. Cyclopoida presnykh vod. [Freshwater Cyclopoida.]. Fauna SSSR, Rakoobraznye, **3** (3): 1-318.

SABATINI, M. E., F. C. RAMIREZ & P. MARTOS, 2000. Distribution pattern and population structure of *Calanus australis* Brodsky, 1959 over the southern Patagonian shelf off Argentina in summer. Journal of Marine Science, **57**: 1856-1866.

SAITO, H. & A. TSUDA, 2000. Egg production and early development of the Subarctic copepods *Neocalanus cristatus*, *N. plumchrus* and *N. flemingeri*. Deep-Sea Research, (I, Oceanographic Research Papers) **47**: 2141-2158.

SAIZ, E. & M. ALCARAZ, 1991. Effects of small-scale turbulence on development time and growth of *Acartia grani* (Copepoda: Calanoida). Journal of Plankton Research, **13**: 873-884.

SAMIA, B., 1993. Contribution à l'étude de *Peroderma cylindricum* Heller, 1865, parasite de la sardine, *Sardina pilchardus* (Walbaum, 1792) des côtes tunisiennes. (Unpubl. Ph.D. Thesis, University of Tunis, Tunis).

SAMEOTO, D. D. & A. W. HERMAN, 1990. Life cycle and distribution of *Calanus finmarchicus* in deep basins on the Nova Scotia shelf and seasonal changes in *Calanus* spp. Marine Ecology Progress Series, **66**: 225-237.

SANCHEZ-VELASCO, L., I. CONTRERAS-ARREDONDO & G. ESQUEDA-ESCARCEGA, 1999. Diet composition of *Euthynnus lineatus* and *Auxis* sp. larvae (Pisces: Scombridae) in the Gulf of California. Bulletin of Marine Science, **65**: 687-698.

SANCHEZ-VELASCO, L. & B. SHIRASAGO, 1999. Distribution and abundance of copepod nauplii in the vicinity of a submarine canyon (NW Mediterranean). Revista de Biología Tropical, **47** (Suppl. 1): 165-173.

SANDOY, S. & J. P. NILSSEN, 1987. Life cycle dynamics and vertical distribution of *Heterocope saliens* (Lillj.) in two anthropogenic acidic lakes in southern Norway. Archiv für Hydrobiologie, **110**: 83-99.

SANTER, B., E. BLOHM-SIEVERS, C. E. CACERES & N. G. HAIRSTON, JR., 2000. Life-history variation in the coexisting freshwater copepods *Eudiaptomus gracilis* and *Eudiaptomus graciloides*. Archiv für Hydrobiologie, **149**: 441-458.

SANTER, B. & F. VAN DEN BOSCH, 1994. Herbivorous nutrition of *Cyclops vicinus*: the effect of a pure algal diet on feeding, development, reproduction and life cycle. Journal of Plankton Research, **16**: 171-195.

SANTER, B. & W. LAMPERT, 1995. Summer diapause in cyclopoid copepods: adaptive response to a food bottleneck? Journal of Animal Ecology, **64**: 600-613.

SANTOS, P. J. P., L. SOUZA-SANTOS & J. CASTEL, 2003. Population dynamics of *Nannopus palustris* Brady, 1880 (Copepoda, Harpacticoida) in the oligo-mesohaline area of the Gironde estuary (France). Cahiers de Biologie Marine, **44**: 329-338.

SARASWATHY, M., 1967. Pelagic copepods from the inshore waters of Trivandrum coast. Proceedings of Symposium on Crustacea, **1**: 74-106.

SARS, M., 1870. Bidrag til Kundskap om Christianiafjordens Fauna. II. Crustacea. Beskrivelse af nye, paa Annelider snyltende Copepodeformer. Nyt Magazin for Naturvidenskaberne, Christiania [Oslo], **17** (2): 113-160, pls. 8-10.

SARVALA, J., 1977a. The naupliar development of *Bryocamptus zschokkei* (Copepoda, Harpacticoida). Annales de Limnologie, **13**: 115-131.

——, 1977b. The naupliar development of six species of freshwater harpacticoid Copepoda. Annales Zoologici Fennici, **14**: 135-161.

——, 1979a. A parthenogenetic life cycle in a population of *Canthocamptus staphylinus* (Copepoda, Harpacticoida). Hydrobiologia, **62**: 113-129.

——, 1979b. Effect of temperature on the duration of egg, nauplius and copepodite development of some freshwater benthic Copepoda. Freshwater Biology, **9**: 515-534.

SAVITZ, J. D., D. A. WRIGHT & R. A. SMUCKER, 1994. Toxic effects of the insecticide diflubenzuron (Dimilin) on survival and development of nauplii of the estuarine copepod, *Eurytemora affinis*. Marine Environmental Research, **37**: 297-312.

SAZHINA, L. I., 1960. Razvitie Chernomorskikh Copepoda. I. Nauplial'nye stadii *Acartia clausi* Giesbr., *Centropages kroyeri* Giesbr., *Oithona minuta* Kritcz. [Development of Black Sea Copepoda I. Nauplii of *Acartia clausi* Giesbr., *Centropages kroyeri* Giesbr.,

Oithona minuta Kritcz.]. Trudy Sevastopol'skoy Biologicheskoy Stantsii, Akademiya Nauk SSSR, **13**: 49-67. [In Russian.]

——, 1961. Razvitie Chernomorskikh Copepoda. II. Nauplial'nye stadii *Calanus helgolandicus* (Claus). [Development of Black Sea Copepoda. II. Nauplius stage of *Calanus helgolandicus* (Claus).]. Trudy Sevastopol'skoy Biologicheskoy Stantsii, Akademiya Nauk SSSR, **14**: 102-108. [In Russian.]

——, 1967. Razvitie Chernomorskikh Copepoda. III. Nauplial'nye stadii *Pontella mediterranea* Claus i *Labidocera brunescens*. [Development of Black Sea Copepoda. III. Nauplii of *Pontella mediterranea* Claus and *Labidocera brunescens* Czern.]. In: Akademiya Nauk SSSR, Okeanograficheskaya Komissiya, Moscow: 99-108, figs. 1-7. [In Russian.]

——, 1969a. Razvitie Chernomorskikh Copepoda. IV. Kopepoditnye stadii *Acartia clausi* Giesbr., *Centropages ponticus* Karavaj., *Oithona minuta* Kritcz. [The development of Black Sea Copepoda. IV. The copepodid stages of *Acartia clausi* Giesbr., *Centropages ponticus* Karavaj, *Oithona minuta* Kritcz.]. Biologiya Morya, Kiev, **17**: 96-143.

——, 1969b. Nauplial'nye lichinki massovykh pelagicheskikh Copepoda Chernogo Morya. Opredelitel' Fauny Chernogo i Azovskogo Morei, **2**: 152-163.

——, 1971. Godichnyi tsikl razvitiya massovykh Copopoda v Chernom More. [Annual cycle of development of mass Copepoda in the Black Sea.]. Gidrobiologicheskii Zhurnal, **7**: 38-46.

——, 1975. Ekologicheskaya izmenchivost' temia razmnozheniya i razvitiya pelagicheskikh kopepod. [The ecological significance of the timing of reproduction and development of pelagic copepods.]. Biologiya Morya, Kiev, **33**: 78-87.

——, 1982. Naupliusy massovykh vidov Copepoda Atlanticheskogo okeana. [Nauplii of mass Copepoda species from the Atlantic Ocean.]. Zoologicheskii Zhurnal, **61**: 1154-1164.

SCHMAUS, P. H., 1917. Die *Rhincalanus* Arten, ihre Systematik, Entwicklung, und Verbreitung. Zoologischer Anzeiger, **48**: 305-319.

SCHMAUS, P. H. & K. LEHNHOFER, 1927. Copepoda 4: *Rhincalanus* Dana 1852 der Deutschen Tiefsee-Expedition. Systematik und Verbreitung der Gattung. Wissenschaftliche Ergebnisse der Deutschen Tiefsee-Expedition auf dem Dampfer "Valdivia" 1898-1899, **23**: 355-400.

SCHMINKE, H. K., 1982. Die Nauplius-Stadien von *Parastenocaris vicesima* Klie, 1935 (Copepoda, Parastenocarididae). Drosera, **82**: 101-108.

SCHNACK-SCHIEL, S. B. & W. HAGEN, 1994. Life cycle strategies and seasonal variations in distribution and population structure of four dominant calanoid copepod species in the eastern Weddell Sea, Antarctica. Journal of Plankton Research, **16**: 1543-1566.

SCHNACK-SCHIEL, S. B., D. THOMAS, G. S. DIECKMANN, H. EICKEN, R. GRADINGER, M. SPINDLER, J. WEISSENBERGER, E. MIZDALSKI & K. BEYER, 1995. Life cycle strategy of the Antarctic calanoid copepod *Stephos longipes*. Progress in Oceanography, **36**: 45-75.

SCHRAM, T., 1993. Supplementary descriptions of the developmental stages of *Lepeophtheirus salmonis* (Krøyer, 1837) (Copepoda: Caligidae). In: G. A. BOXSHALL & D. DEFAYE (eds.), Pathogens of wild and farmed fish: sea lice: 30-47. (Ellis Horwood, London).

SCHRAM, T. & M. ANSTENSRUD, 1985. *Lernaeenicus sprattae* (Sowerby) larvae in the Oslofjord plankton and some laboratory experiments with the nauplius and copepodid (Copepoda, Pennellidae). Sarsia, **70**. 127-134.

SCHULZ, K., 1992. *Kunihulsae arabica*, a new genus and species of calanoid copepod from the Arabian Sea. Mitteilungen aus dem Hamburgischen Zoologischen Museum und Institut, **89**: 175-180.

——, 1996. *Mospicalanus schielae*, a new genus and species of calanoid copepod (Crustacea: Spinocalanidae) from deep Antarctic water. Polar Biology, **16**: 595-600.

SCHUURMANS STEKHOVEN, J. H., 1937. Crustacea Parasitica. I. Parasitica Copepoda. Resultats Scientifiques des Croissières du Navire-Ecole 'Mercator', 1. Mémoires du Musée Royal d'Histoire Naturelle de Belgique, (2) **9**: 11-24.

SCOTT, A., 1901. *Lepeophtheirus* and *Lernaea*. Liverpool Marine Biology Committee Memoirs, **6**: 1-54.

SCOTT, T. & A. SCOTT, 1913. The British parasitic Copepoda. Publications of the Ray Society, **1**: 1-252; **2**, pls. 1-72.

SEBASTIAN, M. J. & K. C. GEORGE, 1964. *Lernaeenicus anchoviellae* n. sp. (Copepoda-Lernaeidae) parasitic on *Anchoviella bataviensis* (Hardenberg), with descriptions of its three post-larval stages. Journal of the Marine Biological Association of India, **6**: 235-240.

SEKIGUCHI, H., I. A. MCLAREN & C. J. CORKETT, 1980. Relationship between growth rate and egg production in the copepod *Acartia clausi hudsonica*. Marine Biology, Berlin, **58**: 133-138.

SELL, A. F., D. VAN KEUREN & L. P. MADIN, 2001. Predation by omnivorous copepods on early developmental stages of *Calanus finmarchicus* and *Pseudocalanus* spp. Limnology and Oceanography, **46**: 953-959.

SHAW, T.-H., 1938. Some observations on the life history of a tide-pool copepod, *Tigriopus fulvus* (Fischer). Bulletin of the Fan Memorial Institute of Biology, (Zoology) **8**: 9-17.

SHEN, C. J. & C. C. CHANG, 1965. Larval development of *Metadiaptomus asiaticus* Ulyanin. Oceanology and Limnology Sinica, **7**: 141-150. [In Chinese, with English summary.]

SHEN, C. J. & S. T. CHEN, 1963. Studies on the larval development of *Sinocalanus mystrophorus* Burckhardt. Acta Hydrobiologica Sinica, **1963** (3): 26-39. [In Chinese, with English summary.]

SHIELDS, R. J. & W. M. TIDD, 1968. Effect of temperature on the development of larval and transformed females of *Lernaea cyprinacea* L. (*Lernaeidae*). Crustaceana, (Supplement) **1**: 87-95.

—— & ——, 1974. Site selection on hosts by copepodids of *Lernaea cyprinacea* L. (Copepoda). Crustaceana, **27**: 225-230.

SHIH, C.-T. & D. C. MACLELLAN, 1977. Descriptions of copepodite stages of *Diaptomus* (*Leptodiaptomus*) *nudus* Marsh, 1904 (Crustacea: Copepoda). Canadian Journal of Zoology, **55**: 912-921.

SHIH, C.-T., L. RAINVILLE & D. [C.] MACLELLAN, 1981. Copepodids of *Bradyidius similis* (Sars, 1902) (Crustacea: Copepoda) in the Saguenay Fjord and the St. Lawrence Estuary. Canadian Journal of Zoology, **59**: 1079-1093.

SHIINO, S. M., 1932. *Ichthyotaces pteroisicola* n. gen. & n. sp. A copepod parasitic on the fish *Pterois lunulata* Temm. & Schl. Annotationes Zoologicae Japonenses, **13**: 417-430, pl. 25.

——, 1959. Ostpazifische parazitierende Copepoden. [Eastern Pacific parasitic copepods.]. Reports of the Faculty of Fisheries, Prefectural University of Mie, **3**: 267-333.

SHIMKEVICH, V., 1893. O pervykh' stadiyakh' razvitiya *Notopterophorus*. [Sur les premiers stades du développement du *Notopterophorus*.]. Vestnik Estestvoznaniya St.-Petersburg, **4**: 200-203. [In Russian, with French summary.]

——, 1896. Sur les premiers stades du développement des Copépodes parasitaires. In: Comptes Rendus du III. Congrès International de Zoologie, Leyde, 16-24 September 1895: 503-504.

——, 1899. Einige Worte über die Entwicklung der parasitischen Copepoden. Zoologischer Anzeiger, **22**: 111-114.

SHMELEVA, A. A. & V. E. ZAIKA, 1973. Vertikal'noe raspredelenie kopepoditnykh stadii veslonogikh rachkov Adriaticheskogo Morya. [Vertical distribution of the copepodid stages of copepods in the Adriatic Sea.]. Okeanologiya, **13**: 872-876. [In Russian, with English summary.]

SHUVAEV, YU. D., 1975. Raspredelenie voztastnykh stadii *Calanus finmarchicus* (Gunn.) v Severnom More vesnoi i osen'yu 1969. [Distribution of growth stages of *Calanus finmarchicus* (Gunn.) in the North Sea in spring and autumn of 1969.]. Trudy Atlanticheskogo Nauchno-issledovatel'skogo Instituta Rybnogo Khozaistva i Okeanographii, **58**: 97-101.

——, 1978. Osobennosti dvizhenii nekotorykh planktonnykh veslonogikh rakoobraznykh. [Aspects of the movements of some planktonic Copepoda.]. Gidrobiologicheskii Zhurnal, **14**: 39-44. [In Russian, with English summary.]

SHODEN, S., T. IKEDA & A. YAMAGUCHI, 2005. Vertical distribution, population structure and life cycle of *Eucalanus bungii* (Copepoda: Calanoida) in the Oyashio region, with notes on its regional variations. Marine Biology, Berlin, **146**: 497-511.

SHOTTER, R. A., 1971. The biology of *Clavella uncinata* (Muller) (Crustacea: Copepoda). Parasitology, **63**: 419-430.

SILEN, L., 1963. *Clionophilus vermicularis* n. gen. n. sp., a copepod infecting the burrowing sponge, *Cliona*. Zoologiska Bidrag, Uppsala, **35**: 269-288.

SKRESLET, S., K. OLSEN, A. MOHUS & K. S. TANDE, 2000. Stage-specific habitats of *Calanus finmarchicus* and *Calanus helgolandicus* in a stratified northern Norwegian fjord. Journal of Marine Science, **57**: 1656-1663.

SLAGSTAD, D. & K. S. TANDE, 1996. The importance of seasonal vertical migration in across shelf transport of *Calanus finmarchicus*. Ophelia, **44**: 189-205.

SLATER, L. M. & R. R. HOPCROFT, 2005. Development, growth, and egg production of *Centropages abdominalis* in the eastern Subarctic Pacific. Journal of Plankton Research, **27**: 71-78.

SLINN, D. J., 1970. An infestation of adult *Lernaeocera* (Copepoda) on wild sole, *Solea solea*, kept under hatchery conditions. Journal of the Marine Biological Association of the United Kingdom, **50**: 787-800.

SMITH, J. A. & P. J. WHITFIELD, 1988. Ultrastructure studies on the early cuticular metamorphosis of adult female *Lernaeocera branchialis* (L.) (Copepoda, Pennellidae). Hydrobiologia, **167-168**: 607-616.

SMITH, S. L. & P. V. Z. LANE, 1987. On the life history of *Centropages typicus*: responses to a fall diatom bloom in the New York Bight. Marine Biology, Berlin, **95**: 305-313.

SMOL, N. & C. HEIP, 1974. The culturing of some harpacticoid copepods from brackish water. Biologisch Jaarboek Dodonaea, **42**: 159-169.

SMYLY, W. J. P., 1957. Observations on the life-history of the harpacticoid copepod, *Canthocamptus staphylinus* (Jurine). Annals and Magazine of Natural History, (12) **10**: 509-512.

——, 1961. The life cycle of the freshwater copepod *Cyclops leuckarti* Claus in Esthwaite Water. Journal of Animal Ecology, **30**: 153-169.

——, 1962. Laboratory experiments with stage V copepodids of the freshwater copepod, *Cyclops leuckarti* Claus, from Windermere and Esthwaite Water. Crustaceana, **4**: 273-280.

——, 1970. Observations on the rate of development, longevity and fecundity of *Acanthocyclops viridis* (Jurine) (Copepoda, Cyclopoida), in relation to type of prey. Crustaceana, **18**: 21-36.

——, 1974. The effect of temperature on the development time of the eggs of three freshwater cyclopoid copepods from the English Lake District. Crustaceana, **27**: 278-284.

SOH, H. Y., H. L. SUH, S. OHTSUKA, Y. H. YOON & S. D. CHOI, 2001. Taxonomic studies on brackish copepods in Korean waters - II. Ontogeny and phylogeny of appendages in copepodid stages of *Tortanus derjugini* Smirnov, 1935 (Copepoda, Calanoida). Journal of Plankton Research, **23**: 1137-1169.

SONG, L. & F. JINCHUAN, 1990. Larval stages of marine planktonic copepods of China: 1-141. (China Ocean Press, Beijing).

SPEIRS, D. C., W. S. C. GURNEY, S. J. HOLMES, M. R. HEATH, S. N. WOOD, E. D. CLARKE, I. H. HARMS, H. J. HIRCHE & E. MACKENZIE, 2004. Understanding demography in an advective environment: modelling *Calanus finmarchicus* in the Norwegian Sea. Journal of Animal Ecology, **73**: 897-910.

SPINDLER, K. D., 1970. Die Bedeutung der Photoperiode für die Entwicklung von *Cyclops vicinus*. Verhandlungen der Deutschen Zoologischen Gesellschaft, Zoologischer Anzeiger, (Supplement) **33** [1969]: 190-195.

——, 1971. Untersuchungen über den Einfluss äusserer Faktoren auf die Dauer der Embryonalentwicklung und den Häutungsrhytmus von *Cyclops vicinus*. Oecologia, **7**: 342-355.

SPROSTON, N. G. & P. H. T. HARTLEY, 1941. Observations on the bionomics and physiology of *Trebius caudatus* and *Lernaeocera branchialis* (Copepoda). Journal of the Marine Biological Association of the United Kingdom, (n. ser.) **25**: 393-417.

SPROSTON, N. G., W. YIN & T. HU, 1950. The genus *Lamproglena* (Copepoda parasitica): the discovery of the life-histories and males of two Chinese species from food fishes, revealing their relationship with *Lernaea*, and of both to the Cyclopoidea. Sinensia, (n. ser.) **1**: 51-84.

STEENSTRUP, J. J. S. & C. F. LÜTKEN, 1861. Bidrag til Kundskab om det aabne Havs Snyltekrebs og Lernaeer samt om andre nue eller hidtil kun ufuldstaendigt kjendte parasitiske Copepoder. Kongelige Danske Videnskabernes Selskabs Skrifter, (Naturhistorisk og Mathematisk Afdeling) (5) **5**: 341-432, pls. 1-15.

STEIN, F., 1852. Über die Beziehung der Gattungen *Caligus* und *Chalimus*. Wiegmanns Archiv, **1**: 91.

STEPANOVA, L. A., 1982. Rodovaya differentsiatsiya lichinok v kopepoditnoi stadii sem. Diaptomidae (Copepoda, Calanoida).[Generic differentiation of 5th copepodite instar larvae of the family Diaptomidae (Copepoda, Calanoida).]. Vestnik Zoologii, **1982**: 29-34. [In Russian.]

STEUER, A., 1935. Postembryonalentwicklung von *Rhincalanus gigas* Brady. Sitzungsberichte der Akademie der Wissenschaften in Wien, (Mathematisch-Naturwissenschaftliche Klasse) **144**: 383-390.

STOCK, J. H., 1959. Copepoda associated with Neapolitan Mollusca. Pubblicazione della Stazione Zoologica di Napoli, **31**: 43-58.

STOHR, S., K. SCHULTZ & H. CH. JOHN, 1996. Population structure and reproduction of *Calanus helgolandicus* (Copepoda, Calanoida) along the Iberian and Moroccan slope. Helgoländer Wissenschaftliche Meeresuntersuchungen, **50**: 457-475.

STOLAROV, V. P., 1936. Nabliudenia nad ciklom razvitia *Lernaea cyprinacea* i ee patogennoe vlanie na Koznye ryby. Trudy Leningradskogo Obshchestva Estestvoispytatelei, **65**: 239-253. [In Russian, with German summary.]

SUÁREZ-MORALES, E. & E. TOVAR, 2004. Postnaupliar stages of a thaumatopsyllid copepod from a reef area of the western Caribbean Sea. Sarsia, **89**: 223-244.

SULLIVAN, B. K., E. BUSKEY, D. C. MILLER & P. J. RITACCO, 1983. Effects of copper and cadmium on growth, swimming and predator avoidance in *Eurytemora affinis* (Copepoda). Marine Biology, Berlin, **77**: 299-306.

SUN, B. & J. W. FLEEGER, 1995. Sustained mass culture of *Amphiascoides atopus* a marine harpacticoid copepod in a recirculating system. Aquaculture, **136**: 313-321.

SWADLING, K. M., 2001. Population structure of two Antarctic ice-associated copepods, *Drescheriella glacialis* and *Paralabidocera antarctica*, in winter sea ice. Marine Biology, Berlin, **139**: 597-603.

SWADLING, K. M. & J. A. E. GIBSON, 2000. Grazing rates of a calanoid copepod (*Paralabidocera antarctica*) in a continental Antarctic lake. Polar Biology, **23**: 301-308.

SWADLING, K. M. & N. H. MARCUS, 1994. Selectivity in the natural diets of *Acartia tonsa* Dana (Copepoda: Calanoida): Comparison of juveniles and adults. Journal of Experimental Marine Biology and Ecology, **181**: 91-103.

SWADLING, K. M., A. D. MCKINNON, G. DE'ATH & J. A. E. GIBSON, 2004. Life cycle plasticity and differential growth and development in marine and lacustrine populations of an Antarctic copepod. Limnology and Oceanography, **49**: 644-655.

SWADLING, K. M., A. D. MCPHEE & A. MCMINN, 2000. Spatial distribution of copepods in fast ice of eastern Antarctica. Polar Bioscience, **13**: 55-65.

SWADLING, K. M., P. D. NICHOLS, J. A. E. GIBSON & D. A. RITZ, 2000. Role of lipid in the life cycles of ice-dependent and ice-independent populations of the copepod *Paralabidocera antarctica*. Marine Ecology Progress Series, **208**: 171-182.

TAKAHASHI, T. & A. OHNO, 1996. The temperature effect on the development of calanoid copepod, *Acartia tsuensis*, with some comments to morphogenesis. Journal of Oceanography, **52**: 125-137.

TAKANO, H., 1971a. Breeding experiments of a marine littoral copepod *Tigriopus japonicus* Mori. Bulletin of the Tokai Regional Fisheries Research Laboratory, **64**: 71-80.

——, 1971b. Notes on the raising of an estuarine copepod *Gladioferens imparipes* Thomson. Bulletin of the Tokai Regional Fisheries Research Laboratory, **64**: 81-88.

TANAKA, O., 1938. Note on *Calanus cristatus* Kröyer. Japanese Journal of Zoology, **7**: 599-601.

TANDE, K. S., 1988a. Aspects of developmental and mortality rates in *Calanus finmarchicus* related to equiproportional development. Marine Ecology Progress Series, **44**: 51-58.

——, 1988b. An evaluation of factors affecting vertical distribution among recruits of *Calanus finmarchicus* in three adjacent high-latitude localities. Hydrobiologia, **167/168**: 115-126.

TANDE, K. S. & R. J. HENDERSON, 1988. Lipid composition of copepodite stages and adult females of *Calanus glacialis* in Arctic waters of the Barents Sea. Polar Biology, **8**: 333-339.

TANDE, K. S. & C. C. E. HOPKINS, 1981. Ecological investigations of the zooplankton community of Balsfjorden, northern Norway: the genital system in *Calanus finmarchius* and the role of gonad development in overwintering strategy. Marine Biology, Berlin, **63**: 159-164.

TANDE, K. S. & D. SLAGSTAD, 1982. Ecological investigation on the zooplankton community of Balsfjorden, northern Norway. Seasonal and short-time variations in enzyme activity in copepodite stage 5 and 6 males and females of *Calanus finmarchicus* (Gunnerus). Sarsia, **67**: 63-68.

TANG, K. W., H. H. JAKOBSEN & A. W. VISSER, 2001. *Phaeocystis globosa* (Prymnesiophyceae) and the planktonic food web: feeding, growth, and trophic interactions among grazers. Limnology and Oceanography, **46**: 1860-1870.

TANIMURA, A., 1992. Postembryonic development of *Paralabidocera antarctica* (I.C. Thompson) (Copepoda, Calanoida) from the fast ice near Syowa Station, Antarctic. Hydrobiologia, **245**: 109-128.

TANIMURA, A., T. HOSHIAI & M. FUKUCHI, 1996. The life cycle strategy of the ice-associated copepod, *Paralabidocera antarctica* (Calanoida, Copepoda), at Syowa Station, Antarctica. Antarctic Science, **8**: 257-266.

TATON, H., 1935. Contribution à l'étude du Copépode gallicole *Mesoglicola delagei* Quidor. Travaux de la Station Biologique de Roscoff, **12**: 53-68.

TAUBE, I., 1966. Embryonalutvecklingens beroende av temperaturen hos *Mesocyclops leuckarti* (Claus) och *Cyclops scutifer* Sars. Skrifter Limnologiska Institutionen, Uppsala, **2** (123): 1-49.

TAUBE, I. & A. NAUWERECK, 1967. Zur Populationsdynamik von *Cyclops scutifer* Sars. I. Die Temperaturabhängigkeit der Embryonalentwicklung von *Cyclops scutifer* Sars in Vergleich zu *Mesocyclops leuckarti* (Claus). Reports of the Institute of Freshwater Research, Drottningholm, **47**: 76-86.

TAW, N., 1974. A new species of *Labidocera* (Copepoda, Calanoida) from Tasmania and its postnaupliar developmental stages. Australian Journal of Marine and Freshwater Ressearch, **25**: 261-272.

TEARE, M., 1978. Postembryonic development of *Tachidius discipes* Giesbrecht, 1881 (Copepoda, Harpacticoida). Cahiers de Biologie Marine, **19**: 343-353.

TESCH, J. J., 1915. Quantitative Untersuchungen über das Vorkommen der Copepoden und ihrer Entwicklungsstadien im Plankton beim Haaks-Feuerschiff 1912, mit einem Anhang: Die Nauplien und Copepoditen von *Euterpina acutifrons* (Dana). Rapporten en Verhandelingen Uitgegeven door het Rijksinstituut voor Visscherijonderzoek, **1**: 269-306.

THOMPSON, B. M., 1982. Growth and development of *Pseudocalanus elongatus* and *Calanus* sp. in the laboratory. Journal of the Marine Biological Association of the United Kingdom, **62**: 359-372.

THORELL, T., 1859. Bidrag till Kännedomen om Krustaceer, som lefva i arter af Slägtet *Ascidia* L. Kongliga Svenska Vetenskaps-Akademiens Handlingar, **3** (8): 1-84.

TIETZE, N. S., P. G. HESTER, K. R. SHAFFER, S. J. PRESCOTT & E. T. SCHREIBER, 1994. Integrated management of waste tire mosquitoes utilizing *Mesocyclops longisetus* (Copepoda: Cyclopidae), *Bacillus thuringiensis* var. *israelensis*, *Bacillus sphaericus*, and methoprene. Journal of the American Mosquito Control Association, **10**: 363-373.

TOKIOKA, T. & R. BIERI, 1966. Juveniles of *Macrosetella gracilis* (Dana) from clumps of *Trichodesmium* in the vicinity of Seto. Publications of the Seto Marine Biological Laboratory, **14**: 177-184.

TOLEDO, J. D., M. S. GOLEZ, M. DOI & A. OHNO, 1999. Use of copepod nauplii during early feeding stage of grouper *Epinephelus coioides*. Fisheries Science, **65**: 390-397.

TORRES-SORANDO, L. J., D. ZACARIAS, E. ZOPPI DE ROA & D. J. RODRÍGUEZ, 2003. Population dynamics of *Oithona hebes* (Copepoda: Cyclopoida) in a coastal estuarine lagoon of Venezuela: a stage-dependent matrix growth model. Ecological Modelling, **161**: 157-166.

TRUJILLO-ORTIZ, A., 1995. Alternative method for the calculation of mean time for the assessment of secondary production by true cohort analysis. Journal of Plankton Research, **17**: 2175-2190.

TSUDA, A., H. SAITO & H. KASAI, 1999. Life histories of *Neocalanus flemingeri* and *Neocalanus plumchrus* (Calanoida: Copepoda) in the western Subarctic Pacific. Marine Biology, Berlin, **135**: 533-544.

——, —— & ——, 2001. Life history strategies of Subarctic copepods *Neocalanus flemingeri* and *N. plumchrus*, especially concerning lipid accumulation patterns. Plankton Biology and Ecology, **48**: 52-58.

TUCKER, C. S., C. SOMMERVILLE & R. WOOTTEN, 2000a. The effect of temperature and salinity on the settlement and survival of copepodids of *Lepeophtheirus salmonis* (Krøyer, 1837) on Atlantic salmon, *Salmo salar* L. Journal of Fish Diseases, **23**: 309-320.

——, —— & ——, 2000b. An investigation into the larval energetics and settlement of the sea louse, *Lepeophtheirus salmonis*, an ectoparasitic copepod of Atlantic salmon, Salmo salar. Fish Pathology, **35**: 137-143.

TULLY, O., 1989. The succession of generations and growth of the caligid copepods *Caligus elongatus* and *Lepeophtheirus salmonis* parasitising farmed Atlantic salmon smolts (*Salmo salar* L.). Journal of the Marine Biological Association of the United Kingdom, **69**: 279-287.

TULLY, O. & D. T. NOLAN, 2002. A review of the population biology and host-parasite interactions of the sea louse *Lepeophtheirus salmonis* (Copepoda: Caligidae). Parasitology, (Supplement) **124**: 165-182.

TULLY, O. & K. F. WHELAN, 1993. Production of nauplii of *Lepeophtheirus salmonis* (Krøyer) (Copepopda: Caligidae) from farmed and wild salmon and its relation to the infestation of wild sea trout (*Salmo rutta* L.) off the west coast of Ireland in 1991. Fisheries Research, **17**: 187-200.

TWOMBLY, S., 1994. Comparative demography and population dynamics of two coexisting copepods in a Venezuelan floodplain lake. Limnology and Oceanography, **39**: 234-247.

——, 1996. Timing of metamorphosis in a freshwater crustacean: comparison with anuran models. Ecology, **77**: 1855-1866.

TWOMBLY, S. & C. W. BURNS, 1996. Exuvium analysis: a nondestructive method of analyzing copepod growth and development. Limnology and Oceanography, **41**: 1324-1329.

TWOMBLY, S. & H. CASWELL, 1991. Demographic analysis of tropical zooplankton populations. Verhandlungen der Internationale Vereinigung für Theoretische und Angewandte Limnologie, **24**: 1183-1187.

TWOMBLY, S. & N. TISCH, 2002. Fitness consequences of the timing of metamorphosis in a freshwater crustacean. Oikos, **97**: 213-222.

UCHIMA, M., 1979. Morphological observation of developmental stages in *Oithona brevicornis* (Copepoda, Cyclopoida). Bulletin of the Plankton Society of Japan, **26**: 59-76.

UMMERKUTTY, A. N., 1960. Studies on Indian copepods 2. An account of the morphology and life history of a harpacticoid copepod *Tisbintra jonesi,* sp. nov. from the Gulf of Mannar. Journal of the Marine Biological Association of India, **2**: 149-164.

URBANOWICZ, F., 1884. Zur Entwicklungsgeschichte der Cyclopiden. Zoologischer Anzeiger, Leipzig, **7**: 615-619.

UYE, S.-I., 1980. Development of neritic copepods *Acartia clausi* and *A. steueri*. II. Isochronal larval development at various temperatures. Bulletin of the Plankton Society of Japan, **27**: 11-18.

——, 1982. Population dynamics and production of *Acartia clausi* Giesbrecht (Copepoda: Calanoida) in inlet waters. Journal of Experimental Marine Biology and Ecology, **57**: 55-83.

——, 1988. Temperature-dependent development and growth of *Calanus sinicus* (Copepoda: Calanoida) in the laboratory. Hydrobiologia, **167-168**: 285-293.

——, 1991. Temperature-dependent development and growth of the planktonic copepod *Paracalanus* sp. in the laboratory. Bulletin of Plankton Society of Japan, (Special Volume): 627-636.

UYE, S.-I., C. HUANG & T. ONBÉ, 1990. Ontogenetic diel vertical migration of the planktonic copepod *Calanus sinicus* in the Inland Sea of Japan. Marine Biology, Berlin, **104**: 389-396.

UYE, S.-I. & S. KASAHARA, 1983. Grazing of various developmental stages of *Pseudodiaptomus marinus* (Copepoda: Calanoida) on naturally occurring particles. Bulletin of the Plankton Society of Japan, **30**: 147-158.

UYE, S.-I. & Y. KAYANO, 1994. Predatory feeding behavior of *Tortanus* (Copepoda: Calanoida): life-stage differences and the predation impact on small planktonic crustaceans. Journal of Crustacean Biology, **14**: 473-483.

VADER, W., 1970. *Antheacheres duebeni* M. Sars, a copepod parasitic in the sea anemone, *Bolocera tuediae* (Johnston). Sarsia, **43**: 99-106.

VALLE, A. DELLA, 1880. Sui Coriceidi parassiti, e sull'anatomia del gen. *Lichomolgus*. Atti della Reale Accademia dei Lincei, Memorie della Classe di Scienze Fisiche, Matematiche e Naturali, Roma, **3**: 107-124.

VARELA SANTIBAÑEZ, C., 1988. Desarrollo post-embrionario de *Diaptomus diabolicus* Brehm (Copepoda, Calanoida) del sur de Chile. Biota de Osorno, Chile, **4**: 31-61.

VARELLA, A., 1985. O ciclo biologico de *Ergasilus bryconis* Thatcher, 1981 (Crustacea: Poecilostomatoida, Ergasilidae) parasitica das branquias do matrinxa, *Brycon erythropterum* (Cope, 1872) e aspectos de sua ecologia: 1-100. (Ph.D. Thesis, Instituto Nacional de Pesquisas da Amazonia e Fundação Universidade do Amazonas, Manaus, Amazonas).

VEGA, M. A. P., 1997a. The functional response of copepodid stages to adult of *Parabroteas sarsi* (Copepoda, Calanoida). Internationale Revue der Gesamten Hydrobiologie, **82**: 95-105.

——, 1997b. The functional response of copepodid stages to adult of *Parabroteas sarsi* (Copepoda, Calanoida). II. Antennae and mouthparts. Crustaceana, **70**: 394-417.

VEJDOVSKY, F., 1877. Untersuchungen über die Anatomie und Metamorphose von *Tracheliastes polycolpus* Nordman. Zeitschrift für Wissenschaftliche Zoologie, **29**: 15-46.

VERVOORT, W., 1946. Biological results of the Snellius Expedition. XV. The bathypelagic Copepoda Calanoida of the Snellius Expedition. I. Families Calanidae, Eucalanidae, Paracalanidae, and Pseudocalanidae. Temminckia, **8**: 1-181.

VIJVERBERG, J., 1977. Population structure, life histories and abundance of copepods in Tjeukemeer, the Netherlands. Freshwater Biology, **7**: 579-597.

——, 1980. Effects of temperature in laboratory studies on development and growth of Cladocera and Copepoda from Tjeukemeer, the Netherlands. Freshwater Biology, **10**: 317-340.

VIJVERBERG, J. & A. F. RICHTER, 1982. Population dynamics and production of *Acanthocyclops robustus* (Sars) and *Mesocyclops leuckarti* (Claus) in Tjeukemeer. Hydrobiologia, **95**: 261-274.

VILELA, M. H., 1969. The life cycle of *Tisbe* sp. (Copepoda, Harpacticoida) under laboratory conditions. Notas e Estudios do Instituto de Biologia Marítima, Lisboa, **36**: 1-16.

——, 1972. The developmental stages of the marine calanoid copepod *Acartia grani* Sars bred in the laboratory. Notas e Estudios do Instituto de Biologia Marítima, Lisboa, **40**: 1-38, 18 pls.

VILLAR-ARGAIZ, M., J. M. MEDINA-SANCHEZ & P. CARRILLO, 2002. Linking life history strategies and ontogeny in crustacean zooplankton: implications for homeostasis. Ecology, **83**: 1899-1914.

VILLATE, F., 1991. Annual cycle of zooplankton community in the Abra harbour (Bay of Biscay): abundance, composition and size spectra. Journal of Plankton Research, **13**: 691-706.

VINCX, M. & C. HEIP, 1979. Larval development and biology of *Canuella perplexa* T. and A. Scott, 1893 (Copepoda, Harpacticoida). Cahiers de Biologie Marine, **20**: 281-299.

VLADIMIRSKAYA, E. V., 1969. Razvitie *Calanus finmarchicus* (Gunner) vesnoĭ v raĭone N'yufaundlenda. [Development of *Calanus finmarchicus* in the Newfoundland area in spring.]. Trudy Vsesoyuznogo Nauchno-Issledovatel'skogo Instituta Morskogo Rybnogo Khozyaistvo i Okeanografii, **65**: 178-197. [In Russian, with English summary.]

——, 1979. Zimnee raspredelenie massovykh vidov kopepod v yuzhnoĭ chasti Morya Skotiya. [Winter distribution of mass species of copepods in the southern part of the Scotia Sea.]. Biologiya Morya, Vladivostok, **1979**: 8-14.

VOPEL, K., J. DEHMLOW, M. JOHANSSON & G. ARLT, 1998. Effects of anoxia and sulphide on populations of *Cletocamptus* confluens (Copepoda, Harpacticoida). Marine Ecology Progress Series, **175**: 121-128.

VOTH, D. R., 1972. Life history of the caligid copepod *Lepeophtheirus hospitalis* Fraser, 1920 (Crustacea: Caligoida): 1-114. (Ph.D. Thesis, Oregon State University, Grand Forks, North Dakota).

WALKER, L. M., 1981. Reproductive biology and development of a marine harpacticoid copepod reared in the laboratory. Journal of Crustacean Biology, **1**: 376-388.

WALTER, E., 1922. Über die Lebensdauer der freilebenden Süsswasser-Cyclopiden und andere Fragen ihrer Biologie. Zoologische Jahrbucher, (Systematik) **44**: 375-420.

WEBBER, M. K. & J. C. ROFF, 1995. Annual structure of the copepod community and its associated pelagic environment of Discovery Bay, Jamaica. Marine Biology, Berlin, **123**: 467-479.

WEGLÉNSKA, T., 1971. The influence of various concentrations of natural food on the development, fecundity and production of planktonic crustacean filtrators. Ekologia Polska, **19**: 427-447.

WERNER, I. & H. AUEL, 2004. Environmental conditions and overwintering strategies of planktonic metazoans in and below coastal fast ice in the Gulf of Finland (Baltic Sea). Sarsia, **89**: 102-116.

WHITE, H. C., 1942. Life history of *Lepeophtheirus salmonis*. Journal of the Fisheries Research Board of Canada, **6**: 24-29.

WHITEHOUSE, J. W. & B. G. LEWIS, 1973. The effect of diet and density on development, size and egg production in *Cyclops abyssorum* Sars, 1863 (Copepoda, Cyclopoida). Crustaceana, **25**: 225-236.

WIERZBICKA, M., 1974. Haemolymph concentration in Cyclopoida copepodids during active and resting stage and the effect of 2,4-D sodium salt. Polskie Archiwum Hydrobiologii, **21**: 269-273.

WIERZEJSKI, A., 1877. Ueber Schmarotzerkrebse von Cephalopoden. Zeitschrift für Wissenschaftliche Zoologie, **29**: 562-583.

WILKES, S. N., 1966. The life history of *Nectobrachia indivisa* Fraser, 1920 (Copepoda: Lernaeopodoida), a parasite of the starry flounder. (Ph.D. Thesis, Oregon State University, Corvallis, Oregon).

WILL, F., 1844. Ueber *Staurosoma*, einem in den Aktinien lebenden Schmarotzer. Archiv für Naturgeschichte, **10**: 337-343.

WILLIAMS, L. W., 1907. List of the Rhode Island Copepoda, Phyllopoda, and Ostracoda with new species of Copepoda. 37th Annual Report of the Commissioners of Inland Fisheries of Rhode Island, **30**: 69-79, pls. 1-3.

WILLIAMS, R., 1974. Vertical distribution and the development of generations of copepods at Ocean Weather Station 'India'. Proceedings of the Challenger Society, **4**: 245-246.

WILLIAMS, R. & D. V. P. CONWAY, 1984. Vertical distribution, and seasonal and diurnal migration of *Calanus helgolandicus* in the Celtic Sea. Marine Biology, Berlin, **79**: 63-73.

—— & ——, 1985. Vertical distribution of *Calanus finmarchicus* and *C. helgolandicus* in relation to the development of the seasonal thermocline in the Celtic Sea. Marine Biology, Berlin, **82**: 145-149.

WILLIAMS, R., D. V. P. CONWAY & N. R. COLLINS, 1987. Vertical distribution of eggs, nauplii and copepodites of *Calanus helgolandicus* (Copepoda: Crustacea) in the Celtic Sea. Marine Biology, Berlin, **96**: 247-252.

WILLIAMS, R. & J. A. LINDLEY, 1980. Plankton of the Fladen Ground during FLEX 76. III. Vertical distribution, population dynamics and production of *Calanus finmarchicus* (Crustacea: Copepoda). Marine Biology, Berlin, **60**: 47-56.

WILLIAMS-HOWZE, J., 1996. The biology and morphology of the marine harpacticoid copepod *Heteropsyllus nunni* Coull, during encystment diapause. Hydrobiologia, **320**: 179-189.

WILLIAMSON, C. E., 1986. The swimming and feeding behavior of *Mesocyclops*. Hydrobiologia, **134**: 11-19.

WILLIAMSON, C. E., N. M. BUTLER & L. FORCINA, 1985. Food limitation in naupliar and adult *Diaptomus pallidus*. Limnology and Oceanography, **35**: 1283-1290.

WILSON, C. B., 1907a. North American parasitic copepods belonging to the family Caligidae. Part 2. - The Trebinae and Euryphorinae. Proceedings of the United States National Museum, **31**: 669-720.

——, 1907b. North American parasitic copepods belonging to the family Caligidae. Parts 3 and 4. - A revision of the Pandarinae and the Cecropinae. Proceedings of the United States National Museum, **33**: 323-490.

——, 1908. North American parasitic copepods: new genera and species of Caliginae. Proceedings of the United States National Museum, **33**: 593-627.

——, 1911. North American parasitic copepods. - Part 9. The Lernaeopodidae: the development of *Achtheres ambloplitis*. Proceedings of the United States National Museum, **39**: 189-226.

——, 1915. North American parasitic copepods belonging to the Lernaeopodidae, with a revision of the entire family. Proceedings of the United States National Museum, **47**: 565-729.

——, 1917. The economic relations, anatomy, and life history of the genus *Lernaea*. Bulletin of the United States Bureau of Fisheries, **35**: 165-198.

——, 1932. The copepods of the Woods Hole region, Massachusetts. Bulletin of the United States National Museum, **158**: 1-635.

WISHNER, K. F., J. R. SCHOENHERR, R. BEARDSLEY & C. CHEN, 1995. Abundance, distribution and population structure of the copepod *Calanus finmarchicus* in a springtime right whale feeding area in the southwestern Gulf of Maine. Continental Shelf Research, **15**: 475-507.

WOODWARD, I. O. & R. W. G. WHITE, 1983. Effects of temperature and food on instar development rates of *Boeckella symmetrica* Sars (Copepoda: Calanoida). Australian Journal of Marine and Freshwater Research, **34**: 927-932.

WOLLENBERGER, L., M. BREITHOLTZ, K. O. KUSK & B. E. BENGTSSON, 2003. Inhibition of larval development of the marine copepod *Acartia tonsa* by four synthetic musk substances. Science of the Total Environment, **305**: 53-64.

WOLLENBERGER, L., L. DINAN & M. BREITHOLTZ, 2005. Brominated flame retardants: activities in a crustacean development test and in an ecdysteroid screening assay. Environmental Toxicology and Chemistry, **24**: 400-407.

WORTHINGTON, E. B., 1931. Vertical movements of fresh-water macroplankton. Internationale Revue der Gesamten Hydrobiologie und Hydrographie, **25**: 394-436.

WORTHINGTON, E. B. & C. K. RICARDO, 1936. Scientific results of the Cambridge Expedition to the East African lakes, 1930-1. No. 17. The vertical distribution and movements of the plankton in Lakes Rudolf, Naivasha, Edward and Bunyoni. Journal of the Linnean Society of London, (Zoology) **40**: 33-69.

WRIGHT, D. A., J. D. SAVITZ, R. DAWSON, J. MAGEE & R. A. SMUCKER, 1996. Effect of diflubenzuron on the maturation and reproductive success of the copepod *Eurytemora affinis*. Ecotoxicology, **5**: 47-58.

WROBLEWSKI, J. S., 1980. A simulation of the distribution of *Acartia clausi* during Oregon upwelling, August, 1973. Journal of Plankton Research, **2**: 43-68.

WUETHRICH, M., 1948. Etude du développement des nauplii de *Diaptomus gracilis* Sars, et *Diaptomus laciniatus* Lilljeborg. Revue Suisse de Zoologie, **55**: 427-445.

WYNGAARD, G. A., 1983. In situ life table of a subtropical copepod. Freshwater Biology, **13**: 275-281.

WYNGAARD, G. A., C. E. GOULDEN & A. NOURBAKHSH, 1994. Life history traits of the tropical freshwater copepod *Mesocyclops longisetus* (Crustacea: Copepoda). Hydrobiologia, **292/293**: 423-427.

XU, Z. & C. W. BURNS, 1991. Development, growth and survivorship of juvenile calanoid copepods on diets of cyanobacteria and algae. Internationale Revue der Gesamten Hydrobiologie, **76**: 73-87.

YAMAGUCHI, A. & T. IKEDA, 2000. Vertical distribution, life cycle, and allometry of two oceanic calanoid copepods (*Pleuromamma scutullata* and *Heterorhabdus tanneri*) in the Oyashio region, western Subarctic Pacific Ocean. Journal of Plankton Research, **22**: 29-46.

—— & ——, 2000. Vertical distribution, life cycle, and developmental characteristics of the mesopelagic calanoid copepod *Gaidius variabilis* (Aetideidae) in the Oyashio region, western north Pacific Ocean. Marine Biology, Berlin, **37**: 99-109.

—— & ——, 2001. Abundance and population structure of three mesopelagic *Paraeuchaeta* species (Copepoda: Calanoida) in the Oyashio region, western Subarctic Pacific Ocean with notes on their carcasses and epizoic ciliates. Plankton Biology and Ecology, **48**: 104-113.

—— & ——, 2002a. Vertical distribution patterns of three mesopelagic *Paraeuchaeta* species (Copepoda: Calanoida) in the Oyashio region, western Subarctic Pacific Ocean. Bulletin of Fisheries Sciences Hokkaido University, **53**: 1-10.

—— & ——, 2002b. Reproductive and developmental characteristics of three mesopelagic *Paraeuchaeta* species (Copepoda: Calanoida) in the Oyashio region, western Subarctic Pacific Ocean. Bulletin of Fisheries Sciences Hokkaido University, **53**: 11-21.

YAMAGUCHI, A., T. IKEDA & K. HIRAKAWA, 1999. Diel vertical migration, population structure and life cycle of the copepod *Scolecithricella minor* (Calanoida: Scolecitrichidae) in Toyama Bay, southern Japan Sea. Plankton Biology and Ecology, **46**: 54-61.

YAMAGUCHI, A., T. IKEDA & N. SHIGA, 1998. Population structure and life cycle of *Pseudocalanus minutus* and *Pseudocalanus newmani* (Copepoda: Calanoida) in Toyama Bay, southern Japan Sea. Plankton Biology and Ecology, **45**: 183-193.

YAMAGUCHI, A. & N. SHIGA, 1997. Vertical distributions and life cycles of *Pseudocalanus minutus* and *P. newmani* (Copepoda; Calanoida) off Cape Esan, southwestern Hokkaido. Bulletin of the Plankton Society of Japan, **44**: 11-20. [In Japanese.]

YAMAGUCHI, A., S. TACHIBANA, K. HIRAKAWA & T. ISHIDA, 2005. Descriptions of the copepodid stages of the mesopelagic copepod, *Gaetanus variabilis* (Brodsky, 1950) (Calanoida, Aetideidae) from the Japan Sea. Crustaceana, **78**: 819-837.

YANG, J. M., 1998. Preliminary study on the feeding of copepodid larvae of the Bohai Sea *Calanus sinicus* Brodsky. Marine Fisheries Research, **19**: 1-4. [In Chinese.]

YASHOUV, A., 1959. On the biology of *Lernaea* in fish ponds. Bamidgeh, Bulletin for Fish Culture in Israel, **11**: 80-89.

YOON, W. D., M. B. SHIM & J. K. CHOI, 1998. Description of the developmental stages in *Acartia bifilosa* Giesbrecht (Copepoda: Calanoida). Journal of Plankton Research, **20**: 923-942.

YOSHIKOSHI, K. & Y. KÔ, 1988. Structure and function of the peritrophic membrane of copepods. Nippon Suisan Gakkaishi, **54**: 1077-1082.

ZACARIAS, D. & E. ZOPPI DE ROA, 1981. Desarrollo larval y postlarval de *Oithona hebes* Giesbrecht (Copepoda: Cyclopoida). Acta Cientifica Venezolana, **11**: 109-123. [In Spanish, with English summary.]

ZANDT, F., 1935. *Achtheres pseudobasanistes* n.n., syn. *Basanistes coregoni* (Neresheimer). Die postembyonale Entwicklung und geographische Verbreitung eines Lernaeopodiden. Zoologische Jahrbücher, (Anatomie und Ontogenie der Tiere) **60**: 289-464.

ZÁNKAI, N. P., 1978. The duration of development of *Eudiaptomus gracilis* (G.O. Sars) (Copepoda) in Lake Balaton. Acta Biologica Debrecina, **15**: 183-198.

——, 1991. Feeding of nauplius stages of *Eudiaptomus gracilis* on mixed plastic beads. Journal of Plankton Research, **13**: 437-454.

——, 1992. On the feeding of *Cyclops vicinus* (Copepoda: Cyclopoida) nauplius larvae. The size limits of the consumable particles. Allattani Kozlemenyek, **78**: 117-125.

——, 1994. Feeding of copepodite and adult stages of *Eudiaptomus gracilis* (G. O. Sars, 1863) (Copepoda, Calanoida) on mixed plastic beads. Crustaceana, **66**: 90-109.

ZARET, T. M. & J. S. SUFFERN, 1976. Vertical migration in zooplankton as a predator avoidance mechanism. Limnology and Oceanography, **21**: 804-813.

ZELDIS, J. R., K. ROBINSON, A. ROSS & B. HAYDEN, 2004. First observations of predation by New Zealand greenshell mussels (*Perna canaliculus*) on zooplankton. Journal of Experimental Marine Biology and Ecology, **311**: 287-299.

ZIEGELMAYER, W., 1925. Metamorphose und Wachstum der Cyclopiden. Zeitschrift für Wissenschaftliche Zoologie, **126**: 493-570.

ZMERZLAYA, E., 1972. *Ergasilus sieboldi* Nordmann, 1832, ego razvitie, biologiya i epizootologicheskoe znachenie. [*Ergasilus sieboldi* Nordmann, 1832, its development, biology and epizootological significance.]. Izvestiya Gosudarstvennogo nauchno-issledovatel'skogo Instituta Ozernogo i Rechnogo Rybnogo Khozyaistva, **80**: 132-177. [In Russian, with English summary.]

ZURLINI, G., I. FERRARI & A. NASSOGNE, 1978. Reproduction and growth of *Euterpina acutifrons* (Copepoda: Harpacticoida) under experimental conditions. Marine Biology, Berlin, **46**: 59-64.

APPENDIX I

List of species and subspecies mentioned in the text with author and date, grouped alphabetically for each higher taxon.

BRANCHIOPODA
Artemia salina Linnaeus, 1758
Leptestheria kawachiensis Uéno, 1927

MYSTACOCARIDA
Derocheilocaris typica Pennak & Zinn, 1943

COPEPODA
Acanthocyclops thomasi (Forbes, 1882)
Acartia clausi Giesbrecht, 1889
Acartia erythraea Giesbrecht, 1889
Acartia longiremis (Lilljeborg, 1853)
Acartia tonsa Dana, 1849
Achtheres percarum Nordmann, 1832
Acrocalanus gibber Giesbrecht, 1888
Alebion lobatus Cressey, 1970
Allantogynus delamarei Changeux, 1960
Alteutha interrupta (Goodsir, 1845)
Amphiascus undosus Lang, 1965
Anchistrotos pleuronichthydis Yamaguti, 1939
Apocyclops dengizicus (Lepeshkin, 1900)
Apocyclops royi (Lindberg, 1940)
Asterocheres lilljeborgi Boeck, 1859
Balaenophilus unisetus Aurivillius, 1879
Benthomisophria palliata G.O. Sars, 1909
Bryocamptus zschokkei alleganiensis Coker, 1934
Bryocyclops caroli Bjornberg, 1985
Calanoides acutus Giesbrecht, 1902
Calanus agulhensis De Decker, Kaczmarak & Marska, 1991
Calanus australis Brodsky, 1959
Calanus chilensis Brodsky, 1959
Calanus finmarchicus (Gunnerus, 1770)
Calanus glacialis Jaschnov, 1955
Calanus helgolandicus (Claus, 1863)
Calanus pacificus Brodsky, 1948
Calanus propinquus Brady, 1883
Caligus centrodonti Baird, 1850
Caligus clemensi Parker & Margolis, 1964

Caligus elongatus Nordmann, 1832
Caligus epidemicus Hewitt, 1971
Caligus multispinosus Shen, 1957
Caligus spinosus Yamaguti, 1939
Canuella perplexa T. Scott & A. Scott, 1893
Canthocamptus mirabilis Sterba, 1968
Cardiodectes medusaeus (Wilson, 1908)
Caribeopsyllus amphiodiae Ho, Dojiri, Hendler & Deets, 2003
Centropages abdominalis Sato, 1913
Centropages typicus Krøyer, 1849
Centropages velificatus (Oliveira, 1947)
Chiridius armatus (Boeck, 1872)
Colobomatus pupa Izawa, 1974
Conchyliurus quintus Tanaka, 1961
Coullana canadensis (Willey, 1923)
Critomolgus anthopleurus Kim, 1996
Cyclopina longifurca Sewell, 1924
Cyclopina schneideri T. Scott, 1904
Cyclops bicuspidatus (Claus, 1857)
Cyclops scutifer G.O. Sars, 1863
Cyclops strenuus Fischer, 1851
Cyclops strenuus strenuus Fischer, 1851
Cyclops vicinus Ulyanin, 1875
Cyclops viridis (Jurine, 1820)
Dermatomyzon nigripes (Brady & Robertson, 1875)
Diacyclops thomasi (Forbes, 1882)
Diaptomus oregonensis Lilljeborg, 1889
Diaptomus siciloides Lilljeborg, 1889
Diarthrodes cystoecus Fahrenbach, 1954
Dioithona oculata (Farran, 1913)
Doridicola longicauda (Claus, 1860)
Doridicola sepiae (Izawa, 1976)
Doropygopsis longicauda (Aurivillius, 1882)
Doropygus bayeri Illg, 1958
Doropygus fernaldi Illg, 1958
Doropygus mohri Illg, 1958
Doropygus seclusus Illg, 1958
Drepanopus forcipatus Giesbrecht, 1888
Drescheriella glacialis Dahms & Dieckmann, 1987
Ectocyclops rubescens Brady, 1904
Ergasilus hypomesi Yamaguti, 1936
Eucalanus attenuatus (Dana, 1849)
Eucalanus crassus Giesbrecht, 1888
Eucalanus elongatus (Dana, 1848)
Eucalanus hyalinus (Claus, 1866)
Eucalanus inermis Giesbrecht, 1893
Eucalanus pileatus Giesbrecht, 1888
Eucalanus subtenuis Giesbrecht, 1888

Euchaeta japonica Marukawa, 1921
Euchirella messinensis (Claus, 1863)
Euchirella rostrata (Claus, 1866)
Eucyclops serrulatus (Fischer, 1851)
Eudiaptomus gracilis (G.O. Sars, 1862)
Eudiaptomus graciloides (Lilljeborg, 1888)
Euryte longicauda Philippi, 1843
Eurytemora affinis (Poppe, 1880)
Eurytemora velox (Lilljeborg, 1853)
Euterpina acutifrons (Dana, 1847)
Gaidius variabilis Brodsky, 1950
Galapalaophonte biarticulata Fiers, 1991
Halicyclops neglectus Kiefer, 1935
Hansenulus trebax Heron & Damkaer, 1986
Harpacticus uniremis Krøyer, 1842
Heliogabalus phascolia Lützen, 1968
Hemicyclops ctenidis Ho & Kim, 1990
Hemicyclops gomsoensis Ho & Kim, 1991
Hemicyclops japonicus Itoh & Nishida, 1993
Herrmannella rostrata Canu, 1891
Lamproglena chinensis Yü, 1938
Lepeophtheirus dissimulatus Wilson, 1905
Lepeophtheirus pectoralis (Müller, 1776)
Lepeophtheirus salmonis (Krøyer, 1837)
Leptinogaster major (Williams, 1907)
Leptodiaptomus novamexicanus (Herrick, 1895)
Lernaea cyprinacea Linnaeus, 1746
Lernaeenicus sprattae (Sowerby, 1806)
Lernaeocera branchialis (Linnaeus, 1767)
Lichomolgus canui G.O. Sars, 1917
Longipedia americana Wells, 1980
Longipedia minor (T. Scott & A. Scott, 1893)
Macrocyclops albidus (Jurine, 1820)
Macrocyclops fuscus (Jurine, 1820)
Macrosetella gracilis (Dana, 1847)
Megacyclops viridis (Jurine, 1820)
Megadiaptomus hebes Kiefer, 1936
Mesocyclops aequatorialis Kiefer, 1929
Mesocyclops aequatorialis similis Van de Velde, 1984
Mesocyclops edax (Forbes, 1890)
Mesocyclops leuckarti (Claus, 1857)
Mesocyclops thermocyclopoides Harada, 1931
Metridia longa (Lubbock, 1854)
Metridia lucens Boeck, 1865
Metridia pacifica Brodsky, 1950
Midicola spinosus (Raffaele & Monticelli, 1885)
Misophria pallida Boeck, 1865
Modiolicola insignis Aurivillius, 1882

Monstrilla hamatapex Grygier & Ohtsuka, 1995
Monstrilla helgolandica Claus, 1863
Mytilicola intestinalis Steuer, 1902
Neanthessius renicollis Izawa, 1976
Neobrachiella robusta (Wilson, 1912)
Neocalanus tonsus (Brady, 1883)
Neoergasilus japonicus (Harada, 1930)
Notodelphys affinis Illg, 1958
Notodelphys ascidicola Allman, 1847
Oithona davisae Ferrari & Orsi, 1984
Oithona ovalis Herbst, 1955
Oithona similis Claus, 1866
Oncaea media Giesbrecht, 1891
Ostrincola koe Tanaka, 1961
Pachypygus gibber (Thorell, 1859)
Panaietis yamagutii Izawa, 1976
Paracyclopina longifurca (Sewell, 1924)
Paracyclops fimbriatus (Fischer, 1853)
Paraeuchaeta norvegica (Boeck, 1872)
Paraleptastacus brevicaudatus Wilson, 1932
Paramphiascella fulvofasciata Rosenfield & Coull, 1974
Paranthessius columbiae Thompson, 1897
Parastenhelia megarostrum Wells, Hicks & Coull, 1982
Parategastes sphaericus (Claus, 1863)
Parathalestris harpactoides (Claus, 1863)
Parkius karenwishnerae Ferrari & Markhaseva, 1996
Peniculisa shiinoi Izawa, 1965
Philoblenna arabica Izawa, 1976
Phyllodiaptomus annae (Apstein, 1907)
Platycopia orientalis Ohtsuka & Boxshall, 1994
Pleuromamma gracilis (Claus, 1863)
Pleuromamma xiphias (Giesbrecht, 1889)
Pontella chierchiae Giesbrecht, 1889
Porcellidium fimbriatum Claus, 1863
Procyclopina feiticeira Lotufo, 1995
Pseudacanthocanthopsis apogonis Yamaguti & Yamasu, 1959
Pseudocalanus elongatus (Boeck, 1865)
Pseudodiaptomus acutus (Dahl, 1894)
Pseudodiaptomus ardjuna Brehm, 1953
Pseudodiaptomus aurivilli Cleve, 1901
Pseudodiaptomus binghami Sewell, 1912
Pseudodiaptomus coronatus Williams, 1906
Pseudodiaptomus euryhalinus Johnson, 1939
Pseudodiaptomus marinus Sato, 1913
Pseudodiaptomus richardi (Dahl, 1894)
Pseudomyicola ostreae Yamaguti, 1936
Pseudomyicola spinosus (Raffaele & Monticelli, 1885)
Pygodelphys aquilonaris Illg, 1958

Rhincalanus gigas Brady, 1883
Ridgewayia klausruetzleri Ferrari, 1995
Sabellacheres illgi Dudley, 1964
Salmincola californiensis (Dana, 1852)
Sarcotaces pacificus Komai, 1924
Scolecodes huntsmani (Henderson, 1931)
Scopalatum vorax (Esterly, 1911)
Scottomyzon gibberum (T. Scott & A. Scott, 1894)
Scutellidium hippolytes (Krøyer, 1863)
Selioides bocqueti Carton, 1963
Speocyclops racovitzai (Chappuis, 1923)
Stenhelia palustris (Brady, 1868)
Stenhelia (Delavalia) palustris Brady, 1868
Tachidius discipes Giesbrecht, 1881
Taeniacanthus lagocephali Pearse, 1952
Taeniastrotos pleuronichthydis (Yamaguti, 1939)
Tegastes clausi G.O. Sars, 1904
Tegobomolochus nasicola Izawa, 1976
Temora longicornis (Müller, 1785)
Temora stylifera (Dana, 1849)
Thalestris longimana Claus, 1863
Thalestris rhodymeniae (Brady, 1894)
Thermocyclops consimilis Kiefer, 1934
Thermocyclops decipiens (Kiefer, 1929)
Thermomesochra reducta Itô & Burton, 1980
Tigriopus japonicus Mori, 1938
Tisbe gracilis (T. Scott, 1895)
Tracheliastes polycolpus Nordmann, 1832
Zaus robustus Itô, 1974
Zygomolgus poucheti (Canu, 1891)

ISOPODA
Asellus aquaticus (Linnaeus, 1758)
Lirceus macrourus Garman, 1890

AMPHIPODA
Gammarus chevreuxi Sexton, 1924

DECAPODA
Panulirus argus (Latreille, 1804)

GLOSSARY

Abdomen – that part of the body of a copepod posterior to the somite bearing the genital opening. The abdomen includes four somites, three of which do not bear a paired appendage; the posterior somite bears the caudal rami and is the first abdominal somite to appear during development.

Aesthetasc – a transformed seta of antenna 1 or another oral appendage of copepodids. An aesthetasc usually has a sclerotized base but otherwise is poorly sclerotized; it often is considered to have a chemosensory function.

Anal somite – the posterior somite and abdominal somite onto which the anus opens. The anal somite bears a paired appendage, collectively called the caudal rami. The anal somite is herein considered the first abdominal somite, despite its posterior position, because it is the first abdominal somite to appear during post-embryonic development.

Anameric – the addition of only one somite to the body at each molt during development.

Antenna 1 – the anterior limb of the cephalon; it is uniramous in copepods.

Antenna 2 – the limb of the cephalon posterior to antenna 1; in most naupliar stages, it bears a distinct protopodal masticatory arthrite that is not present in copepodid stages or in non-feeding naupliar stages, especially not in those of species with lecithotrophic embryos.

Anteriad – toward the anterior end of the body.

Anterioposterior axis – an imaginary line through the rostral area of the head and the anal somite.

Appendage – paired extension of a somite along a proximodistal axis and usually composed of serially repeated elements. Appendages of copepods include the limbs of the five cephalic somites, the limbs of the seven thoracic somites, and the caudal ramus of the anal segment (see also Limb and Swimming leg).

Architecture – the morphological organization of the body.

Arthrite – a ventrally articulating, sclerotized extension of a protopodal segment that is moved by muscles.

Arthrodial membrane – an unsclerotized, flexible section of the exoskeleton between the sclerotized parts of two somites or two segments.

Basis – the distal segment of the protopod; it bears no more than two ventral, setose endites. The rami, exopod and endopod of a limb, originate on the basis (see Coxa and Praecoxa).

Bud – the earliest step of a developing limb; a limb bud does not articulate with its somite and bears setae including at least the crown group of terminal setae. A limb bud often is considered functionless on immature stages although limb 6 of podopleans is a bud that covers the genital opening of the adult male (see Transformed limb and Secondary bud).

Caudal ramus – the appendage of the posterior abdominal somite of a copepod. It does not have a propodal/ramal configuration and its homologies to serially repeated limbs of the cephalon and thorax have not been determined. The caudal ramus bears setae in a pattern similar to an exopod. Its axial orientation is not known.

Cephalon – a complex of all of the somites of the head.

Cephalothorax – a complex of the cephalic somites plus at least one thoracic somite.

Chalimus – one of up to four stages in the copepodid phase of development of caligid-like parasitic copepods; the chalimus usually is attached to the host, often by a frontal filament held by maxilla 2. The first chalimus is molted from the first copepodid stage; the four chalimus stages correspond to the second to fifth copepodid stages.

Complex – two or more unarticulated somites or segments resulting from the failure of an arthrodial membrane to form between the somites or segments comprising the complex.

Configuration – the morphological organization of an appendage.

Copepodid – [alternative spelling: copepodite] a developmental stage without a naupliar arthrite on antenna 2, usually with articulating thoracic somites and more than three transformed limbs, and often with articulating abdominal somites. Copepodid stages are designated by Roman numerals.

Coxa – the middle segment of the protopod, proximal to the basis and distal to the praecoxa, with a single setose endite; the mandibular gnathobase of copepods includes the coxal endite with its single seta (see Basis and Praecoxa).

Denticle – a solid extension of the epicuticle of a segment (see Seta and Setule).

Diapause – a significant period of quiescence during development.

Distad – toward the distal end of a limb.

Dorsad – toward the dorsal aspect of the body.

Dorsoventral axis – an imaginary line from the surface of the body opposite the limbs to the surface bearing the limbs; the terms 'lateral' and 'medial' often are used in place of 'dorsal' and 'ventral' in descriptions of appendages.

Ecdysis – the process of shedding the exoskeleton during molting.

Element – a serially repeated component or part of a segment, e.g., the dorsal formation seta, the ventral formation seta, or the finishing arthrodial membrane.

Endite – a non-articulating, ventral attenuation of a protopodal segment.

Endopod – a ventral extension of the proximodistal axis of a limb originating on the basis of the protopod and usually segmented. Dorsal setae are absent from endopodal segments except for the penultimate and the antepenultimate segments. An endopodal segment may bear more than one ventral seta.

Exite – a non-articulating, dorsal attenuation of a protopodal segment.

Exopod – a dorsal extension of the proximodistal axis of a limb originating on the basis of the protopod and usually segmented. Segments of the exopod bear a dorsal seta and often a ventral seta, but usually there is only one of each kind of seta on a segment.

Finishing arthrodial membrane – an arthrodial membrane that completes a segment and whose formation defines the location of the segment along the proximodistal axis of a limb. The finishing arthrodial membrane usually is the distal arthrodial membrane of a segment that forms proximal to the source segment, and is the proximal arthrodial membrane of a segment that forms distal to the source segment.

Flagellomere – a segment of a flagellum that is part of the ramus of antenna 1 or antenna 2 of many crustaceans. Flagellomeres are not moved by muscles and are formed distad to the first flagellomere, which is the source segment for the flagellomeres. Copepods do not have a flagellum on either antenna 1 or antenna 2.

Flagellum – the distal section of either ramus of antenna 1 or antenna 2 of crustaceans that is made up of articulating segments lacking intrinsic muscles; the articulating segments are known as flagellomeres.

Formation seta – the first dorsal and/or first ventral seta that forms during development of a segment. The first dorsal seta of an exopodal segment often forms before the first ventral seta, and the first ventral seta of an endopodal segment often forms before the first dorsal seta.

Gnathobase – the ventral extension of the mandibular coxa, including its setose endite.

Gymnopleans – copepods in which the major body articulation of the adult is between the sixth and the seventh thoracic somites; there is also a significant difference in size between these two somites (see also Podopleans and Thaumatopsylloids).

Homolog – a corresponding part in different copepods that has been inherited from the common ancestor of those different copepods; "homologous" refers to these corresponding parts.

Interpodal bar – a ventral exoskeletal structure uniting the contralateral pair of thoracic limbs; an interpodal bar may unite thoracic limb pairs 2-6 of most copepods.

Labium – see Paragnaths.

Labrum – a lobe-like flap originating near the anterior margin of the head, between the bases of the first antennae, and extending posteriorly across the ventral surface of the body to the mouth area.

Limb – the paired appendages of the five cephalic and seven thoracic somites. A limb has three axes, anterioposterior, proximodistal, and dorsoventral (the latter often called mediolateral). A limb develops in steps and may be composed of up to three protopodal segments (praecoxa, coxa and basis), and usually an endopod and an exopod both of which may be segmented (see also Appendage and Swimming leg).

Mandible – the paired limb of the somite of the cephalon posterior to antenna 2. The mandible bears a coxal gnathobase during the copepodid phase of development; it may (the fourth to sixth naupliar stages of calanoid copepods) or may not (all other copepods) bear a coxal gnathobase during the naupliar phase of development.

Maxilla 1 – the paired limb of the somite of the cephalon posterior to the mandible.

Maxilla 2 – the paired limb of the somite of the cephalon posterior to the maxilla 1.

Maxilliped – the paired limb of the first thoracic somite of copepods.

Molt – the transition from one developmental stage to the next.

Naupliar shield – the expanse of the exoskeleton of an unspecified number of somites, uninterrupted by arthrodial membranes, and covering the dorsal and lateral part of the naupliar body.

Nauplius – a developmental stage whose somites do not articulate, with only three transformed limbs, and with an arthrite on the coxa of antenna 2. A naupliar arthrite may not be present on nauplii that do not feed. Naupliar stages are designated by Roman numerals.

Palp – the basis plus rami of a limb. The palp may be reduced to a poorly-sclerotized extension of the exoskeleton bearing a crown of small setae corresponding to the terminal setae of one or both rami.

Paragnath(s) – a pair of articulating lobes often bearing setae and located posterior to the mandible and anterior to maxilla 1. Homologies of the paragnath have not been determined, but it may be the praecoxa of the mandible or it may be that section of the praecoxa of maxilla 1 bearing a second lobe. Also collectively known as Labium.

Pattern – the order in which somites are added to the body relative to other somites, or that elements of segments are added to a limb relative to other elements; also the process of adding in a fixed order during development.

Phylotypic stage – the stage of development, relative to other stages, that has diverged least during the evolutionary history of a monophyletic group of organisms. The phylotypic stage of copepods is the first copepodid.

Polyarthrans – copepods of the families Canuellidae and Longipediidae, traditionally classified as Harpacticoida, now opposed to oligoarthran harpacticoids and allegedly representing a separate clade in the Copepoda [cf., e.g., Tieman, 1984, Crustaceana, (Suppl.) **7**: 47-59; 2004, Invertebrate Zoology, **1** (1): 29-51].

Podopleans – copepods on which the major body articulation of the adult is between the fifth and the sixth thoracic somites; often there is a significant difference in size between these two somites (see also Thaumatopsylloids and Gymnopleans).

Posteriad – toward the posterior end of the body.

Post-formation seta – one or more setae added to a segment after the initial dorsal and/or ventral formation setae have been added. Examples of post-formation setae can be found on the endopod of the maxilliped of calanoids and polyarthrans, and on the proximal exopodal segment of platycopiids.

Post-mandibular appendages – the cephalic limbs, maxilla 1 and maxilla 2; plus the thoracic limbs, maxilliped, swimming legs 1-4, and limbs 5-6; and the caudal ramus. Except for the caudal ramus, these appendages are added in strict anterioposterior order during post-embryonic development.

Praecoxa – the proximal segment of the protopod; it is proximal to the coxa and bears up to three setose ventral endites (see Coxa and Basis).

Presentation – the first appearance of limb; a limb may initially appear as either a limb bud or a transformed limb.

Prosome – that part of the adult copepod body anterior to the major body articulation.

Proximad – toward the proximal end of a limb.

Proximodistal axis – an imaginary line through the insertion of a limb on its somite, and the tip of the limb; on biramous limbs, the proximodistal axis is duplicated through each ramus.

Ramus – a group of serially repeated segmental elements along a proximodistal axis and originating on the basis; the exopod and the endopod are rami.

Reorganized limb – a transformed limb or secondary bud that has been reconfigured from the limb bud.

Rule of Serial Homologs – if a serial homolog that is formed late during the normal course of development is present, then serial homologs that are formed earlier during the normal course of development also are expected to be present. This rule is derived when the body or the limbs are patterned by truncation, so that the last formed of a set of serially homologous elements will be the first to fail to form as a result of truncated development.

Secondary bud – a small, poorly differentiated limb that has been reconfigured from a limb bud or a transformed limb of a copepodid; a secondary bud usually is found on a chalimus (see Bud and Transformed limb).

Segment – a composite group of elements that are serially repeated components of a limb; these elements usually include formation and post-formation setae, muscles, and a finishing arthrodial membrane.

Serial homologs – corresponding elements on serially repeated somites of the body or on serially repeated segments of a limb.

Seta – an articulating extension of a segment, usually not directly along the proximodistal axis (see Denticle and Setule).

Setal precedence – a process by which the formation setae of the presumptive proximal and middle segments of the rami of swimming legs initially appear on the distal segment complex. These setae are allocated to the proximal or to the middle segment after a distal, finishing arthrodial membrane is formed, which separates segments of the complex later in development.

Setule – a solid extension of the epicuticle of a seta (see Denticle and Seta).

Somite – a composite group of elements, usually exoskeletal, musculature, and nerve, which makes up a serially repeated component of the body.

Source segment – a segment from which a limb is patterned by the formation of new segment elements; a source segment is homologous to the formative zone (Fuller, 1920) or the meriston (Henson, 1947) of the antenna 1 of hemimetabolous insects. A source segment is located between the oldest and youngest element of a limb that is patterned either proximally or distally, or between the youngest elements of a limb that is patterned both proximally and distally.

Stage – a period of development between two molts in which the exoskeleton does not change; synonymous terms are instar and stadium.

Swimming leg(s) – a contralateral pair of thoracic limbs that are flattened anterioposteriorly and united by an interpodal bar; these attributes are shared only in the transformed limb and later steps of limb development (see also Appendage and Limb).

Syncoxa – a segment complex of the praecoxa and the coxa of a protopod.

Thaumatopsylloids – copepods for which there is a significant difference in size between the fourth and the fifth thoracic somites of the adult (see also Podopleans and Gymnopleans).

Thorax – that part of the body of a copepod posterior to the cephalic somite bearing maxilla 2, anterior to the abdomen, and including the somite bearing the genital opening.

Transformed Limb – a limb that is similar in configuration to the limb of the adult; a transformed limb is reconfigured from the limb bud. The protopodal segmentation of a transformed limb of copepods is complete but often the setation is not; the rami of a transformed limb are present but often not completely patterned (see Bud and Secondary bud).

Urosome – that part of the adult copepod body posterior to the major body articulation.

Ventrad – toward the ventral aspect of the body.

Von Vaupel Klein's Organ – on swimming leg 1 of calanoid copepods: the ventral seta of the basis, which is curved and often recurved, over a sensory area of pores and/or denticles on the proximal anterior face of the endopod.

TAXONOMIC INDEX

SUBJECT INDEX

www.ingramcontent.com/pod-product-compliance
Lightning Source LLC
LaVergne TN
LVHW011127300926
849160LV00008B/546

* 9 7 8 9 0 0 4 1 5 7 1 3 2 *